AF213969

Nanodiamonds in Analytical and Biological Sciences

Nanodiamonds in Analytical and Biological Sciences

Principles and Applications

Edited by

Wesley Wei-Wen Hsiao
Department of Chemical Engineering, National Taiwan University of Science and Technology, Taiwan

Trong-Nghia Le
Institute of Atomic and Molecular Sciences, Academia Sinica, Taiwan

WILEY

This edition first published 2023
© 2023 John Wiley & Sons Ltd

All rights reserved. No part of this publication may be reproduced, stored in a retrieval system, or transmitted, in any form or by any means, electronic, mechanical, photocopying, recording or otherwise, except as permitted by law. Advice on how to obtain permission to reuse material from this title is available at http://www.wiley.com/go/permissions.

The right of Wesley Wei-Wen Hsiao and Trong-Nghia Le to be identified as the authors of the editorial material in this work has been asserted in accordance with law.

Registered Office
John Wiley & Sons Ltd, The Atrium, Southern Gate, Chichester, West Sussex, PO19 8SQ, UK

For details of our global editorial offices, customer services, and more information about Wiley products visit us at www.wiley.com.

Wiley also publishes its books in a variety of electronic formats and by print-on-demand. Some content that appears in standard print versions of this book may not be available in other formats.

Trademarks: Wiley and the Wiley logo are trademarks or registered trademarks of John Wiley & Sons, Inc. and/or its affiliates in the United States and other countries and may not be used without written permission. All other trademarks are the property of their respective owners. John Wiley & Sons, Inc. is not associated with any product or vendor mentioned in this book.

Limit of Liability/Disclaimer of Warranty
While the publisher and authors have used their best efforts in preparing this work, they make no representations or warranties with respect to the accuracy or completeness of the contents of this work and specifically disclaim all warranties, including without limitation any implied warranties of merchantability or fitness for a particular purpose. No warranty may be created or extended by sales representatives, written sales materials or promotional statements for this work. This work is sold with the understanding that the publisher is not engaged in rendering professional services. The advice and strategies contained herein may not be suitable for your situation. You should consult with a specialist where appropriate. The fact that an organization, website, or product is referred to in this work as a citation and/or potential source of further information does not mean that the publisher and authors endorse the information or services the organization, website, or product may provide or recommendations it may make. Further, readers should be aware that websites listed in this work may have changed or disappeared between when this work was written and when it is read. Neither the publisher nor authors shall be liable for any loss of profit or any other commercial damages, including but not limited to special, incidental, consequential, or other damages.

A catalogue record for this book is available from the Library of Congress

Hardback ISBN: 9781394202133; ePub ISBN: 9781394202157; ePDF ISBN: 9781394202140; oBook ISBN: 9781394202164

Cover image: © Vink Fan/Shutterstock
Cover design by Wiley

Set in 9.5/12pt STIXTwoText by Integra Software Services Pvt. Ltd, Pondicherry, India
Printed and bound by CPI Group (UK) Ltd, Croydon, CR0 4YY

C9781394202133_061023

To our family
Wu-Hua Hsiao, Yu-Chu Lai, Pei-Fen Lu, *and* ***Yu-Ching Hsiao***
Nguyen Thi Ngan, Le Thi Hong Nhung
and ***Le Thi Hong Hanh***
for their constant and dedicated support
and to
Prof. Huan-Cheng Chang
for his valuable advice and guidance

Contents

Contributors

Haroon Aman
School of Mathematics and Physics
The University of Queensland
Australia
National Institute of Lasers and Optronics College, Pakistan Institute of Engineering and Applied Sciences
Pakistan

Stefanny Angela
Department of Chemical Engineering
National Taiwan University of Science and Technology
Taiwan

Chris Chen-Hua Chang
Department of Chemical Engineering
National Taiwan University of Science and Technology
Taiwan

Shu-Wei Chang
Department of Computer Science and Information Engineering, National Taiwan University
Taiwan

Marvin Chen
College of Letter and Science, the University of California, Berkeley
USA

Yun-Yu Chen
Global Health and Public Policy, Duke Kunshan University
China

Wei-Hung Chiang
Department of Chemical Engineering
National Taiwan University of Science and Technology
Taiwan

Yen-Tse Chiang
Faculty of Life Science and Medicine
King's College London
UK

Ryu Endo
Department of Biomedical Engineering, The Ohio State University
USA

Raymond Hsin
Department of Biomedical Engineering, Washington University in St. Louis
USA

Richard Hsin
Department of Computer Engineering
Rose-Hulman Institute of Technology
USA

Yuen Yung Hui
Institute of Atomic and Molecular Sciences, Academia Sinica
Taipei Taiwan

Tan-Thanh Huynh
School of Applied Chemistry, Tra Vinh University
Vietnam

Ming-Wei Jen
College of Education and Human Ecology, The Ohio State University
USA

Joshua Ko
Department of Chemical Engineering
National Taiwan University of Science and Technology
Taiwan

Xuan Mai Lam
Department of Chemical Engineering
National Taiwan University of Science and Technology
Taiwan

Thi-Hong-Hanh Le
Department of Chemical Engineering
National Taiwan University of Science and Technology Taiwan

Hsuan-Yi Lin
Department of Power Mechanical Engineering, National Tsing Hua University
Taiwan

Yu–Chen Lin
College of Management, National Taiwan University
Taiwan

Yu-Jou Lin
Department of Chemical Engineering
National Taiwan University of Science and Technology
Taiwan

Yu-Teng Lin
Department of Chemical Engineering
National Taiwan University of Science and Technology
Taiwan

Miranda Liu
Department of Biomedical Engineering, Duke University
USA

Steven Che-Wei Lu
Department of Chemical Engineering
National Taiwan University of Science and Technology
Taiwan

Dinh Minh Pham
Institute of Biotechnology, Vietnam Academy of Science and Technology
Vietnam

Shahzad Ahmad Qureshi
Department of Computer and Information Sciences, Pakistan

Institute of Engineering and Applied Sciences, Pakistan
Centre for Mathematical Sciences
Pakistan Institute of Engineering and Applied Sciences
Pakistan

Neha Sharma
Department of Chemical Engineering
National Taiwan University of Science and Technology
Taiwan

Yan-Kai Tzeng
Department of Physics, Stanford University
USA

Duc-Thang Vo
College of Engineering, National Taiwan University of Science and Technology
Taiwan

Jo-Yu Wang
Department of Chemical Engineering
National Taiwan University of Science and Technology
Taiwan

Syun-Ying Wang
School of Environmental and Biological Sciences, Rutgers University
USA

Kartika Wardhani
School of Medicine, Stanford University
USA

Tzu-Chun You
College of Letter and Science, the University of California, Berkeley
USA

Tzu-Syuan You
Faculty of Arts and Science, University of Toronto
Canada

Editor's Preface: Nanodiamonds – A Rising Star in Nanotechnology

Nanotechnology: From Big to Small

Delivered to the American Physical Society, Richard Feynman's famous 1959 Caltech lecture, "There's Plenty of Room at the Bottom," is today recognized as one of the first and most influential meditations on the topic known today as *nanotechnology* (6). Nanotechnology is exactly what Feynman spoke of when he "theorized" the compression of massive amounts of information into tiny spaces. Today, nanotechnology is on the brink of becoming precisely what Feynman conceived: A world in which machines are too tiny to be seen.

Though many overemphasize the "nano" of nanotechnology, size is not the most important aspect of this field. Rather, emphasis is on the concept of building machines from a bottom-up perspective. As nanotechnology practitioners, we operate at the atomic level to create improved machines, materials, medicines, energy sources, and commodities. This bottom-up approach has already led to many advances in the world of science and promises to lead to many more over the next century.

Nanotechnology relies on the use of nanomaterials, which are defined as anything on a scale of 1 to 100 nm. Just as Feynman found quantum mechanics to operate at a different level than classic Newtonian physics, scientists are finding that materials often operate on the *nanoscale* differently than they do on the *macroscale*. Using these unique nano-properties, scientists develop new technologies that would have been impossible to create in the past.

Many hold the misconception that nanotechnology is a niche field, which it is not. On the contrary, nanotechnology is a multipurpose field, as its impact reaches virtually all industries and research areas.

The Nanodiamond – A Scientist's Best Friend

A true diamond never overshadows.
It's the brilliance of the subtle shine.
That's most attractive.
– Dena Tyson, Xceptance, 2008

Humans have loved diamonds throughout history. On the surface, the reason is simple – they are sparkly. But the more philosophical of our ancestors gave more tangible reasons: diamonds are both rare and unbreakable. Today, researchers can synthesize diamonds, recreating the physical construction of what was once thought to be a singularity of nature.

To synthesize a diamond, we simply copy nature. In the laboratory, researchers convert graphite to physically and chemically identical diamonds found in nature through applying a high transient pressure under a high adiabatic temperature. The first proven synthetic diamond was created in 1954 by General Electric (GE) using the high-pressure, high-temperature (HPHT) method. However, as the size of this synthetic diamond was too small, with the largest size of 0.15 mm, it was applicable for the industrial field instead of gem use. Later, the first nanosized diamonds were developed by USSR in 1963. Today, these substances are known as *nanodiamonds* (NDs), as the commercial market (which began in the 1990s) calls it.

Modern analytical tools, such as scanning electron microscopy, infrared spectroscopy, optical microscopy, the X-ray diffraction technique, and Raman spectroscopy, help us understand the construction of NDs. The ND is unique in that it is both tiny and tough; traits that common sense often tells us are opposites. Not only are NDs used in the production of purposefully hard materials and polishing media, but they are also used in biotechnology and have the potential to play a major part in saving human lives.

As an allotrope (sp^3 hybridization) of carbon, diamond has several important superlative physical properties:

- It is the hardest material known to date.
- It has the greatest thermal conductivity of any bulk material.
- It possesses by far the highest refractive index of all dielectric materials.

Nanoscale diamonds preserve some of these properties. However, the technicalities of ND production have traditionally been expensive, time-consuming, and crude. Since 1963, NDs have been prepared by detonation to generate "detonation nanodiamonds" (DNDs). Today, manufacturers crush or ball mill micrometer or millimeter diamonds obtained by HPHT and chemical vapor deposition (CVD) to reduce production costs. Despite some differences in surface structure, the core of

HPHT NDs and CVD NDs are generally similar to DNDs. DNDs are particulates (with an average diameter of 5 nm). Their primary particles have a highly uniform size distribution. Due to the large number of disordered (sp^2) carbon atoms on DNDs' surface during the synthesis period, they tend to bind together and form agglomerates. DND clusters can serve as a drug delivery system. Owing to the presence of less sp^2 carbon on the surface, HPHT NDs exhibit reduced agglomeration. Therefore, even though their size distribution is broader than DNDs, HPHT NDs are amenable to many bioimaging applications. CVD NDs are thin films formed on substrates. Several industries have used these NDs as surface coating for a variety of cutting and dressing instruments.

DNDs are now taking on new roles. Biologists currently use them as diagnostic and therapeutic tools in biology and medicine, especially for nanometer-sized diamonds (1–4). Thanks to the excellent properties of biocompatibility, high loading capacity, and unique surface function, ND particles are well suited for many drug delivery applications. Another distinguishing attribute of ND is their bright, stable fluorescence, which comes from their crystals' defects. This discovery was made in 2005 (5), and from that, a wide array of studies have been done, marking the beginning of a new era in analytical and biological applications. For instance, fluorescent nanodiamond (FND) enables the use of cell labeling, imaging, and tracking. These applications directly result from FND's exceptionally high biocompatibility and unique optical properties (5). Recent research in the field has focused on applying surface-functionalized FNDs in bioimaging, quantum sensing, and drug or gene delivery. Results have shown that FND, with their surfaces' function group, can conveniently be immobilized with protein or nucleic acid (5). Additionally, when exposed to green–yellow light, FND containing nitrogen-vacancy (NV) color centers can emit bright, nonphotobleaching, nonphotoblinking, tissue-penetrating red photons.

Book Structure

The design and working principles of NDs are the focus of this book. A practical understanding of this field necessitates a multidisciplinary background: Physics, chemistry, materials science, electrical and optoelectronics engineering, as well as bioengineering are all viewpoints that may be added to the ND.

This book contains 13 chapters:

- Chapter 1: Nanodiamonds
- Chapter 2: Fluorescent Nanodiamonds
- Chapter 3: Fabrication of Fluorescent Nanodiamonds
- Chapter 4: Surface Modification of Nanodiamonds
- Chapter 5: Toxicity Assessments of Nanodiamonds
- Chapter 6: *In Vitro* Bioimaging of Fluorescent Nanodiamonds

- Chapter 7: *In Vivo* Bioimaging of Fluorescent Nanodiamonds
- Chapter 8: Quantum Sensing of Fluorescent Nanodiamonds
- Chapter 9: Nanoscale Thermometry with Fluorescent Nanodiamonds
- Chapter 10: Nanodiamond-Enabled Drug Delivery
- Chapter 11: Nanodiamond for Mass-Spectrometry-Based Analysis of Peptides, Proteins, and Proteomes
- Chapter 12: Emerging Roles of Artificial Intelligence in Nanodiamond Sensing
- Chapter 13: Perspective and Outlook on Nanodiamond Research

Research on the unique properties of ND/FND will likely lead scientists to find new uses for these materials, especially as particles in a multimodal imaging/therapy platform. This book touches on four major areas of ND/FND: surface modification, bioimaging, drug delivery, and quantum sensing. We hope this book can help readers choose an appropriate approach in successive ND research.

Surface Modification

As a carbon-based nanomaterial family member, NDs appear to be the most biocompatible compared with other nanoscale carbon-based materials. For this reason, biologists have begun researching their potential applications in nanomedicine (7). However, NDs or any materials practically used in biology face the following challenges: difficulties in directly modifying the NDs surface, low colloidal stability due to its aggregation in physiological media, and nonspecific protein adsorption.

In addition, colloidal stability is one of the most important features affecting the application of NDs. Like other nanoparticles (NPs), NDs' colloidal stability is not static and can decline in solution with high ionic strength. This poses a challenge for the usage of NDs in biological systems. All physiological buffers and culture mediums have high salt content. Therefore, prior to developing analytical and biological applications, researchers must first understand the surface features and aspects of different environments. This is because these environmental interactions influence the colloidal stability of NDs.

The book examines three strategies for modifying the surface of ND: functionalization, encapsulation, and bioconjugation. We also examine the toxicity of these nanomaterials in biological systems (including both *in vitro* and *in vivo*). In this book, we focus particularly on the latest developments of surface-modified NDs in analytical chemistry, biological sciences and nanoscale medicine.

Bioimaging

Bioimaging has emerged as a major factor in mainstream life science research. Not only does it grant researchers the ability to visualize biological processes in cells and organisms, but it also allows for the real-time monitoring of molecules and cells in three dimensions – making it an incredibly useful tool in disease identification and treatment. Indeed, its noninvasiveness, sensitivity, and quantifiability have positioned bioimaging as a bastion in the field of medical research. Thanks to the tremendous research, we are now able to make a connection between the complex biochemical processes and the functioning of proteins in living cells. This stems from the decades-long effort to create molecular tags in the form of fluorescent probes. Researchers in the 2000s developed fluorescent NPs, allowing modern-day scientists to select from an array of bioimaging methods such as dye-doped nanospheres (8), quantum dots (QDs) (9), and metallic nanoclusters (10). Thanks to their unique surface properties, NDs can flexibly form bonds with many kinds of biomolecules, including peptides, antibodies, nucleic acids, or small-molecule ligands. Thus, NDs can be effectively used as fluorescent cellular markers. However, there are certain challenges associated with fluorescence bioimaging applications, including the color fading of imaging probes, photodamage to testing samples, cell autofluorescence, and strong light scattering from tissue. QDs, semiconductor NPs typically made of cadmium or selenium, are one of the most promising candidates for bioimaging. Compared with organic fluorophores, QDs have larger absorption coefficients and much higher levels of brightness and photostability. However, their inherent toxicity has hampered QDs' use for biolabeling living cells and organisms (11). Upconversion nanoparticles (UCNPs), another attractive alternative, face similar biocompatibility issues since these inorganic nanocrystallites are doped with a transition metal, lanthanide, or actinide ions (12).

FND is the "brightest" member of the nanocarbon family. It was first shown to be useful in cellular fluorescence imaging in 2005 (13) and in *in vivo* stem cell tracking in 2013 (14). Surface-functionalized FNDs are gaining traction as potential optical nanoprobes for *in vivo* imaging, and diagnostics as technological and biomedical demands grow. FNDs allow researchers to observe, monitor, and quantify molecules and cells at a high spatial resolution – a capability that is needed in order to understand complex biological systems fully. FND-based technology has developed in tandem with the existing modern imaging modalities.

After providing a brief background of the optical properties of FNDs, this book will discuss a broad range of FND-based bioimaging technologies, including multiphoton excitation, super-resolution, cathodoluminescence, and magnetic resonance imaging. Though important, NDs as light scattering biolabels and photoacoustic contrast agents has also been discussed (15, 16).

Drug Delivery

Epidemics evolve in lockstep with human progress. At present, countries around the world are wrestling with the COVID-19 epidemic. Pandemics have serious consequences for human life and health and severe economic and world peace implications. To help us contend with disease outbreaks, identification, testing, therapy approaches, and novel vaccinations are continually being studied and presented. COVID-19 has shifted the demand for interventions directed against a single pathogen to vaccinations and therapies that can boost immune responses to groups of infections. The desire for faster vaccine development and increased worldwide access to these inoculations has also motivated the growth of drug delivery systems. Aside from COVID-19, medication and drug delivery techniques must be further developed to assist patients in overcoming various diseases, such as cancer, chronic illnesses, and other infectious diseases.

Nanotechnology has been mobilized in healthcare, particularly as medication carriers. NPs-based treatments are being researched to see if these materials can overcome the constraints of biological barriers, microenvironmental and cellular, that are heterogeneous across patient populations and diseases. The NPs' delivery system allows the transport of hydrophobic drugs and biologics, as well as direct targeting of disease sites. In addition, NPs, with their unique physicochemical properties such as substantial loading capacities, stable quantum properties, and broad functional surface areas, enable the transport of a vast number and variety of molecules such as drugs, proteins, and genes (17).

Many NPs have been studied and used as drug delivery and monitoring carriers. They include lipid-based vehicles (e.g., liposomes, solid lipid NPs, and micelles), polymer carriers (e.g., hydrogels, polymersome, dendrimer, and nanofiber), metallic and semiconducting NPs (e.g., nanogold and QDs), carbon structures (e.g., fullerene, nanotube, graphene, and ND), and inorganic NPs (e.g., silica and metal oxides) (18, 19). However, NPs cannot work without any flaws. Due to their tiny dimensions, NPs react easily to biomolecules found in cells or in the extracellular environment, leading to unexpected or undesired biological effects. Hence, despite its potential, the use of NPs in clinical settings remains a developing research area that requires further study (20, 21).

Due to their excellent biocompatibility and nontoxicity, carbon-based nanomaterials have received more attention from biologists than metal- or semiconductor-based nanomaterials. NDs, a new class of carbon nanomaterials, have been particularly singled out as they exhibit a number of distinctive properties such as low toxicity, stable fluorescence, facile functionalization, intrinsic biocompatibility, and other fundamental properties of bulk diamonds (22). Compared with other members of the nanocarbon family, such as carbon black, carbon nanotubes, and graphene, NDs distinguish themselves by having the highest biocompatibility and lowest cytotoxicity when applied to living cells (23). Therefore, there is no doubt about the advantages of using NDs in biomedical applications. NDs of varied shapes and sizes have been created using various synthesis techniques. DNDs and HPHT NDs are commonly employed in biological studies (24). Both have excellent biocompatibility since they can remain in cells without impacting their *in vitro* and *in vivo* activities (25). DND, with a size of several nanometers, can quickly enter cell membranes in a noninvasive manner, thus preserving the cells' integrity during the treatment period (26). This makes them ideal candidates for small chemical drug transport. In addition to their small size, NDs are notable for their abnormal reactive surface and capacity to form hydrogels, making them significant active moiety carriers in molecular biotechnologies (27).

Surface modification or functionalization of NDs with biological molecules (e.g., small molecules, peptides and proteins, genes, as well as small interfering RNA [siRNA]) is an emerging research topic in nanomedicine. These biomolecules can be flexibly attached to the surface of NDs to form the ND-X modality. ND-X modalities have been tested in both *in vitro* and *in vivo* models and can be used to target specific cells (28–30). They can also detect, monitor, and even change biochemical processes within cells. Therefore, NDs can be used as both a monitoring tool (e.g., detecting precancerous alterations) and a therapy (e.g., carrying drugs). Because of their possession of many features required for clinical applications (e.g., stability and compatibility in biological conditions, as well as scalability in manufacturing), NDs have emerged as a viable choice in many biomedical applications (5, 7, 31, 32).

The delivery of drugs is a multidisciplinary area that includes chemistry, biology, and medicine. Due to their intricacy, using NDs necessitates a vital understanding of advanced drug and gene delivery methods. Despite considerable accomplishments in recent years, ND drug delivery monitoring is still in its infancy. In this book, we will discuss the essential elements of NDs, as these materials will certainly play a crucial role in developing enhanced targeted drug delivery and monitoring systems going forward.

Quantum Sensing

Exploiting flaws in ND fragments has allowed researchers to construct precise quantum sensors in biocompatible materials (33). Quantum sensing has begun applying FNDs that contain negatively-charged nitrogen-vacancy (NV^-) centers, allowing the display of quantum behavior up to room temperature.

The quantum properties of FND stem from the NV^- center's unique triplet ground state, the spin of which is susceptible to both optical detection and manipulation (34). Locked within a diamond matrix, the NV^- center has a spin coherence time of up to 10 ms, even in ambient environments (35). In light of the ND's chemical inertness and biocompatibility, FND is the perfect bridge between quantum physics and biology, unlocking new opportunities in the field of bioimaging (16). By combining the optically detected magnetic resonance (ODMR) technique and multiphoton microscopy, researchers have arrived at a possible method of creating high-resolution optical images under the skin (36). Other color centers in NDs, besides the NV center, also gain great attraction in bioimaging. The silicon-vacancy center shows particular potential, as it has a narrow near-infrared emission at 739 nm with full width at half maximum of 4 nm (37). Its magnetic features are currently the focus of research (38).

As a result of the ODMR spectrum's sensitivity to changes in magnetic field, electric field, temperature, and pressure, ND's NV^- center is likely to be adopted in quantum sensing. The NV^- center has already been operated with microwave stimulation and developed as a nanoprobe for sensing local temperature and magnetic fields. However, to broaden the range of FND applications in biomedicine, it must be developed into an all-optical nanoprobe that does not require microwave excitation (39).

This book aims to elucidate the biophysical groundwork for nanoscale – or quantum – sensing. FNDs both reveal and enhance our understanding of a wide array of topics – chemical reactions within single cells, signaling in neural networks, and the origin of magnetism – thus forming the core of this new field. Here, we begin by highlighting various "electrifying" quantum sensor applications, including magnetic field sensing, orientation tracking, nanothermometry, nanoscopic spin probes, and nuclear spin sensing.

Closing: Last but Never the Least

During the last two decades, FNDs have gained significant interest from biologists, chemists, physicists, and material scientists due to their unique structural, electrical, optical, chemical, and mechanical properties. Recently, there has been increased demand for creative industrial applications using FND. This book discusses the latest research on ND/FND synthesis and their wider applications.

This book is, undeniably, the most comprehensive reference book on FNDs. Containing plenty of illustrations, case studies, practical examples, and historical perspectives, this text will help strengthen your understanding of a variety of core concepts. Each chapter addresses, in-depth, a single integral topic. In addition to covering FND's fundamental properties, synthesis, mechanisms, and functionalization, we also introduce a broad range of methodological concepts and techniques. Though its focus is on what has already proved workable, this book also presents a wealth of laboratory achievements and offers a profound, precise, and accessible analysis of the impacts and current applications of FNDs, up to 2022. As a result, this book is relevant to undergraduate and graduate students alike, as well as to academic researchers, industrial personnel, and instructors in biology, chemistry, physics, materials science, and other relevant fields.

The discovery of NDs/FNDs has sparked a flood of international scientific interest, and this novel material in the field of materials science and biochemical physics has revealed a cornucopia of new physics and prospective applications. Finally, we hope this book will give our readers ideas and inspiration and serve as a tremendous resource in this emerging field.

Wesley Wei-Wen Hsiao
Trong-Nghia Le
Taipei, Taiwan, August 2023

References

1 Chang H-C, Hsiao WW-W, Su M-C. *Fluorescent Nanodiamonds*. Hoboken, NJ, USA: John Wiley & Sons; 2018.

2 Hsiao WW-W, Le T-N, Chang H-C. Applications of fluorescent nanodiamond in biology. In: Meyers, R.A. (Ed.), *Encyclopedia of Analytical Chemistry*. Hoboken, NJ, USA: John Wiley & Sons; 2022. pp. 1–43, https://doi.org/10.1002/9780470027318.a9776.

3 Ho DE. *Nanodiamonds: Applications in Biology and Nanoscale Medicine*. Morwell: Springer; 2009.

4 Schrand AM, Hens SAC, Shenderova OA. Nanodiamond particles: Properties and perspectives for bioapplications. *Crit Rev Solid State Mater Sci*. 2009;34(1–2):18–74.

5 Vaijayanthimala V, Chang HC. Functionalized fluorescent nanodiamonds for biomedical applications. *Nanomedicine-Uk*. 2009;4(1):47–55.

6 Feynman RP. There's plenty of room at the bottom. *Eng Sci*. 1960;23(5):22–36.

7 Xing Y, Dai LM. Nanodiamonds for nanomedicine. *Nanomedicine-Uk*. 2009;4(2):207–18.

8 Yan JL, Estevez MC, Smith JE, Wang KM., He XX, Wang L, *et al*. Dye-doped nanoparticles for bioanalysis. *Nano Today*. 2007;2(3):44–50.

9 Medintz IL, Uyeda HT, Goldman ER, Mattoussi H. Quantum dot bioconjugates for imaging, labelling and sensing. *Nat Mater*. 2005;4(6):435–46.

10 Zheng J, Nicovich PR, Dickson RM. Highly fluorescent noble-metal quantum dots. *Annu Rev Phys Chem*. 2007;58:409–31.

11 Derfus AM, Chan WCW, Bhatia SN. Probing the cytotoxicity of semiconductor quantum dots. *Nano Lett*. 2004;4(1):11–18.

12 Ang LY, Lim ME, Ong LC, Zhang Y. Applications of upconversion nanoparticles in imaging, detection and therapy. *Nanomedicine-Uk*. 2011;6(7):1273–88.

13 Yu S-J, Kang M-W, Chang H-C, Chen K-M, Yu Y-C. Bright fluorescent nanodiamonds: No photobleaching and low cytotoxicity. *J Am Chem Soc*. 2005;127(50):17604–5.

14 Wu T-J, Tzeng Y-K, Chang W-W, Cheng C-A, Kuo Y, Chien C-H, *et al*. Tracking the engraftment and regenerative capabilities of transplanted lung stem cells using fluorescent nanodiamonds. *Nat Nanotechnol*. 2013;8(9):682–9.

15 Hui YY, Cheng CL, Chang HC. Nanodiamonds for optical bioimaging. *J Phys D: Appl Phys*. 2010;43(37).

16 Hui YY, Chang HC. Recent developments and applications of nanodiamonds as versatile bioimaging agents. *J Chin Chem Soc*. 2014;61(1):67–76.

17 Niemeyer CM. Nanoparticles, proteins, and nucleic acids: Biotechnology meets materials science. *Angew Chem Int Ed*. 2001;40(22):4128–58.

18 Chow EKH, Ho D. Cancer nanomedicine: From drug delivery to imaging. *Sci Transl Med*. 2013;5(216):216rv4.

19 Koo OM, Rubinstein I, Onyuksel H. Role of nanotechnology in targeted drug delivery and imaging: A concise review. *Nanomedicine-Uk*. 2005;1(3):193–212.

20 Bae YH, Park K. Targeted drug delivery to tumors: Myths, reality and possibility. *J Control Release*. 2011;153(3):198–205.

21 Desai N. Challenges in development of nanoparticle-based therapeutics. *AAPS J*. 2012;14(2):282–95.

22 Qin J-X, Yang X-G, Lv C-F, Li Y-Z, Liu K-K, Zang J-H, *et al*. Nanodiamonds: Synthesis, properties, and applications in nanomedicine. *Mater Des*. 2021;210:110091.

23 Vaijayanthimala V, Cheng P-Y, Yeh S-H, Liu K-K, Hsiao C-H, Chao J-I, *et al*. The long-term stability and biocompatibility of fluorescent nanodiamond as an in vivo contrast agent. *Biomaterials*. 2012;33(31):7794–802.

24 Mochalin VN, Shenderova O, Ho D, Gogotsi Y. The properties and applications of nanodiamonds. *Nat Nanotechnol*. 2012;7(1):11–23.

25 Zhu Y, Li J, Li WX, Zhang Y, Yang XF, Chen N, *et al*. The biocompatibility of nanodiamonds and their application in drug delivery systems. *Theranostics*. 2012;2(3):302–12.

26 Man HB, Ho D. Diamond as a nanomedical agent for versatile applications in drug delivery, imaging, and sensing. *Phys Status Solidi A*. 2012;209(9):1609–18.

27 Man HB, Ho D. Nanodiamonds as platforms for biology and medicine. *J Lab Autom*. 2013;18(1):12–18.

28 Kaur R, Badea I. Nanodiamonds as novel nanomaterials for biomedical applications: Drug delivery and imaging systems. *Int J Nanomedicine*. 2013;8:203–20.

29 Moosa B, Fhayli K, Li S, Julfakyan K, Ezzeddine A, Khashab NM. Applications of nanodiamonds in drug delivery and catalysis. *J Nanosci Nanotechno*. 2014;14(1):332–43.

30 Perevedentseva E, Lin Y-C, Cheng C-L. A review of recent advances in nanodiamond-mediated drug delivery in cancer. *Expert Opin Drug Del*. 2021;18(3):369–82.

31 Vaijayanthimala V, Lee DK, Kim SV, Yen A, Tsai N, Ho D, *et al*. Nanodiamond-mediated drug delivery and imaging: Challenges and opportunities. *Expert Opin Drug Del*. 2014;12(5):735–49.

32 Lam R, Ho D. Nanodiamonds as vehicles for systemic and localized drug delivery. *Expert Opin Drug Del.* 2009;6(9):883–95.

33 Schirhagl R, Chang K, Loretz M, Degen CL. Nitrogen-vacancy centers in diamond: Nanoscale sensors for physics and biology. *Annu Rev Phys Chem.* 2014;65:83–105.

34 Bradac C, Gaebel T, Pakes CI, Say JM, Zvyagin AV, Rabeau JR. Effect of the nanodiamond host on a nitrogen-vacancy color-centre emission state. *Small.* 2013;9(1):132–9.

35 Rondin L, Tetienne JP, Hingant T, Roch JF, Maletinsky P, Jacques V. Magnetometry with nitrogen-vacancy defects in diamond. *Rep Prog Phys.* 2014;77(5):056503.

36 Hegyi A, Yablonovitch E. Nanodiamond molecular imaging with enhanced contrast and expanded field of view. *J Biomed Opt.* 2014;19(1):011015.

37 Aharonovich I. Diamond nanocrystals for photonics and sensing. *Jpn J Appl Phys.* 2014;53(5):05FA01.

38 Hepp C, Muller T, Waselowski V, Becker JN, Pingault B, Sternschulte H, *et al.* Electronic structure of the silicon vacancy color center in diamond. *Phys Rev Lett.* 2014;112(3):036405.

39 Plakhotnik T, Doherty MW, Cole JH, Chapman R, Manson NB. All-optical thermometry and thermal properties of the optically detected spin resonances of the NV^- center in nanodiamond. *Nano Lett.* 2014;14(9):4989–96.

1

Nanodiamonds

*Tan-Thanh Huynh, Marvin Chen, Thi-Hong-Hanh Le, Xuan Mai Lam, Kartika Wardhani, and Wesley Wei-Wen Hsiao**

* Corresponding author

Diamonds, as a carbon allotrope (sp^3 hybridization), possess exceptional physical properties, such as (i) being the hardest substance known, (ii) having the highest thermal conductivity of any material, and (iii) exhibiting an extraordinarily large refractive index among dielectric materials. Some of these properties are retained, even on the nanoscale, making nanodiamonds highly versatile. Consequently, diamonds have found a new role in analytical chemistry and biological sciences. Nanodiamonds (NDs) on the nanoscale are now being used in various diagnostic and therapeutic applications. Just as diamonds are famously known as a "girl's best friend," NDs have rapidly become one of the most highly promising materials for scientists. Before delving into NDs' research capabilities, this chapter summarizes the fundamental features of NDs, including structure, classification, preparation, and physical and chemical characteristics.

1.1 Structure and Classification of Nanodiamonds

NDs, also known as ultradisperse diamonds, are allotropes of nanocarbon materials, such as fullerenes, tubes, onions, rods, platelets, and graphene, with sizes in the range of several nanometers (1, 2). In the 1960s, diamond nanoparticles, or NDs, were initially manufactured in the former Soviet Union using detonation (3). Although this discovery was kept secret for nearly 20 years as a result of the

Nanodiamonds in Analytical and Biological Sciences: Principles and Applications, First Edition.
Edited by Wesley Wei-Wen Hsiao and Trong-Nghia Le.
© 2023 John Wiley & Sons Ltd. Published 2023 by John Wiley & Sons Ltd.

Cold War, NDs created by explosive destruction have found widespread applicability in modern science and technology (4). In addition to detonation, NDs can also be synthesized by crushing or ball milling diamonds that are in the size of micrometers or millimeters, created either through high-pressure high-temperature (HPHT) or chemical vapor deposition (CVD) (5). While the lattice structure of HPHT and CVD-NDs is similar to that of NDs created via detonation (DNDs), their sizes and surface characteristics are significantly different. DNDs have a highly uniform size distribution of 5 nm. During synthesis, covalent bonding between disordered (sp^2) carbon atoms on their surfaces results in densely bonded agglomerates. Biologists have been using these DND clusters as drug delivery vehicles. In contrast, HPHT-NDs are monocrystalline crystallites with exceptional optical transparency. Despite their wide size range, their lower surface sp^2 carbon content results in less aggregation, thus making them more appropriate for bioimaging applications. CVD-NDs can be deposited on substrates to form thin films. Industries use these materials to form the surface coating of cutting and dressing tools.

NDs can be sorted by size into nanocrystalline (<100 nm) or ultrananocrystalline (<10 nm) (6). These substances have existed for millions of years and can be found in the earth's sediment layers, in extraterrestrial materials, and in natural petroleum (7). ND particles are characteristically polyhedral and consist of an sp^3-hybridized diamond inner core covered with an sp^2-hybridized graphite outer shell or amorphous carbon that has dangling bonds with terminated functional groups on the surface. Thanks to unstable electrons around the curved surface, both graphitic (sp^2) and diamonded (sp^3) bonds are interchangeable. As a result, one-of-a-kind flexible templates are created. ND structure is reportedly built of two interpenetrating face-centered cubic lattices that are tightly packed (Figure 1.1). Along with the elemental cube space diagonal, one lattice is shifted by one-quarter of its length relative to the other (8).

Diamonds are classified into different types, as shown in Figure 1.2. Type I and type II of natural diamonds are categorized by the presence or lack of nitrogen impurities in the crystal lattice structure. Type I contains 0.001–0.3% nitrogen, while type II contains less than 1 ppm nitrogen. The distribution of nitrogen impurities further separates type I diamonds into type Ia and type Ib. Type Ia diamonds have an aggregated form of nitrogen, whereas type Ib diamonds have atomically scattered, single substitutional nitrogen impurities (C-centers). Usually, natural diamonds contain more than 1000 ppm of nitrogen in aggregate form and are classified as type Ia. Synthetic diamonds, made through the HPHT technique, with about 100 ppm of atomically isolated nitrogen, are grouped as type Ib. Type Ia is classified as type IaA, which has two aggregated nitrogen atoms (A-center); type IaB, which has platelets and four aggregated nitrogen atoms around the closed lattice vacancy (B-center); and type IaA/B, possessing

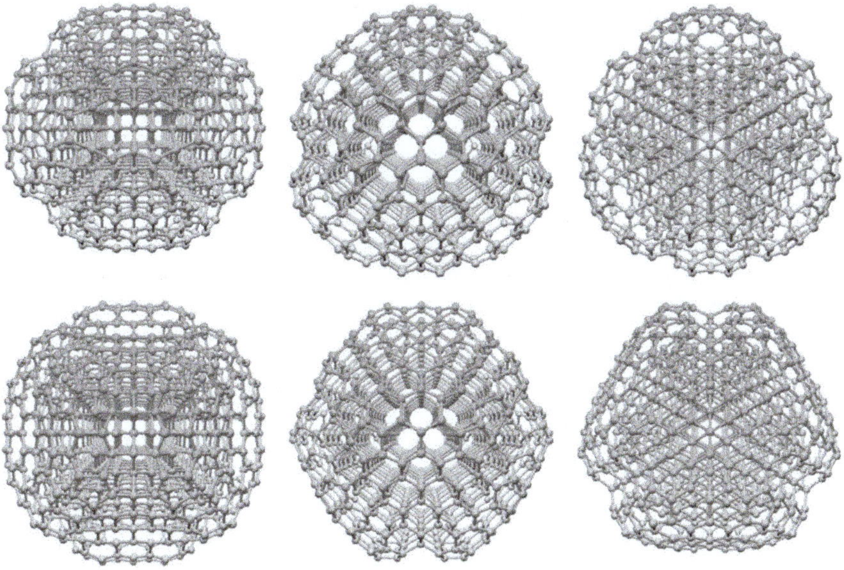

Figure 1.1 Structures of nanodiamond from various crystallographic axes. (9) / Elsevier. Lai, Lin, and Amanda S. Barnard. Surface phase diagram and thermodynamic stability of functionalisation of nanodiamonds. Journal of Materials Chemistry 2012;22(33): 16774–16780.

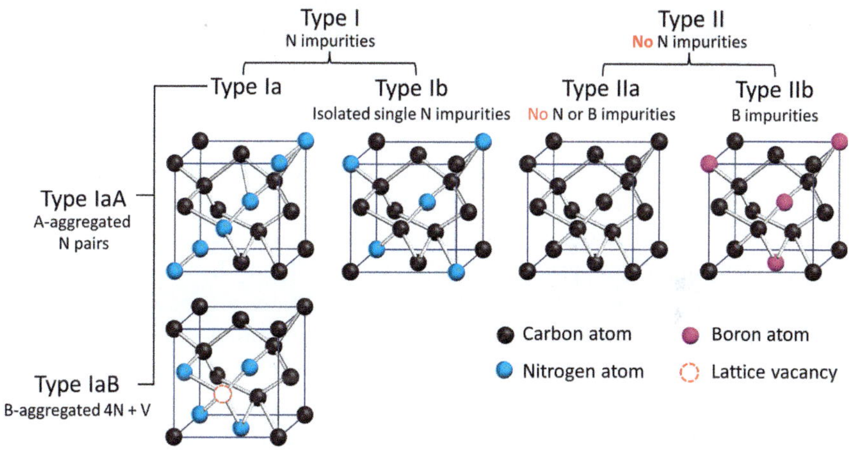

Figure 1.2 Classification of diamonds.

the feature of both types IaA and IaB (10). On the other hand, type II diamonds continue to be categorized into type IIa, in which no impurities are present, and type IIb, which is characterized by the presence of boron atoms in the lattice structure. Type IIa is considered the purest diamond, with only 1–2% abundance. Characteristics of different types of diamonds are described in Table 1.1.

Table 1.1 Characteristics of different types of diamond.

Type	Feature	Color	Natural Abundance	Note
Ia	Nitrogen atom concentration up to 0.3% clustered together within the carbon lattice	Colorless, yellow	98%	Most natural diamonds
Ib	Nitrogen atom concentration up to 500 ppm, spread evenly throughout the carbon lattice	Yellow, orange, brown	0.1%	Almost all synthetic diamonds
IIa	No or few nitrogen atoms	Colorless, yellow, brown, pink, purple	1–2%	The "purest" diamonds
IIb	No or little nitrogen but boron contained in the p-type semiconductor	Blue, gray	0.1%	

1.2 Preparation of Nanodiamonds

Diamonds have a metastable structure at standard temperature and pressure and are much less thermodynamically stable than its common allotrope, graphite. Therefore, diamonds are generally formed in nature in the lithospheric mantle at more than 140 km depth. In industrial production, diamonds can be synthesized via the HPHT growth technique, detonation synthesis, or the CVD method. HPHT NDs (Figure 1.3a) are gained through a "top-down" process by crushing an HPHT diamond and subjecting it to a levigation procedure under high pressure and temperature in a hydraulic press using metal catalysts. HPHT NDs consist of substitutional nitrogen impurities, which lead to highly fluorescent nitrogen-vacancy (NV) centers upon appropriate treatment. As a result, they have been found to be particularly useful biomarkers. DNDs (Figure 1.3b) are monodisperse particles formed by the controlled detonation of carbon-containing suitable precursors such as combined trinitrotoluene (TNT) and hexogen (RDX) in a closed chamber in the absence of oxygen. CVD NDs can be produced on distinct substrates by forming nanocrystalline and ultracrystalline films. CVD (Figure 1.3c) grows diamonds by employing low-pressure carbon-containing gasses. The melting temperature (at the process pressure) of the substrate must be greater than that of the necessary deposition temperature for diamond growth (usually higher than 700 °C) to facilitate the formation of the diamond rather than amorphous carbon. CVD NDs are particularly suitable for manufacturing biocompatible coatings with excellent mechanical and wear-resistant properties for biosensor applications after the appropriate surface modification with specific biochemicals.

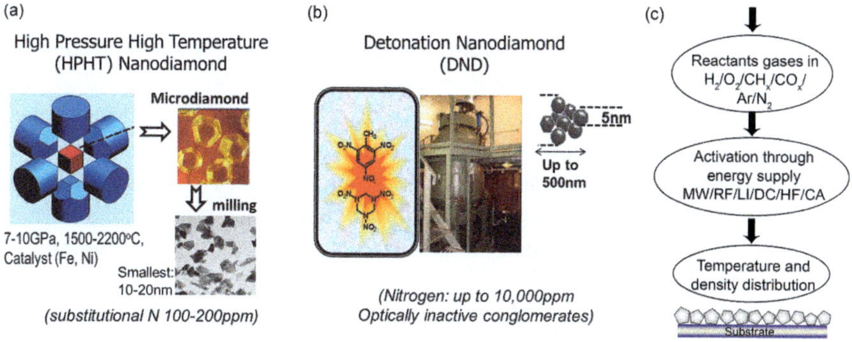

Figure 1.3 Schematic synthesized diagram of (a) HPHT ND, (b) DND, and (c) CVD processes. *Source:* Reproduced from ref. (9) / AIP Publishing LLC.

Historically, the first reported NDs to be used were DNDs (11). In comparison, CVD and HPHT synthesis under frequent conditions of 1400 °C and 5–6 GPa are the predominant industrial methods for current synthetic ND production. These approaches are often used to form micron-sized or bigger diamonds that can be processed to yield NDs. NDs, on the other hand, can also be produced by the detonation of explosives in sealed containers, which produces DNDs with a diameter ranging from 4 to 7 nm (12, 13). Diamondoids, such as adamantane, which are naturally found in crude oil, are the tiniest conceivable diamonds. Many other approaches can be used to obtain bigger NDs with a diameter of up to 100 nm.

1.2.1 High-Pressure, High-Temperature

The synthesis of "man-made diamonds" using the HPHT transformation of graphite was first described in 1955 by researchers from General Electric (9). In this method, a hydraulic press is used to create a pressure of 5–11 GPa. While keeping a temperature in the range of 1200–2200 °C, a metal catalyst, typically group VIII-transition metals, i.e., Fe, Co, Ni, and their alloys, is added to accelerate the process of converting graphite powder into diamond (14, 15). Recent studies have expanded the catalyst class to include Mg-based or 15 rare-earth metals (16, 17). The diamond crystals obtained by the HPHT technique possess blocky shapes with the primary particle size ranging from 10 nm to about 1 cm. One limitation of the HPHT method is that most diamonds produced by this technique are type Ib. Their single substitutional nitrogen atom concentration in the crystal lattice (100–200 ppm), as natural impurities, results in a color change to yellowish or yellow–orange tones leading to the limiting the applications (18, 19). These "man-made diamonds" can be differentiated from natural type Ia diamonds by determining the concentration of nitrogen impurities (Table 1.1). Irradiation of type Ib HPHT diamonds with high-energy electrons, protons, and helium ions, followed by thermal annealing, results in the formation of negatively charged NV centers with red emissions, leading to the production of fluorescent NDs (FNDs) (15). HPHT diamonds are monocrystalline and are of high enough quality to be used for various industrial operations, such as the cutting and machining of mechanical components, as well as the polishing and grinding of optics (larger HPHT diamonds can also be used as gemstones). Type Ib HPHT diamonds contain only diamond carbon structures (e.g., diamond-like carbon [DLC], graphite, and amorphous carbon).

1.2.2 Detonation Technique

In 1963, a group of Soviet Union scientists discovered single crystals of diamond particles in soot-formed and so-called detonation diamonds (DNDs) by detonating an oxygen-deficient TNT/RDX mixture composition in an inert media (3, 1). From the 1980s onward, DNDs have attracted the interest of a growing number of researchers (20). DNDs have the smallest particle sizes among particulate synthetic diamonds, with an average primary size of roughly 5 nm (21). As a result of shock-wave propagation, the carbon is squeezed into crystalline grains of diamond, which is produced by an intense explosion within only a fraction of a microsecond in an oxygen-deficient atmosphere. However, until recently, the applications of DNDs remained relatively unexplored due to USSR security restrictions and a lack of contemporary interest in nanotechnology. Over the past decade, there has been a noticeable interest in the use of DND particles for various commercial applications, such as quantum computing (22), catalysis (23), lubrication (24), microelectronics (25), electrochemistry (26), and biomedicine (27, 28). Two other examples of DND applications are the formation of hard coating composites (29, 30) and the polishing and seeding of substrates for CVD diamond growth (31, 32).

DND particles can have a size of up to 200 nm due to harsh explosion states in the reactor that inevitably induce significant particle aggregation. In addition to electrostatic interaction, particles are linked by covalent bonds between surface functional groups and soot structures around each primary particle (Figure 1.4). To generate a graphite-free diamond, rigorous purification of the soot is necessary. Figure 1.4b depicts a structural model of the unusually strong DND agglomeration. The primary diamond particles are surrounded by graphitic and soot-like material. To refine the primary particles of DND with a size of 4–5 nm and minimal amounts of sp^2 carbon impurities, intricate postprocessing deagglomeration and purification can be applied to break the disordered sp-sp^2 bonded carbon atoms on the surfaces. Methods including wet ball milling with ZrO_2 microbeads (33) and annealing in different atmospheres are then introduced (34). As a result, the primary particles of DND are readily available in kilogram amounts in the forms of colloidal sols, gels, and assemblies, which are suitable for a variety of applications.

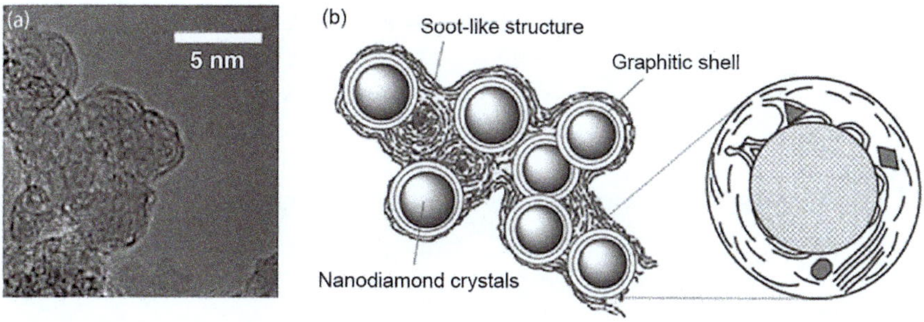

Figure 1.4 (a) High-resolution TEM image of typical clusters of DNDs. The particles are surrounded by graphitic and soot-like material. (b) Structure model of the diamond agglomerates. *Source:* Reproduced from ref. (35) / John Wiley & Sons.

1.2.3 Chemical Vapor Deposition

Diamond growth through CVD emerged in the 1950s. CVD is one of the most versatile thin-film deposition techniques employed for nanocrystalline diamond film synthesis. However, in early CVD experimental setups, the growth rate was low due to the deposition of graphite, which led to mixed sp^3/sp^2 phases (36). This would change with the breakthrough discovery of Angus *et al.*, who found that the presence of a large proportion of atomic hydrogen in the gas mixtures leads to the preferential etching of graphite, resulting in increased diamond growth (37, 38). As a result, CVD diamond growth has become an active and extensive area of research that has, since the early 1980s, led to the industrial manufacture and the utilization of diamond materials in various applications.

The CVD process begins with the carbon-containing precursors' dissociation above a solid substrate surface, followed by the reaction of the decomposition products with the production of a diamond layer, which causes diamond deposition onto that surface. Methane is a common carbon source for diamond formation, although other carbon-containing gasses can also be used. CVD diamond growth requires a means of activation for the precursor molecules (methane or other hydrocarbons) diluted in a surplus of hydrogen, which is typically carried out using thermal methods (e.g., a hot filament), electric discharge (e.g., direct current plasma jet, radio frequency, or microwave plasma-enhanced CVD), or a combustion flame (e.g., oxyacetylene torch) (39). The typical methane concentration in the gas mixture ranges from 0.5% to 2%. During diamond growth, the decomposed methane or other hydrocarbon molecules are exclusively responsible for carbon supply, while atomic hydrogen is the most crucial component of the gas phase mixture, which terminates the "dangling" carbon bonds on the surface of the growing diamond or diamond nuclei to prevent the formation of graphite (sp^2 bonds) and promote diamond growth (sp^3 bonds). Scanning electron

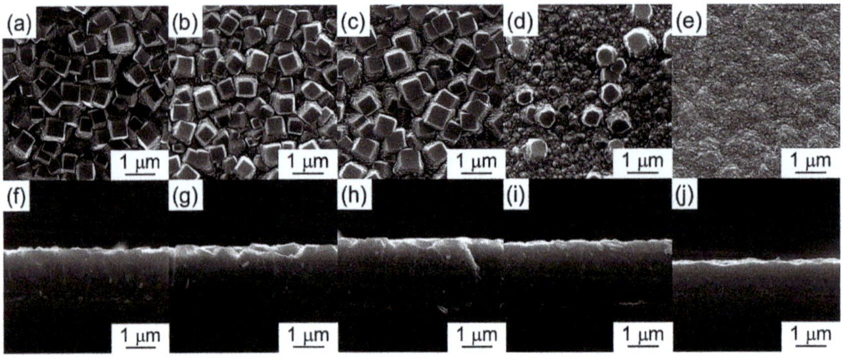

Figure 1.5 SEM images CVD diamond film deposited at various CH_4/H_2 ratios (a and f) 0.5, (b and g) 1.0, (c and h) 1.5, (d and i) 2.0, and (e and j) 2.5 vol% at 1073 K. *Source:* Reproduced from ref. (40) / Royal Society of Chemistry.

microscopic (SEM) images CVD diamond film deposited at various CH_4/H_2 ratios are shown in Figure 1.5.

To generate diamond nuclei, the CVD method requires a substrate preparation on the surface of substrates that can be enlarged prior to deposition. Growth pretreatment processes are necessary to introduce nucleation sites on the surface of the substrate. Mechanical abrading, ultrasonic seeding, and ion bombardment are generally applied in these processes (36). Various deposition conditions yield films of various parameters, wherein CVD technology can produce thin (100 cm in diameter) polycrystalline diamond films. The CVD process is less cost-effective than HPHT. However, the CVD process is attractive in that it can manufacture diamond films of various grades. For instance, the addition of other gasses (e.g., trimethyl boron or diborane) to the gas mixture during growth (the doping of diamond with boron, phosphorus, or other dopants) results in the creation of certain defects, which can be used for the precise tuning of the diamond's properties (e.g., electrical conductivity) (41–44). CVD films can also be milled to gain ND particles of different sizes and shapes. The CVD method makes synthesizing diamond films with differing isotopic forms of ^{12}C and ^{13}C possible (45).

1.2.4 Alternative Methods of Producing Nanodiamonds

Besides the methods mentioned above, many other techniques have been reported and utilized to synthesize NDs, such as laser ablation (46), chlorination of carbides (47), autoclave synthesis from supercritical acids (48), laser-induced high temperature and high pressure from carbon onions (49), ion irradiation of graphite (50), femtosecond laser irradiation of ethanol (51), and ultrasound cavitation (52), as depicted in Figure 1.6.

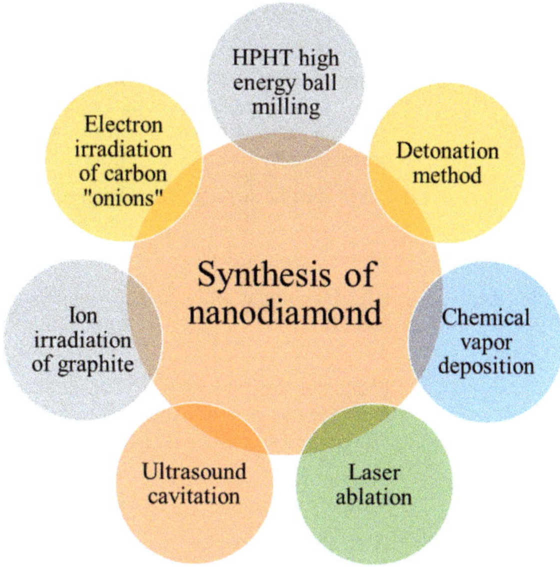

Figure 1.6 Methods for preparation of NDs, including HPHT, detonation method, CVD, laser ablation, electron irradiation of carbon onions, and ion irradiation of graphite, cold synthesis.

1.3 Physical and Chemical Properties of Nanodiamonds

In brief, since the synthesized NDs have the same core structure, they exhibit comparable optical and mechanical properties as bulk diamonds. NDs are typically distinguished by a high Young's modulus and stiffness, superior thermal conductivity, and a high refractive index along with stability under harsh conditions. Furthermore, NDs possess versatile surface chemistry that can be loaded with any functional groups. This means that the macroscopic and microscopic characteristics of the as-produced NDs, including their hydrophilicity, colloidal stability, and reactivity, can be modified, allowing them to be applied in various fields. A detailed description of the exceptional properties of NDs in detail can be found in Table 1.2.

a) ***Hardness***
Table 1.3 compares some of the physical properties of ND to those of titanium and stainless steel. An ND is nearly 50 times harder than titanium or stainless steel. The high toughness of NDs makes them highly useful in certain biomedical arenas. For instance, NDs can be used for implants as well as for surgical cutting implements (53, 54).
b) ***Size***
The capacity of NDs to control the constancy of their tiny nano size and the shape of agglutinates is highly beneficial. The effect of different ND sizes on biocompatibility and other applications is discussed in the following chapters.

Table 1.2 Physicochemical properties of nanodiamond for biomedical applications.

Property	Characteristics	Application
Structural	Variable sizes with low polydispersity	Multiple uses, advancing cellular internalization
	Large specific surface area	High adsorption capacity and affinity proteins and enzymes absorption; purification
	High specific gravity	Dense structure for solid phase support
Biological	High biocompatibility, low toxicity	Crucial features for biomedical applications, such as investigations into cells, tissues, and other biological systems
	Facilitates the binding of bioactive compounds while retaining their functionality	Targeted therapeutics; incorporation of molecules, labels, hormones, inhibitors, antigens, drugs
Chemical	Chemically inert/corrosion-resistant, pH stable	High stability; long-term applications, including implants, coatings, scaffolds
	High chemical purity	Biocompatible interface
	Surface oxygen-containing groups	Water-dispersible for further modification
	Ease of surface functionalization	Conjugation of polymer, metal composite, active agents (drugs and biomolecules)
	A high number of surface unpaired electrons	X-rays, protective coatings and surfaces, detection devices
	Radiation/Ozone resistance	Free radical scavenger/multiple radical donors; electrochemistry
Optical (FND)	Photoluminescence: nonphotobleaching, nonblinking, originates from NV defects	Fluorescent probe and imaging tool for biolabeling
Mechanical	High tenacity and toughness	Composite additive; possible cell lysis; ballistic delivery to tissues and cells; autoclaving
	Fine abrasive	Homogenization of composites/cosmetics, skin polishing
Electrochemical	Electrochemical metal plating	Improves durability, the life of medical instruments/implants
Thermal	Capable of withstanding extremely high/low temperatures	Stable during sterilization, manufacturing, and storage
Technical	Mass production (i.e., detonation synthesis)	Commercial availability
	Exists naturally in meteorites and can be synthesized in a lab via diverse techniques	Availability of various quality/purity samples

Source: (55) Hsiao et al., 2022 / John Wiley & Sons.

Table 1.3 Comparison of properties of CVD ND, titanium, and stainless steel (316 grade).

Properties	CVD ND	Titanium	Stainless steel
Hardness (kg/mm^2)	10,000	230	210
Young's modulus (GPa)	1000	120.2	215.3
Bulk modulus (GPa)	442	108.6	166
Thermal conductivity (Watts/cm.°C)	20	0.21	0.16 3.8 (Cu)
Thermal expansion ($\times 10^{-8}$ K^{-1})	1.1	8.8	17.2
Coefficient of friction	0.05	0.1 (*Graphite*)	
Refractive index (in the near IR)	2.41–2.44	3.42–3.5 (Si)	

c) ***Chemical Inertness***

ND particles are one of the hardest materials and demonstrate chemical stability, even in corrosive and other harsh environments. This characteristic makes ND a particularly valuable tool in biology since the biological environment generates corrosive effects (56, 57).

d) ***Chemical Surface Modification***

There are significant variations in the characteristics and reactivities of HPHT NDs, DNDs, and CVD NDs, owing to the different dimensions, shapes, and chemical compositions of the created ND particles. Modification of the ND surface has a pivotal influence on the characteristics and behavior of NDs, particularly when these particles are employed as biosensors, biochips, or as a substrate to attached biological molecules. Before each use, the ND surface is subjected to a 13.5-MHz inductively coupled hydrogen plasma (15 torrs) at 800 °C to yield a hydrogen-terminated surface (58). The chemical approach to linking DNA to a diamonds' surface takes place by means of hydrogen-terminated nanocrystalline diamond films (59). NDs with a diameter size range of 5–100 nm can be successfully carboxylated (60). This results in great physical absorption capacities (e.g., hydrophobic and hydrophilic interaction), which allows them to immobilize biomolecules (61). More detailed information about NDs and surface modification can be found in Chapter 4.

e) ***Biocompatibility***

Biocompatibility plays an important role in how NDs can be applied to a biological system. Research in cell and animal models has demonstrated the excellent biocompatibility of NDs (62). Prepared FND powder with a size of 100 nm has been shown to have good biocompatibility and low cytotoxicity in cultured kidney cells (63). In addition, ND particles with a size of 2–10 nm exhibit nontoxicity to many kinds of cells of different origins through 3-[4,5-dimethylthiazol-2-yl]-2,5-diphenyltetrazolium bromide (MTT) assay and luminescent ATP production assays (64). Cell morphology is not disturbed after incubating cells with NDs, even though NDs surround a cell's borders and attach to neurite extensions. Further details about ND's biocompatibility as well as toxicity assessment are discussed in Chapter 5.

f) ***Optical and Photoluminescence Properties***
The feature of NDs that allows them to be used in optical biolabeling and bioimaging is their outstanding photoluminescence (16). Different kinds of impurity sites inside the diamond's core, as well as defects in the diamond lattice or sp^2 clusters on the ND surface, give NDs the ability to emit light of varying frequencies through light excitation (61). The negatively charged NV center, NV^-, is a point defect comprising a substitutional nitrogen atom close to a carbon atom vacancy with C_{3v} symmetry. It has a zero-phonon line (ZPL) at the wavelength of 638 nm and a broad phonon sideband at 560 nm for the $3A \rightarrow 3E$ transition (65). When stimulated by the green–yellow light, the center emits far-red fluorescence at 690 nm with a near-unity quantum yield. Furthermore, even when subjected to continuous high-power laser stimulation at room temperature, its fluorescence is exceptionally stable, with no evidence of photoblinking or photobleaching (66, 67). Thanks to this abnormal property, the NV^- center has been used as a single-photon source for quantum information applications. Additional applications of NDs related to their optical and photoluminescence properties are described in Chapters 6–10.

g) ***Large Surface Area and Relatively High Adsorption Capacity***
Diamond particles have an extremely large relative surface area. This leads to the spontaneous surface adsorption of water or hydrocarbon molecules in ambient conditions (64). This has important implications for drug delivery, as more drugs can be loaded onto ND particles (68). A clean ND particle surface can be achieved by a suitable handling system (69). Surface properties with high adsorption capacity are superior for proteomic research, as will be discussed in greater detail in Chapter 11.

1.4 Colloidal Properties of Nanodiamonds

The formation of a stable colloidal dispersion is of great importance to the handling and processing of biomedical and biophysical applications. For instance, agglomerates or aggregates of ND particles can be detrimental to systemic drug delivery, fluorescent and imaging probes, as well as other applications. The colloidal behavior of NDs varies considerably according to several aspects of NDs, specifically the synthetic methods, size, origin, as well as the residual chemical groups on the diamonds' surface. If no countermeasures are adopted, HPHT NDs and CVD NDs agglomerate mildly, but DNDs form highly bound agglomerates.

Colloidal stabilities in an aqueous medium and their tendency toward agglomeration are determined by dynamic light scattering (70) and zeta potential analyses, which are strongly correlated with the hydrodynamic particle size change in solution and the surface charge of the particle, respectively. A colloid is reported to be stable if its zeta potential is less than −30 mV or more than +30 mV. ND particles can form colloids with zeta potentials in the range of −50 and +50 mV, depending on their surface termination groups and preceding treatments. Primary DND particles possess a highly positive zeta potential of around 40 mV, possibly because of the dense hydroxylation of their surface. The treatment of DND aggregates with oxidizing substances is generally carried out to partially remove sp^2 carbon atoms and stabilize the colloid by forming various oxidic groups on the DND surface. NDs produced by the HPHT synthesis, followed by a milling process, can exist as stable colloids in an aqueous solution by treating their surfaces using a mixture of oxidative mineral acids. The oxidative acid-treated HPHT ND surfaces are strongly acidic.

ND particles must have good colloidal stability in biological liquids and electrolyte solutions for their broad utility in bioapplications. However, the repulsive Coulomb forces occurring between the charged colloidal particles do not balance with the van der Waals interactions due to the presence of electrolytes. This causes the undesired aggregation of ND colloids to rise, demonstrating the necessity of improving ND stability in buffers and biological liquids (71). Liang and coworkers were able to strengthen electrostatic forces by modifying or coating the particles' surfaces with charged groups (72). However, the particles were substantively charged in this instance, thereby compromising their utility for some applications. Thus, using high-molecular-weight compounds, such as polymers (73) or proteins (71), for the surface functionalization of NDs is a preferable way to bring up steric stabilization (Figure 1.7). These molecules can be conveniently covalently attached (74) or adsorbed (75) to the surface of ND particles.

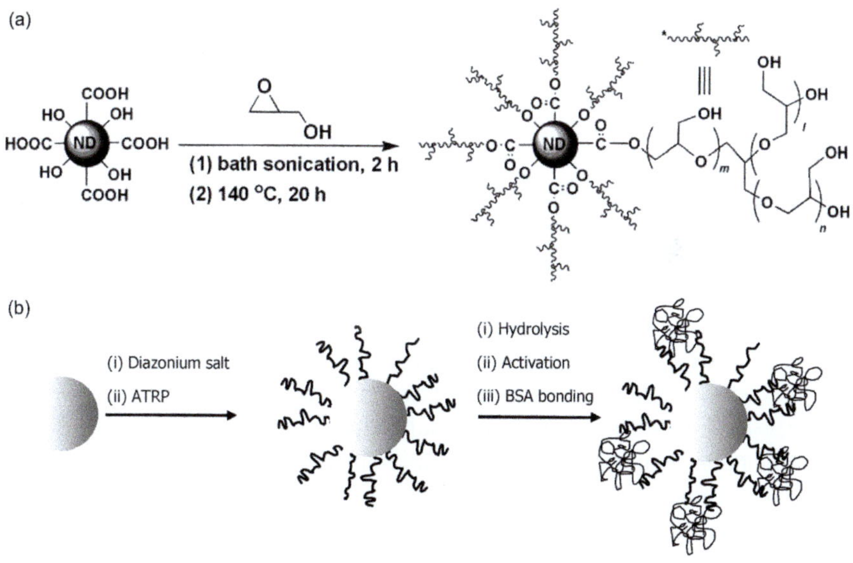

Figure 1.7 Enhanced colloidal stability of NDs by surface functionalization with (a) hyperbranched polyglycerol via ring-opening polymerization and (b) bovine serum albumin-grafted poly(methacrylic acid) via atom transfer radical polymerization. *Source:* Figure (a) is adapted from ref. (73). Figure (b) is adapted from ref. (71).

Acknowledgments

This work was supported by the National Science and Technology Council (NSTC), Taiwan, under Grant No. 112-2221-E-011-031.

References

1 Shenderova O, Zhirnov V, Brenner D. Carbon nanostructures. *Crit Rev Solid State Mater Sci*. 2002;27(3–4):227–356.
2 Geim AK, Novoselov KS. The rise of graphene. *Nat Mater*. 2007;6(3):183–91.
3 Danilenko VV. On the history of the discovery of nanodiamond synthesis. *Phys Solid State*. 2004;46(4):595–9.
4 Mochalin VN, Shenderova O, Ho D, Gogotsi Y. The properties and applications of nanodiamonds. *Nat Nanotechnol*. 2012;7(1):11–23.
5 Schrand AM, Hens SAC, Shenderova OA. Nanodiamond particles: Properties and perspectives for bioapplications. *Crit Rev Solid State Mater Sci*. 2009;34(1–2):18–74.
6 Williams OA. Nanocrystalline diamond. *Diam Relat Mater*. 2011;20(5–6):621–40.
7 Kennett DJ, Kennett JP, West A, West GJ, Bunch TE, Culleton BJ, *et al*. Shock-synthesized hexagonal diamonds in Younger Dryas boundary sediments. *Proc Natl Acad Sci USA*. 2009;106(31):12623–8.
8 Bhosale RR, Osmani RA, Ghodake PP, Harkare BR, Shaikh SM, Chavan SR. Nanodiamonds: A new-fangled drug delivery system. *J Pharm Res*. 2013;3 (12):1395–1403.
9 Bundy F, Hall HT, Strong H, Wentorf R. Man-made diamonds. *Nature*. 1955;176(4471):51–5.
10 Jones R, Goss J. A5. 1 Theory of aggregation of nitrogen in diamond. In: Nazare MH, Neves, AJ (Eds.), *Properties, Growth and Applications of Diamond*. Stevenage, United Kingdom: The Institution of Engineering and Technology; 2001, 26.
11 Shang SY, Tong Y, Zhang SX, Huang FL. Study on phase transformation mechanism of various carbon sources in detonation synthesis of diamond. *Fuller Nanotub Carbon Nanostructures*. 2020;28(11):877–85.
12 Mochalin VN, Shenderova O, Ho D, Gogotsi Y. The properties and applications of nanodiamonds. *Nat Nanotechnol*. 2011;7(1):11–23.
13 Bagge-Hansen M, Bastea S, Hammons J, Nielsen M, Lauderbach L, Hodgin R, *et al*. Detonation synthesis of carbon nano-onions via liquid carbon condensation. *Nat Commun*. 2019;10(1):1–8.
14 Bovenkerk HP, Bundy FP, Hall HT, Strong HM, Wentorf RH. Preparation of diamond. *Nature*. 1959;184(4693):1094–8.

15 Chang Y-R, Lee H-Y, Chen K, Chang C-C, Tsai D-S, Fu C-C, *et al.* Mass production and dynamic imaging of fluorescent nanodiamonds. *Nat Nanotechnol.* 2008;3(5):284–8.

16 Palyanov YN, Kupriyanov IN, Khokhryakov AF, Borzdov YM. High-pressure crystallization and properties of diamond from magnesium-based catalysts. *CrystEngComm.* 2017;19(31):4459–75.

17 Palyanov YN, Borzdov YM, Kupriyanov IN, Khohkhryakov AF, Nechaev DV. Rare-earth metal catalysts for high-pressure synthesis of rare diamonds. *Sci Rep.* 2021;11(1):1–11.

18 Vaijayanthimala V, Chang HC. Functionalized fluorescent nanodiamonds for biomedical applications. *Nanomedicine-Uk.* 2009;4(1):47–55.

19 Tisler J, Balasubramanian G, Naydenov B, Kolesov R, Grotz B, Reuter R, *et al.* Fluorescence and spin properties of defects in single digit nanodiamonds. *ACS Nano.* 2009;3(7):1959–65.

20 Greiner NR, Phillips DS, Johnson JD, Volk F. Diamonds in detonation soot. *Nature.* 1988;333(6172):440–2.

21 Ozawa M, Inaguma M, Takahashi M, Kataoka F, Krüger A, Ōsawa E. Preparation and behavior of brownish, clear nanodiamond colloids. *Adv Mater.* 2007;19(9):1201–6.

22 Petráková V. *Interactions of nitrogen-vacancy centers with charged surfaces of functionalized nanodiamond particles for the detection of cellular processes.* 2013.

23 Schwartz V, Overbury SH, Liang C. Carbon-mediated catalysis: Oxidative dehydrogenation on graphitic carbon. *Novel Materials for Catalysis and Fuels Processing. ACS Symposium Series. 1132: American Chemical Society*; 2013. pp. 247–58.

24 Berman D, Narayanan B, Cherukara MJ, Sankaranarayanan SK, Erdemir A, Zinovev A, *et al.* Operando tribochemical formation of onion-like-carbon leads to macroscale superlubricity. *Nat Commun.* 2018;9(1):1–9.

25 Terranova ML, Orlanducci S, Rossi M, Tamburri E. Nanodiamonds for field emission: State of the art. *Nanoscale.* 2015;7(12):5094–114.

26 Holt KB, Caruana DJ, Millán-Barrios EJ. Electrochemistry of undoped diamond nanoparticles: Accessing surface redox states. *J Am Chem Soc.* 2009;131(32): 11272–3.

27 Zhu Y, Li J, Li WX, Zhang Y, Yang XF, Chen N, *et al.* The biocompatibility of nanodiamonds and their application in drug delivery systems. *Theranostics.* 2012;2(3):302–12.

28 Lee D-K, Kee T, Liang Z, Hsiou D, Miya D, Wu B, *et al.* Clinical validation of a nanodiamond-embedded thermoplastic biomaterial. *Proc Natl Acad Sci USA.* 2017;114(45):E9445–54.

29 Naragino H, Egiza M, Tominaga A, Murasawa K, Gonda H, Sakurai M, *et al.* Hard coating of ultrananocrystalline diamond/nonhydrogenated amorphous carbon composite films on cemented tungsten carbide by coaxial arc plasma deposition. *Appl Phys A*. 2016;122(8):1–10.

30 Chayeuski V, Zhylinski V, Cernashejus O, Visniakov N, Mikalauskas G. Structural and mechanical properties of the ZrC/Ni-nanodiamond coating synthesized by the PVD and electroplating processes for the cutting knifes. *J Mater Eng Perform*. 2019;28(3):1278–85.

31 Mandal S. Nucleation of diamond films on heterogeneous substrates: A review. *RSC Adv*. 2021;11(17):10159–82.

32 Ralchenko V, Saveliev A, Voronina S, Dementjev A, Maslakov K, Salerno M, *et al.*, editors. *Nanodiamond Seeding for Nucleation and Growth of CVD Diamond Films*. Dordrecht: Springer Netherlands; 2005.

33 Osawa E. Monodisperse single nanodiamond particulates. *Pure Appl Chem*. 2008;80(7):1365–79.

34 Dideikin A, Aleksenskii A, Baidakova M, Brunkov P, Brzhezinskaya M, Davydov VY, *et al*. Rehybridization of carbon on facets of detonation diamond nanocrystals and forming hydrosols of individual particles. *Carbon*. 2017;122: 737–45.

35 Krueger A. Diamond nanoparticles: Jewels for chemistry and physics. *Adv Mater*. 2008;20(12):2445–9.

36 May PW. Diamond thin films: A 21st-century material. *Philos Transact A Math Phys Eng Sci*. 2000;358(1766):473–95.

37 Angus JC, Wang Y, Sunkara M. Metastable growth of diamond and diamond-like phases. *Annu Rev Mater Sci*. 1991;21(1):221–48.

38 Angus JC, Hayman CC. Low-pressure, metastable growth of diamond and "diamondlike" phases. *Science*. 1988;241(4868):913–21.

39 Gruen DM. Nanocrystalline diamond films. *Annu Rev Mater Sci*. 1999;29(1):211–59.

40 Tu R, Xu T, Li D, Zhang S, Yang M, Li Q, *et al*. Morphology and mechanical behavior of diamond films fabricated by IH-MPCVD. *RSC Adv*, 2018;8(29):16061–8.

41 Poferl DJ, Gardner NC, Angus JC. Growth of boron-doped diamond seed crystals by vapor deposition. *J Appl Phys*. 1973;44(4):1428–34.

42 Macpherson JV. A practical guide to using boron doped diamond in electrochemical research. *Phys Chem Chem Phys*. 2015;17(5):2935–49.

43 Katamune Y, Mori D, Arikawa D, Izumi A, Shimaoka T, Ichikawa K, *et al*. n-Type doping of diamond by hot-filament chemical vapor deposition growth with phosphorus incorporation. *Appl Phys A*. 2020;126(11):1–6.

44 Pinault-Thaury M-A, Temgoua S, Gillet R, Bensalah H, Stenger I, Jomard F, *et al*. Phosphorus-doped (113) CVD diamond: A breakthrough towards bipolar diamond devices. *Appl Phys Lett*. 2019;114(11):112106.

45 Anthony TR, Banholzer WF. Properties of diamond with varying isotopic composition. *Diam Relat Mater*. 1992;1(5):717–26.

46 Amans D, Diouf M, Lam J, Ledoux G, Dujardin C. Origin of the nanocarbon allotropes in pulsed laser ablation in liquids synthesis. *J Colloid Interface Sci.* 2017;489:114–25.

47 Osswald S, Chmiola J, Gogotsi Y. Structural evolution of carbide-derived carbons upon vacuum annealing. *Carbon.* 2012;50(13):4880–6.

48 Galimov ÉM, Kudin AM, Skorobogatskii VN, Plotnichenko VG, Bondarev OL, Zarubin BG, *et al.* Experimental corroboration of the synthesis of diamond in the cavitation process. *Dokl Phys.* 2004;49(3):150–3.

49 Xiao J, Ouyang G, Liu P, Wang CX, Yang GW. Reversible nanodiamond-carbon onion phase transformations. *Nano Lett.* 2014;14(6):3645–52.

50 Gorrini F, Cazzanelli M, Bazzanella N, Edla R, Gemmi M, Cappello V, *et al.* On the thermodynamic path enabling a room-temperature, laser-assisted graphite to nanodiamond transformation. *Sci Rep.* 2016;6(1):35244.

51 Nee C-H, Yap S-L, Tou T-Y, Chang H-C, Yap S-S. Direct synthesis of nanodiamonds by femtosecond laser irradiation of ethanol. *Sci Rep.* 2016;6 (1):33966.

52 Khachatryan AK, Aloyan SG, May PW, Sargsyan R, Khachatryan VA, Baghdasaryan VS. Graphite-to-diamond transformation induced by ultrasound cavitation. *Diam Relat Mater.* 2008;17(6):931–6.

53 Huang Q, Yu D, Xu B, Hu W, Ma Y, Wang Y, *et al.* Nanotwinned diamond with unprecedented hardness and stability. *Nature.* 2014;510(7504):250–3.

54 Rifai A, Houshyar S, Fox K. Progress towards 3D-printing diamond for medical implants: A review. *Ann 3D Print Med.* 2021;1:100002.

55 Hsiao WW-W, Le T-N, Chang H-C. Applications of fluorescent nanodiamond in biology. In: Meyers, RA (Ed.), *Encyclopedia of Analytical Chemistry.* Hoboken, NJ, USA: John Wiley & Sons; 2022. pp. 1–43.

56 Azevedo AF, Corat EJ, Ferreira NG, Trava-Airoldi VJ. Wettability and corrosion tests of diamond films grown on Ti6Al4V alloy. *Surf Coat Technol.* 2005;194(2):271–5.

57 Rahmani P, Shojaei A, Tavandashti NP. Nanodiamond loaded with corrosion inhibitor as efficient nanocarrier to improve anticorrosion behavior of epoxy coating. *J Ind Eng Chem.* 2020;83:153–63.

58 Thoms BD, Owens MS, Butler JE, Spiro C. Production and characterization of smooth, hydrogen-terminated diamond C (100). *Appl Phys Lett.* 1994;65(23):2957–9.

59 Yang W, Auciello O, Butler JE, Cai W, Carlisle JA, Gerbi JE, *et al.* DNA-modified nanocrystalline diamond thin-films as stable, biologically active substrates. *Nat Mater.* 2002;1(4):253–7.

60 Chang H-C, Han C-C *Diamond crystallites for biotechnological applications.* Google Patents; 2006.

61 Holt KB. Diamond at the nanoscale: Applications of diamond nanoparticles from cellular biomarkers to quantum computing. *Philos Transact A Math Phys Eng Sci.* 2007;365(1861):2845–61.

62 Chauhan S, Jain N, Nagaich U. Nanodiamonds with powerful ability for drug delivery and biomedical applications: Recent updates on in vivo study and patents. *J Pharm Anal.* 2020;10(1):1–12.

63 Yu S-J, Kang M-W, Chang H-C, Chen K-M, Yu Y-C. Bright fluorescent nanodiamonds: No photobleaching and low cytotoxicity. *J Am Chem Soc.* 2005;127(50):17604–5.

64 Schrand AM, Huang HJ, Carlson C, Schlager JJ, Osawa E, Hussain SM, *et al.* Are diamond nanoparticles cytotoxic? *J Phys Chem B.* 2007;111(1):2–7.

65 Tetienne J-P, Lombard A, Simpson DA, Ritchie C, Lu J, Mulvaney P, *et al.* Scanning nanospin ensemble microscope for nanoscale magnetic and thermal imaging. *Nano Lett.* 2016;16(1):326–33.

66 Reineck P, Gibson BC. Near-infrared fluorescent nanomaterials for bioimaging and sensing. *Adv Opt Mater.* 2017;5(2):1600446.

67 Schirhagl R, Chang K, Loretz M, Degen CL. Nitrogen-vacancy centers in diamond: Nanoscale sensors for physics and biology. *Annu Rev Phys Chem.* 2014;65:83–105.

68 Avvaru SP, Noolvi MN, Aminbhavi TM, Chkraborty S, Dash A, Shukla SS. Aromatase inhibitors evolution as potential class of drugs in the treatment of postmenopausal breast cancer women. *Mini Rev Med Chem.* 2018;18(7):609–21.

69 Kanyuk M. Ultrafine fluorescent diamonds in nanotechnology. *Biotechnol Acta.* 2014;7(4):9.

70 Wilson ER, Parker LM, Orth A, Nunn N, Torelli M, Shenderova O, *et al.* The effect of particle size on nanodiamond fluorescence and colloidal properties in biological media. *Nanotechnology*, 2019;30(38):385704.

71 Dahoumane SA, Nguyen MN, Thorel A, Boudou JP, Chehimi MM, Mangeney C. Protein-functionalized hairy diamond nanoparticles. *Langmuir.* 2009;25(17):9633–8.

72 Liang Y, Meinhardt T, Jarre G, Ozawa M, Vrdoljak P, Schöll A, *et al.* Deagglomeration and surface modification of thermally annealed nanoscale diamond. *J Colloid Interface Sci.* 2011;354(1):23–30.

73 Zhao L, Takimoto T, Ito M, Kitagawa N, Kimura T, Komatsu N. Chromatographic separation of highly soluble diamond nanoparticles prepared by polyglycerol grafting. *Angew Chem Int Ed.* 2011;50(6):1388–92.

74 Sreenivasan VK, Ivukina EA, Deng W, Kelf TA, Zdobnova TA, Lukash SV, *et al.* Barstar: Barnase—a versatile platform for colloidal diamond bioconjugation. *J Mater Chem.* 2011;21(1):65–8.

75 Maitra U, Gomathi A, Rao C. Covalent and noncovalent functionalisation and solubilisation of nanodiamond. *J Exp Nanosci.* 2008;3(4):271–8.

2

Fluorescent Nanodiamonds

Wesley Wei-Wen Hsiao, Hsuan-Yi Lin, Ming-Wei Jen, Tan-Thanh Huynh, Thi-Hong-Hanh Le, and Yan-Kai Tzeng*

* *Corresponding author*

A fluorescent nanodiamond (FND) refers to a type of nanodiamond (ND) that contains a high-density ensemble of fluorescent color centers. The color centers in diamonds that fluoresce in the visible light region are mainly made of nitrogen atom (N) and carbon atom vacancy (V). These two elements (or crystallographic defects) can combine in different ratios to form a variety of complexes. Four types of vacancy-related centers in NDs are the most noteworthy: V^0, NV, H3, and N3. For practical use as bioimaging agents, FND particles should be smaller than 100 nm and contain at least 1 ppm of fluorescent centers. This means that the number of defects in each FND of 100-nm diameter is greater than 100. This chapter focuses on the properties of FND.

2.1 Introduction to Fluorescent Nanodiamonds

FNDs have attracted much interest in recent years. Their nearly infinite stable photoluminescence without photobleaching and photoblinking enables long-term imaging and tracking in complex cellular milieus (1–3). FND's magneto-optical properties also make them valuable quantum nanosensors and biological nanothermometers for different characteristics in the microenvironment (4–6). The quantum sensing capabilities of FNDs are related to the magnetic states of the nitrogen-vacancy (NV) defect centers inside FND particles. The quantum spin properties of NV defect centers make them useful for many physiochemical studies, such as magnetic fields (7, 8), temperature (9–12), electric potentials (13–15),

Nanodiamonds in Analytical and Biological Sciences: Principles and Applications, First Edition.
Edited by Wesley Wei-Wen Hsiao and Trong-Nghia Le.
© 2023 John Wiley & Sons Ltd. Published 2023 by John Wiley & Sons Ltd.

and pH (16). Other features of FNDs make them ideal multimodal nanoprobes for performing biological research. For instance, FNDs have remarkable biocompatibility (17), are relatively simple to mass manufacture (18), are inherently nontoxic (19), and have outstanding inertness with facile surface groups for biofunctionalization (20, 21). Fluorescence in NDs is the result of electronic transitions involving radiative relaxation upon light absorption by hundreds of different crystallographic defects (color centers) existing within ND's crystal lattice. A plethora of optically active defect centers in NDs are complexes composed of impurity atoms combined with inherent structural defects, such as lattice vacancies. The NV defect centers in the sp^3 carbon lattice core of NDs make these particles fluorescent when illuminated. Within a single ND, NV centers can be neutral (NV^0) or negatively charged (NV^-), both of which exhibit a photostable luminescence and fluorescence profile that allows for an acceptable contrast to the autofluorescence characteristic of cells – such as fluorescence lifetimes up to 25 ns and over 30 ns for NV^- and NV^0, respectively. The fluorescence lifetime of NV centers is significantly greater than their autofluorescence emission lifetime (typically <4 ns). The fluorescent NV centers provide a wide emission peak with a maximum of approximately 680 nm, emitting infrared/near-infrared (IR/NIR) region. FNDs, which are monocrystalline particles, hold exceptionally photo-stable NV^- defect centers that produce intrinsic bright far-red fluorescence emissions up to a maximum of 700-nm wavelength under green–yellow excitation light. Around 70% of this emission efficiently falls within the NIR window of biological tissue (1, 22).

The emission wavelength region of FND-NV lies above the complex background autofluorescence bands of 450–670 nm. This allows for the use of particular emission filters to restrict the accumulation of NV emissions to longer wavelengths (700 nm), distinct from any background autofluorescence. In FNDs, negatively charged NV defect centers, NV^-, are of primary interest because they are magneto-optically active and have long spin coherence times at room temperature. This means that it is possible to optically polarize these centers. This allows signal separation for selective imaging with high background noise (10). The quantum spin characteristics of NV defects in FNDs can also be used to enhance image contrast by selectively modulating FND emissions with microwaves, magnetic or radio-frequency fields, and NIR light (23, 24). Compared with other carbon allotropes, FNDs present a high refractive index and high density due to their ultrastable diamond structure. As a result of the unique spin features of the NV^- fluorescent defects, their high quantum yield, and other features described above, FNDs can be used as fluorescent tags in biological imaging and target delivery applications (25) or as a qubit in real-time quantum computing (26), and quantum sensing (27, 28). Using microwave-based spin manipulation, FNDs can also be applied as ultrasensitive fluorescent labels for quantum diagnostics platforms, resulting in enhanced sensitivity and an optical readout approach for the early diagnosis of diseases (29).

2.2 Optical Properties of Fluorescent Nanodiamonds

The photoluminescent characteristics of FNDs are attributed to color centers, also known as local defects, present in ND crystals. These color centers act similarly to individual atoms or molecules within the solid host of the NDs. The most noticeable emitter in diamonds is the NV center, which has been employed in several proof-of-concept studies (1). In addition to NV, silicon-vacancy (SiV), germanium-vacancy (GeV), tin-vacancy (SnV), and nickel (Ni) color centers, other color centers have also been discovered. SiV centers have been the subject of much research, in part because of their limited emission bandwidth. Other color centers based on heavy impurities (nickel, chromium, and rare-earth elements) show a lot of promising prospects; however, the production of such color centers is still challenging, and additional study is needed to understand all of their photophysical properties. The adsorption and emission characteristics of the different color centers in diamonds are compiled in Figure 2.1.

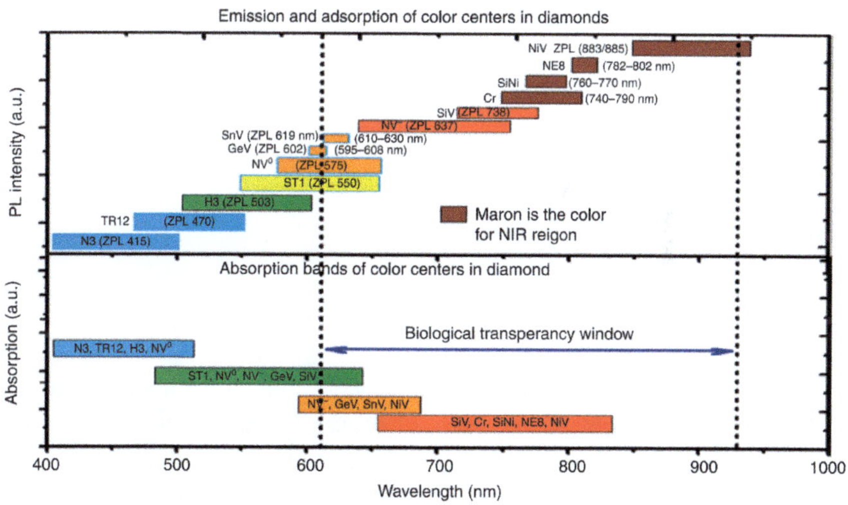

Figure 2.1 Absorption and emission bands of color centers in diamond. *Source:* (30) Alkahtani et al., 2018 / De Gruyter.

2.2.1 Vacancy-Related Color Centers

Diamond is a wide-bandgap material ($E_g = 5.5$ eV), transparent from the ultraviolet to the far-infrared. Structural defects and impurities result in its optical absorption and emission in the crystal matrix. Over the last 50 years, numerous spectroscopic approaches have been used to thoroughly study the characteristics of vacancy-related color centers in both natural and manufactured diamonds (31). Table 2.1 briefly describes the optical properties of V^0, NV^0, NV^-, H3, and N3 centers in bulk diamonds. Structures of vacancy-related color centers in diamonds are described in Figure 2.2.

The neutral vacancy (V^0) is a great candidate among diamond's aforementioned well-characterized color centers. It can be created in either natural or synthetic diamonds by radiation damage. V^0 center has a zero-phonon line (ZPL) at 741 nm and a phonon sideband peaking at ~620 nm in the absorption spectrum. Being excited with red photons, the vacancy generates NIR fluorescence with a quantum yield of only ~1% at room temperature (Table 2.1). Despite NDs poor yield, a

Table 2.1 Optical properties of vacancy-related defect centers in diamond[a].

Centers	Point group	ZPL/nm	λ_{em}/nm	τ/ns	Quantum yield
V^0 (GR1)	T_d	741.2	898	2.55	0.014
NV^-	C_{3v}	637.6	685	11.6	0.99
NV^0	C_{3v}	575.4	600	19[b]	–
N-V-N (H3)	C_{2v}	503.5	531	16	0.95
N_3+V (N3)	C_{3v}	415.4	445	41	0.29

[a] Listed is zero-phonon lines (ZPLs), emission maxima (λ_{em}), and emission lifetimes (τ). Data from ref. (32).
[b] Data from ref. (33).

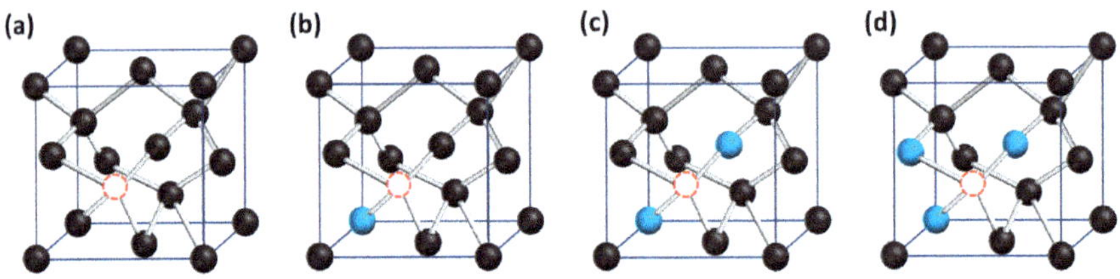

Figure 2.2 Structures of vacancy-related color centers in diamond: (a) V^0, (b) NV, (c) H3, and (d) N3. The carbon atoms, nitrogen atoms, and vacancies are denoted by black spheres, blue spheres, and red dashed circles, respectively.

high-density ensemble (up to 1000 ppm) of centers can be created by ion beam irradiation to compensate for this deficiency (34).

The NV center is a point defect composed of a substitutional nitrogen atom close to a carbon atom vacancy with C_{3v} symmetry. It exists in two different forms: NV^0 and NV^-. The neutral charge form NV^0 has a distinct ZPL at 575 nm, which allows researchers to quickly differentiate it from the negative form NV^- with a ZPL at 638 nm. Both the ZPLs exhibit broad phonon sidebands by ~50-nm red shifted in their emission spectra. The coexistence of these two centers in FNDs gives rise to an emission band peaking at 685 nm. Compared with NV^-, NV^0 is lower in density because of the electron transfer reaction with the atomically distributed nitrogen in its neutral charge state (N^0) as $NV^0 + N^0 \rightarrow NV^- + N^+$, which often occurs in type Ib diamond (35). Despite its low abundance, this center is more valuable than its negative counterpart in cathodoluminescence imaging due to the conversion of $NV^- \rightarrow NV^0$ during electron irradiation (36). In contrast, the NV^- center plays a more critical role in photoluminescence imaging.

The third type of vacancy-related color center exists, usually in natural diamonds. The H3 center is formed by a nitrogen-vacancy-nitrogen complex, N-V-N, in NDs with greater nitrogen content. At the wavelength of 470 nm, blue light excites the H3 core, leading it to generate green fluorescence at 531 nm, with a fluorescence quantum yield proximity 1. The single H3 center is believed to emit steady fluorescence without either photobleaching or photoblinking, which is similar to NV^- (37). They are useful for both photoluminescence and cathodoluminescence imaging (36). Lastly, blue fluorescence can also be emitted from the ND containing an N3 center, which is made of three nitrogen atoms encircling a vacancy in a type Ia diamond. This type of center (with a ZPL at 415 nm) is frequently generated alongside the H3 center in natural diamond. However, its blue emission with the phonon sidebands peaking at ~445 nm easily overlaps with cell and tissue autofluorescence, making the center less attractive for photoluminescence imaging applications.

2.2.2 Negative Nitrogen-Vacancy Center

Of the five types of fluorescent centers listed in Table 2.1, the negative NV (NV^-) center deserves special attention owing to its extraordinary optical features. First, its absorption center is at the wavelength of 560 nm with a fluorescence quantum yield close to 1. Second, the emission band peaks at 685 nm, and over 70% of the released photons are in the NIR window of biological tissue (38). Third, it has a far longer fluorescence lifetime than autofluorescence in cells and tissues (39), enabling *in vitro* and *in vivo* time-gated fluorescence imaging. Fourth, the fluorescence emission is perfectly stable, undergoing neither photoblinking nor photobleaching, which also holds under continuous high-power laser stimulation at ambient temperature (40). As a result, FNDs with high-density ensembles of NV^- have been increasingly used by researchers as bioimaging contrast agents due to their above-mentioned properties.

The NV^- center also possesses two unpaired electrons, which is another important characteristic. The unperturbed electronic energy structure of the center comprises a triplet ground state ^3A with spin quantum numbers of $m_s = 0$ and ± 1 (Figure 2.3) and a triplet excited state ^3E. The two $m_s = \pm 1$ sublevels are degenerate at zero external magnetic field and are energetically higher than the $m_s = 0$ sublevel by 2.87 GHz due to crystal field splitting. Optical transitions between the triplet ground state and the excited state mostly conserve their spin states. However, intersystem crossing (ISC) from the excited state to the metastable singlet state (^1A) is possible, depending on the sublevel of the excited state. The ISC rate from the $m_s = \pm 1$ sublevel of the excited state the singlet state is higher than the transition rate from $m_s = 0$ sublevel of the excited state. The singlet state then decays to the $m_s = 0$ sublevel of the ground state by nonradiative processes. Hence, optical pumping preferentially populates the $m_s = 0$ sublevel of the ground state, which is also called spin-polarized. Furthermore, when applying a resonant microwave of 2.87 GHz, the electron spin transits from the $m_s = 0$ sublevel to the $m_s = \pm 1$ sublevel of the ground state. This spin transition causes a reduction of NV's fluorescence intensity by approximately 30% because the $m_s = \pm 1$ sublevel of the excited state prefers nonradiative decay, as shown in Figure 2.4. Hence, the

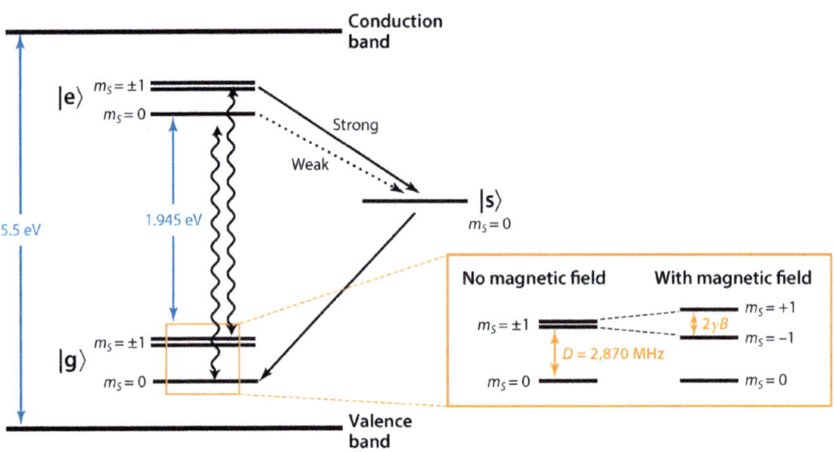

Figure 2.3 Energy level diagram of the NV^- center in diamond. |g⟩ denotes the electronic ground state (^3A), |e⟩ the electronic excited state (^3E), and |s⟩ the metastable singlet state. Wiggly arrows indicate the radiative transition and black solid and dotted arrows indicate strong and weak non-radiative decays to the singlet state, respectively. Inset: The three spin sublevels with $m_s = 0$ and $m_s = \pm 1$ at zero and nonzero magnetic field. *Source:* (4) Schirhagl et al., 2014 / Annual Review of Physical Chemistry.

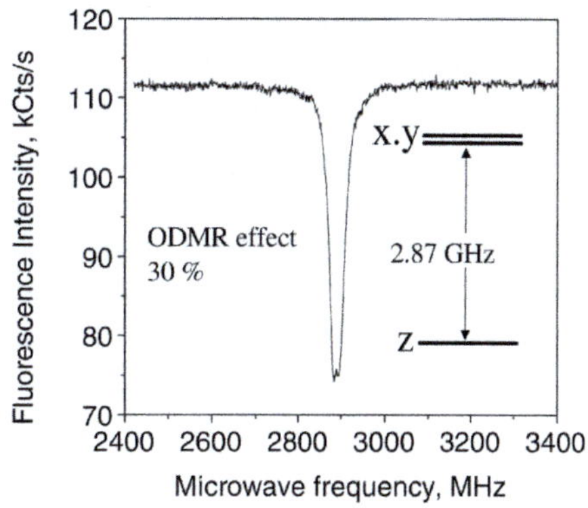

Figure 2.4 Optically detected magnetic resonance spectrum of a single defect. The spectrum was obtained at room temperature with 514-nm irradiation and without an external magnetic field. *Source:* (42) Jelezko et al., 2006 / John Wiley & Sons.

fluorescence intensity conveys information about the spin state of the NV center. We can read out the spin state of the NV center optically. This is called optically detected magnetic resonance (ODMR) (41). The spin state of the NV center may likewise be investigated and modified using the continuous microwave and pulsed microwave, respectively.

ODMR's effect has been observed in a handful of atoms and molecules. The effect of ODMR on an NV center is unique because it is observable at the single spin level even in ambient temperature. Moreover, the NV^- center has an exceptionally long coherence time. Therefore, the spin state remains coherent for longer than 1 ms in isotope-free bulk diamond, thus rendering FND a powerful tool for taking ultrasensitive measurements of temperature, magnetic fields, and even mechanical stress at the nanometer level. Furthermore, background-free detection of FNDs in cells and tissues can be easily realized by magnetic or microwave modulation of NV^- fluorescence (43).

2.2.3 Silicon-Vacancy Center

Besides the NV center, which generally could be the most investigated diamond color center, the SiV center currently has drawn much attention as another promising optical defect center. Since the silicon atom is bigger than carbon, the SiV center forms by replacing two adjacent carbon atoms with a silicon atom, leaving a single silicon atom between two vacancies in the diamond lattice. Two common methods used to synthesize SiV-NDs are adding Si gas during diamond formation by chemical vapor deposition (CVD) and implanting silicon ions into pure diamonds (44).

Due to their divacancy structure, the negative SiV center possesses the following outstanding optical characteristics, including high brightness, stable and narrow zero-photon line, and minimal spectral diffusion (45, 46). It is known as the brightest solid-phonon source for single-photon emission at room temperature and can emit up to 4.8×10^6 photons/s (47). In addition, its stable photoluminescence feature is less sensitive to the nanodiamond size than the NV center (48). In particular, the photoluminescence of the SiV center forms a small phonon sideband, with 70% of its fluorescent dominated in the NIR region (ZPL=738 nm) (49). Moreover, the minimum nanodiamond size to maintain a stable color center in NV was 24 nm, while the value of the SiV center was only 8 nm (48, 50). These unique features made SiV centers in nanodiamonds become a potential candidate for biomarkers in the infrared region. In 2017, Nguyen et al. investigated the temperature-dependent in ZPL of SiV in diamonds (51). Their finding suggested that the SiV ZPL shifted linearly in the range of temperature of 295 ± 5 K with subkelvin sensitivity (51). The combination of outstanding optical properties, chemical stability, and dependence on the temperature of ZPL made SiV become more and more valuable in bioimaging and thermometry fields.

2.3 Introduction to Förster Resonance Energy Transfer

For cell biologists, Förster resonance energy transfer (FRET)–based sensors are important tools for gathering information about fluorescent proteins, interactions between proteins, and the conformational biology of components such as reporters. Scientists can design sensors to get information about phenomena in living cells. FRET sensors have had a significant impact on many fields. Most biologists use organic dye molecules or green fluorescent protein (GFP) labeling to resolve biological questions. However, photobleaching precludes the observation of long-time signals. Quantum dots (QDs) were developed to overcome the photobleaching problem, but they present a more serious biological problem: cytotoxicity (52). As for photophysical properties, QDs exhibit the photoblinking effect. Therefore, developing a new fluorescence nanomaterial is crucial. One source of alternative nanomaterial is nanocrystals of diamonds containing NV centers. FNDs are excellent fluorescent probes for biological systems, such as in fluorescence imaging or magnetic resonance fields (42, 53). These nanocrystals have perfect photostability, demonstrating no bleaching or blinking. In addition, they have a brightness similar to a single GFP (54). FND can be easily functionalized on its surface, e.g., it can bond with IRDye molecules as an acceptor for fluorescence resonance energy transfer (55, 56). The efficiency of the energy transfer can reach 7% with a 19-nm FND. From a biophysical point of view, FRET can not only provide dynamic information in real time, but it can also help shed light on the mechanisms of protein folding and unfolding. The FND can act as a permanent FRET donor, comparable with an organic dye or fluorescent protein, but without the low photostability or photoblinking that characterizes these other alternatives (57).

2.3.1 Theory

An appropriate pair of materials must contain one energy pair: a donor and an acceptor. The acceptor can be fluorescent material or absorb light without any emission, also known as a "quencher." NV^- charged centers in FNDs work as energy donor systems in the Förster resonance energy transfer (FRET) process. For FRET, the emission band of the donor molecule must overlap with the absorption band of the acceptor molecule, which allows energy to transfer from the donor to the acceptor molecule. When both the donor and acceptor are within 10 nm of each other, FRET occurs on the overlapping region between the spectra of excitation and emission. The donor molecules are brought to an excited state, and the donor and acceptor interact with each other by the dipole–dipole coupling effect. The donor can transfer *hv* energy to the acceptor molecules in a nonradiative form. The acceptor releases the energy in the form of heat or emits the photons by fluorescence, as shown in Figure 2.5.

For FRET to work, three conditions must be met: (i) a higher quantum yield of the donor; (ii) an overlap between the donor and the acceptor; and (iii) a distance restriction of $1 \pm 0.5R_0$ between the donor and acceptor (58). When FRET works, the fluorescent intensity of the donor decreases as the fluorescent intensity of the acceptor increases. At the same time, the lifetime of the donor shortens, while the lifetime of the acceptor expands.

The calculation of the efficiency (E) of energy transfer between the donor and the acceptor is important in quantifying the interaction between these materials. The efficiency of the energy transfer depends upon the percentage of overlap, the dipole–dipole interaction, and the distance between the donor and the acceptor. The formula is

$$E = R_0^6 \Big/ \left(r^6 + R_0^6\right)$$

$$R_0^6 = 8.79 \times 10^{-25} \kappa^2 \, n^{-4} \, \Phi_D \, J_{DA}.$$

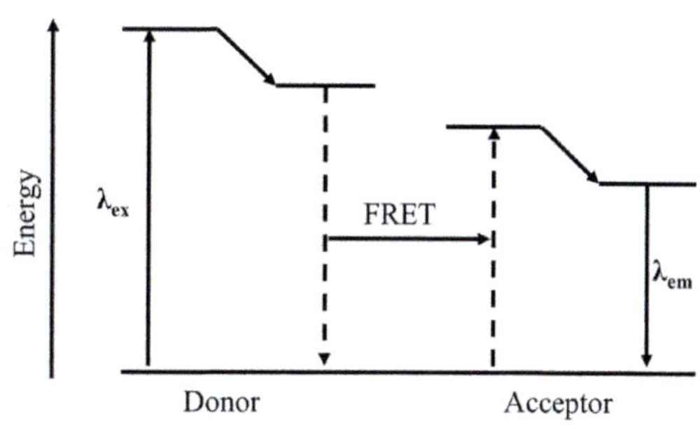

Figure 2.5 Principle of Förster resonance energy transfer.

When the efficiency of energy transfer is 50% for specific energy pairs, R_0 is called the radius of Förster, which is the distance between the donor and the acceptor. R_0 is a constant; r is the distance between the donor and acceptor; k^2 is the dipole orientation factor based the dipole on the donor and the acceptor, which is usually assumed to be $2/3$. Φ_D is the fluorescence quantum yield of the donor under the absence of the acceptor; η is the refractive index of the medium between the donor and the acceptor. In an experiment, FRET efficiency is based on the presence and absence of an acceptor as well as the fluorescence lifetime and intensity of a donor

$$E = 1 - F_{DA}/F_D = 1 - \tau_{DA}/\tau_D$$

where F_{DA} and τ_{DA} are the donor fluorescence intensities and lifetime with an acceptor, respectively, and F_D and τ_D are the donor fluorescence intensities and lifetime without an acceptor, respectively.

The Förster radius and the experimental energy transfer efficiency can reveal the distance between the donor and the acceptor, r, as well as the related molecular structure

$$r = R_0 \left(\frac{1 - E}{E} \right)^{1/6}.$$

2.3.2 FRET Efficiency

FRET rate can be obtained by measuring the donor's lifetime reduction of photoluminescence in the presence of an acceptor. According to Förster theory, FRET efficiency is denoted as E and expressed as

$$E = \frac{F_D - F_{DA}}{F_D} = 1 - \frac{F_{DA}}{F_D}$$

E: the efficiency of fluorescence resonance energy transfer
F_D: the fluorescent intensity before fluorescence resonance energy transfer
F_{DA}: the fluorescent intensity after fluorescence resonance energy transfer.

The fluorescent efficiency, before and after FRET, can be calculated with different donor-to-acceptor ratios. For example, Figure 2.6 shows a FRET efficiency of 20-nm FNDs with various IRDye/FND ratios. Using ethylenediamine to link FND and IRDye, the FRET efficiency of 20-nm FNDs as a donor is approximately 7%, which indicates that FND can transfer the fluorescence energy to IRDye through resonance energy transfer.

Compared to the prior linking procedure, a considerable improvement in the FRET experiment can be observed using 7-nm FNDs as donors. When the ratio of acceptors to donors is 3, the efficiency reaches 42.7% due to the FNDs becoming smaller and the linkage reagent changing to cysteamine (Figure 2.7).

It is natural that, because of this trend, with FND size decreasing at the same time that their fluorescent intensity increases, FNDs have great potential for many applications. Experimental and theoretical data on energy transfer efficiency are summarized in Figure 2.8 (56). The energy transfer efficiency calculated from intensity was plotted versus the energy transfer efficiency calculated from lifetime. The curve in the figure shows the expected theoretical result, which approximately fits with the experimental data.

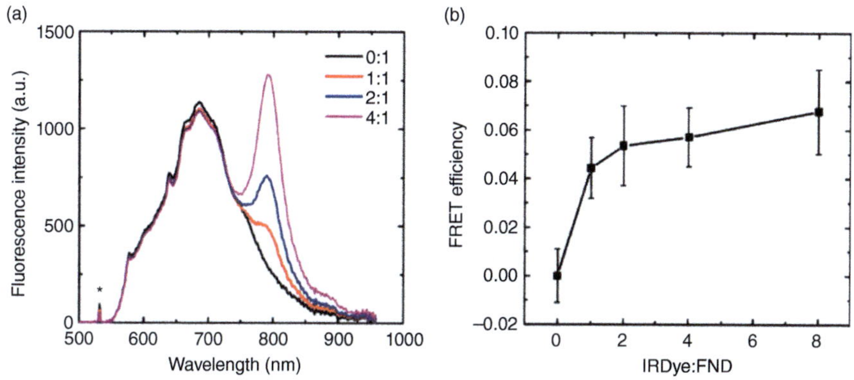

Figure 2.6 (a) Fluorescent spectra and (b) FRET efficiency of 20-nm FNDs with different IRDye/FND ratios. *Source:* (55) Mohan et al., 2010 / John Wiley & Sons.

Figure 2.7 FRET efficiency of 7-nm FNDs with different IRDye/FND ratios.

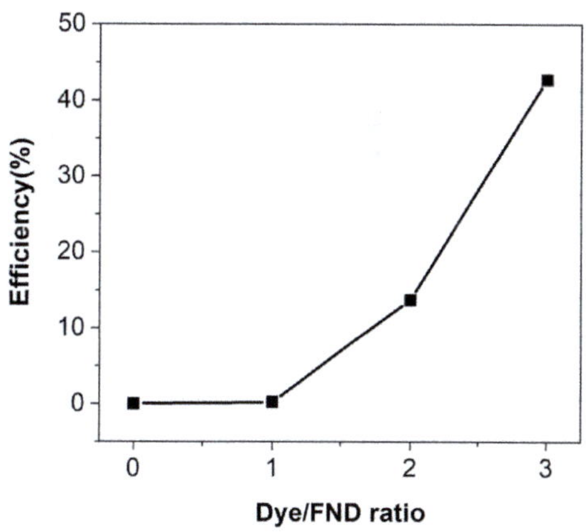

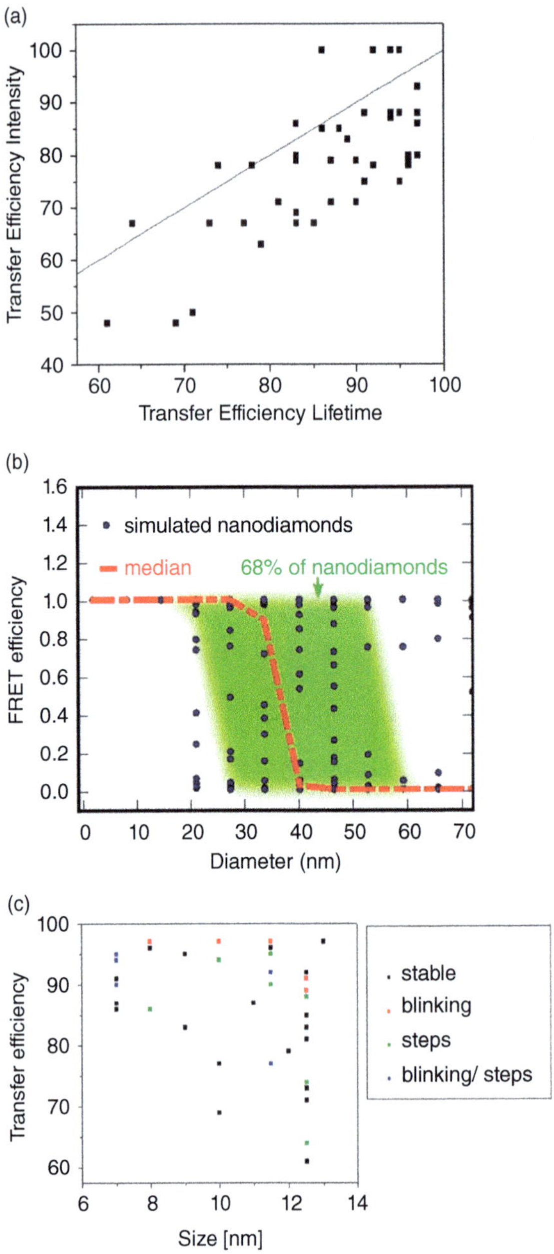

Figure 2.8 Energy transfer efficiency between NV center and Black Hole Quencher 3 (BHQ3). (a) Comparison between the transfer efficiency based on experimental lifetime and intensity data and the theoretical curve. (b) Theoretical FRET efficiency as a function of NDs size. The black points denote simulated nanodiamonds. For each diameter, an ensemble of nanodiamonds was simulated and used to compute the quenching efficiency's median (red) and standard deviation (green). (c) Experimental FRET efficiency as a function of the size of the nanodiamond, including indication for steps and blinking particles. *Source:* (56) Tisler et al., 2011 / American Chemical Society.

Acknowledgments

This work was supported by the National Science and Technology Council (NSTC), Taiwan, under Grant No. 112-2221-E-011-031.

References

1 Hsiao WW-W, Hui YY, Tsai P-C, Chang H-C. Fluorescent nanodiamond: A versatile tool for long-term cell tracking, super-resolution imaging, and nanoscale temperature sensing. *Acc Chem Res*. 2016;49(3):400–7.

2 Waddington DE, Sarracanie M, Zhang H, Salameh N, Glenn DR, Rej E, et al. Nanodiamond-enhanced MRI via in situ hyperpolarization. *Nat Commun*. 2017;8(1):1–8.

3 Hsiao WW-W, Le T-N, Chang H-C. Applications of fluorescent nanodiamond in biology. In: Meyers, R.A. (Ed.), *Encyclopedia of Analytical Chemistry*. Hoboken, NJ, USA: John Wiley & Sons; 2022. pp. 1–43.

4 Schirhagl R, Chang K, Loretz M, Degen CL. Nitrogen-vacancy centers in diamond: Nanoscale sensors for physics and biology. *Annu Rev Phys Chem*. 2014;65:83–105.

5 Sotoma S, Terada D, Segawa TF, Igarashi R, Harada Y, Shirakawa M. Enrichment of ODMR-active nitrogen-vacancy centres in five-nanometre-sized detonation-synthesized nanodiamonds: Nanoprobes for temperature, angle and position. *Sci Rep*. 2018;8(1):1–8.

6 Sotoma S, Epperla CP, Chang H-C. Diamond nanothermometry. *ChemNanoMat*. 2018;4(1):15–27.

7 Jones ZR, Niemuth NJ, Robinson ME, Shenderova OA, Klaper RD, Hamers RJ. Selective imaging of diamond nanoparticles within complex matrices using magnetically induced fluorescence contrast. *Environ Sci Nano*. 2020;7(2):525–34.

8 Le Sage D, Arai K, Glenn DR, DeVience SJ, Pham LM, Rahn-Lee L, et al. Optical magnetic imaging of living cells. *Nature*. 2013;496(7446):486–9.

9 Tsai P-C, Epperla CP, Huang J-S, Chen OY, Wu C-C, Chang H-C. Measuring nanoscale thermostability of cell membranes with single gold–diamond nanohybrids. *Angew Chem Int Ed*. 2017;56(11):3025–30.

10 Simpson DA, Morrisroe E, McCoey JM, Lombard AH, Mendis DC, Treussart F, et al. Non-neurotoxic nanodiamond probes for intraneuronal temperature mapping. *Acs Nano*. 2017;11(12):12077–86.

11 Zhang T, Liu G-Q, Leong W-H, Liu C-F, Kwok M-H, Ngai T, et al. Hybrid nanodiamond quantum sensors enabled by volume phase transitions of hydrogels. *Nat Commun*. 2018;9(1):1–8.

12 Wu Y, Alam MNA, Balasubramanian P, Ermakova A, Fischer S, Barth H, et al. Nanodiamond theranostic for light-controlled intracellular heating and nanoscale temperature sensing. *Nano Lett*. 2021;21(9):3780–8.

13 Karaveli S, Gaathon O, Wolcott A, Sakakibara R, Shemesh OA, Peterka DS, et al. Modulation of nitrogen vacancy charge state and fluorescence in nanodiamonds using electrochemical potential. *Proc Natl Acad Sci USA*. 2016;113(15):3938–43.

14 Dolde F, Fedder H, Doherty MW, Nöbauer T, Rempp F, Balasubramanian G, et al. Electric-field sensing using single diamond spins. *Nat Phys*. 2011;7(6):459–63.

15 Iwasaki T, Naruki W, Tahara K, Makino T, Kato H, Ogura M, et al. Direct nanoscale sensing of the internal electric field in operating semiconductor devices using single electron spins. *Acs Nano*. 2017;11(2):1238–45.

16 Rendler T, Neburkova J, Zemek O, Kotek J, Zappe A, Chu Z, et al. Optical imaging of localized chemical events using programmable diamond quantum nanosensors. *Nat Commun*. 2017;8(1):1–9.

17 Jung H-S, Cho K-J, Ryu S-J, Takagi Y, Roche PA, Neuman KC. Biocompatible fluorescent nanodiamonds as multifunctional optical probes for latent fingerprint detection. *ACS Appl Mater Interfaces*. 2020;12(5):6641–50.

18 Chang Y-R, Lee H-Y, Chen K, Chang C-C, Tsai D-S, Fu C-C, et al. Mass production and dynamic imaging of fluorescent nanodiamonds. *Nat Nanotechnol*. 2008;3(5):284–8.

19 Prabhakar N, Khan MH, Peurla M, Chang H-C, Hänninen PE, Rosenholm JM. Intracellular trafficking of fluorescent nanodiamonds and regulation of their cellular toxicity. *ACS Omega*. 2017;2(6):2689–93.

20 Hemelaar SR, Nagl A, Bigot F, Rodríguez-García MM, de Vries MP, Chipaux M, et al. The interaction of fluorescent nanodiamond probes with cellular media. *Microchimica Acta*. 2017;184(4):1001–9.

21 Shenderova OA, McGuire GE. Science and engineering of nanodiamond particle surfaces for biological applications. *Biointerphases*. 2015;10(3):030802.

22 Hui YY, Hsiao WW-W, Haziza S, Simonneau M, Treussart F, Chang H-C. Single particle tracking of fluorescent nanodiamonds in cells and organisms. *Curr Opin Solid State Mater Sci*. 2017;21(1):35–42.

23 Abe E, Sasaki K. Tutorial: Magnetic resonance with nitrogen-vacancy centers in diamond—microwave engineering, materials science, and magnetometry. *J Appl Phys*. 2018;123(16):161101.

24 Laube C, Oeckinghaus T, Lehnert J, Griebel J, Knolle W, Denisenko A, et al. Controlling the fluorescence properties of nitrogen vacancy centers in nanodiamonds. *Nanoscale*. 2019;11(4):1770–83.

25 Chang B-M, Lin H-H, Su L-J, Lin W-D, Lin R-J, Tzeng Y-K, et al. Highly fluorescent nanodiamonds protein-functionalized for cell labeling and targeting. *Adv Funct Mater*. 2013;23(46):5737–45.

26 Childress L, Hanson R. Diamond NV centers for quantum computing and quantum networks. *MRS Bull*. 2013;38(2):134–8.

27 Chang H-C, Hsiao WW-W, Su M-C *Fluorescent Nanodiamonds*. Hoboken, NJ, USA: John Wiley & Sons; 2018.

28 Degen CL, Reinhard F, Cappellaro P. Quantum sensing. *Rev Mod Phys*. 2017;89(3):035002.

29 Miller BS, Bezinge L, Gliddon HD, Huang D, Dold G, Gray ER, et al. Spin-enhanced nanodiamond biosensing for ultrasensitive diagnostics. *Nature.* 2020;587(7835):588–93.

30 Alkahtani MH, Alghannam F, Jiang L, Almethen A, Rampersaud AA, Brick R, et al. Fluorescent nanodiamonds: Past, present, and future. *Nanophotonics.* 2018;7(8):1423–53.

31 Zaitsev AM. *Optical Properties of Diamond: A Data Handbook.* Morwell: Springer-Verlag Berlin Heidelberg GmbH; 2001.

32 Hui YY, Chang HC. Recent developments and applications of nanodiamonds as versatile bioimaging agents. *J Chin Chem Soc.* 2014;61(1):67–76.

33 Liaugaudas G, Davies G, Suhling K, Khan RUA, Evans DJF. Luminescence lifetimes of neutral nitrogen-vacancy centres in synthetic diamond containing nitrogen. *J Phys Condens Matter.* 2012;24(43):435503 (5pp).

34 Fang CY, Chang CC, Mou CY, Chang HC. Preparation and characterization of ion-irradiated nanodiamonds as photoacoustic contrast agents. *J Nanosci Nanotechno.* 2015;15(2):1037–44.

35 Rondin L, Dantelle G, Slablab A, Grosshans F, Treussart F, Bergonzo P, et al. Surface-induced charge state conversion of nitrogen-vacancy defects in nanodiamonds. *Phys Rev B.* 2010;82(11):115449.

36 Tizei LHG, Meuret S, Nagarajan S, Treussart F, Fang CY, Chang HC, et al. Spatially and spectrally resolved cathodoluminescence with fast electrons: A tool for background subtraction in luminescence intensity second-order correlation measurements applied to subwavelength inhomogeneous diamond nanocrystals. *Phys Status Solidi A.* 2013;210(10):2060–5.

37 Hsu JH, Su WD, Yang KL, Tzeng YK, Chang HC. Nonblinking green emission from single H3 color centers in nanodiamonds. *Appl Phys Lett.* 2011;98(19):193116.

38 Vaijayanthimala V, Cheng P-Y, Yeh S-H, Liu K-K, Hsiao C-H, Chao J-I, et al. The long-term stability and biocompatibility of fluorescent nanodiamond as an in vivo contrast agent. *Biomaterials.* 2012;33(31):7794–802.

39 Billinton N, Knight AW. Seeing the wood through the trees: A review of techniques for distinguishing green fluorescent protein from endogenous autofluorescence. *Anal Biochem.* 2001;291(2):175–97.

40 Fu CC, Lee HY, Chen K, Lim TS, Wu HY, Lin PK, et al. Characterization and application of single fluorescent nanodiamonds as cellular biomarkers. *Proc Natl Acad Sci USA.* 2007;104(3):727–32.

41 Doherty MW, Manson NB, Delaney P, Jelezko F, Wrachtrup J, Hollenberg LCL. The nitrogen-vacancy colour centre in diamond. *Phys Rep.* 2013;528(1):1–45.

42 Jelezko F, Wrachtrup J. Single defect centres in diamond: A review. *Phys Status Solidi A.* 2006;203(13):3207–25.

43 Igarashi R, Yoshinari Y, Yokota H, Sugi T, Sugihara F, Ikeda K, et al. Real-time background-free selective imaging of fluorescent nanodiamonds in vivo. *Nano Lett.* 2012;12(11):5726–32.

44 Zhang H, Aharonovich I, Glenn DR, Schalek R, Magyar AP, Lichtman JW, et al. Silicon-vacancy color centers in nanodiamonds: Cathodoluminescence imaging markers in the near infrared. *Small*. 2014;10(10):1908–13.

45 Choi S, Leong V, Davydov VA, Agafonov VN, Cheong MWO, Kalashnikov DA, et al. Varying temperature and silicon content in nanodiamond growth: Effects on silicon-vacancy centres. *Sci Rep*. 2018;8(1):3792.

46 Lagomarsino S, Flatae AM, Sciortino S, Gorelli F, Santoro M, Tantussi F, et al. Optical properties of silicon-vacancy color centers in diamond created by ion implantation and post-annealing. *Diam Relat Mater*. 2018;84:196–203.

47 Neu E, Steinmetz D, Riedrich-Möller J, Gsell S, Fischer M, Schreck M, et al. Single photon emission from silicon-vacancy colour centres in chemical vapour deposition nano-diamonds on iridium. *New J Phys*. 2011;13(2):025012.

48 Nunn N, Torelli M, McGuire G, Shenderova O. Nanodiamond: A high impact nanomaterial. *Curr Opin Solid State Mater Sci*. 2017;21(1):1–9.

49 Vlasov II, Shiryaev AA, Rendler T, Steinert S, Lee S-Y, Antonov D, et al. Molecular-sized fluorescent nanodiamonds. *Nat Nanotechnol*. 2014;9(1):54–8.

50 Shershulin VA, Sedov VS, Ermakova A, Jantzen U, Rogers L, Huhlina AA, et al. Size-dependent luminescence of color centers in composite nanodiamonds. *Phys Status Solidi A*. 2015;212(11):2600–5.

51 Nguyen CT, Evans RE, Sipahigil A, Bhaskar MK, Sukachev DD, Agafonov VN, et al. All-optical nanoscale thermometry with silicon-vacancy centers in diamond. *Appl Phys Lett*. 2018;112(20):203102.

52 Hardman R. A toxicologic review of quantum dots: Toxicity depends on physicochemical and environmental factors. *Environ Health Perspect*. 2006;114(2):165–72.

53 Schoenfeld RS, Harneit W. Real time magnetic field sensing and imaging using a single spin in diamond. *Phys Rev Lett*. 2011;106(3):030802.

54 Castellani CM, Torres-Ocampo AP, Breffke J, White AB, Chambers JJ, Stratton MM, et al. Live-cell FLIM-FRET using a commercially available system. *Methods Cell Biol*. 2020;158:63–89.

55 Mohan N, Tzeng YK, Yang L, Chen YY, Hui YY, Fang CY, et al. Sub-20-nm fluorescent nanodiamonds as photo-stable biolabels and fluorescence resonance energy transfer donors. *Adv Mater*. 2010;22(7):843–7.

56 Tisler J, Reuter R, Lammle A, Jelezko F, Balasubramanian G, Hemmer PR, et al. Highly efficient FRET from a single nitrogen-vacancy center in nanodiamonds to a single organic molecule. *Acs Nano*. 2011;5(10):7893–8.

57 Yu S-J, Kang M-W, Chang H-C, Chen K-M, Yu Y-C. Bright fluorescent nanodiamonds: No photobleaching and low cytotoxicity. *J Am Chem Soc*. 2005;127(50):17604–5.

58 Periasamy A. Fluorescence resonance energy transfer microscopy: A mini review. *J Biomed Opt*. 2001;6(3):287–91.

3

Fabrication of Fluorescent Nanodiamonds

Wesley Wei-Wen Hsiao, Jo-Yu Wang, Syun-Ying Wang, Stefanny Angela, and Yan-Kai Tzeng*

* *Corresponding author*

This chapter focuses on the fabrication methods and strategies used to generate a mass scale quantity of fluorescent nanodiamonds (FNDs), which is crucial for their applications in analytical and biological applications. Additionally, we delve into topics such as size reduction and the spectroscopic characterization of FNDs at the ensemble level with various optical methods.

3.1 Introduction to FND Fabrication

FND has recently generated much interest for their prospective use as fluorescent tags in biological systems and the sensing applications enabled by the nitrogen-vacancy (NV) fluorescent defects' distinctive spin characteristics (1–3). Diamonds have a high refractive index and density compared with other carbon allotropes due to their unique structure. While diamonds are not the most thermodynamically favorable allotrope under normal conditions, they may be created under extreme heat and pressure.

FND has gained popularity due to its unique properties of electron spin and stable fluorescence (4–6). These properties come from a unique structure within the diamond: the negative charge state NV^- centers. In the 1970s, high-speed ions bombarded diamonds, generating internal defects (7), which found that these lattice defects helped FND release bright, stable fluorescence. Conventional nanodiamond (ND) does not have such a structure and must undergo a unique process to gain NV^- centers. FND must be bombarded with high-energy particles and quenched in a high vacuum to acquire this property.

Nanodiamonds in Analytical and Biological Sciences: Principles and Applications, First Edition.
Edited by Wesley Wei-Wen Hsiao and Trong-Nghia Le.
© 2023 John Wiley & Sons Ltd. Published 2023 by John Wiley & Sons Ltd.

The production of an "acceptable number" of internal defects in the diamond is crucial. Many techniques allow for the bombarding of a diamond: proton beams, ion beams, neutron beams, and electron beams (8–10). However, the specific energy dose is critical for each of these techniques. Furthermore, hyperactivity is time-consuming and uneconomical. Theoretical simulation models are a better approach to experimenting with different particles at different energies (11). These simulations assist us in obtaining post-collision data, including the target's penetration depth and estimates of the number of flaws.

3.2 Theoretical Simulations

In theoretical simulations, ion stopping and range tables, as well as Transport of Ions in Matter (TRIM), are commonly utilized in Stopping and Range of Ions in Matter (SRIM). In 1983, Ziegler and Biersack established software that could employ quantum mechanics to investigate interactions between ions and matter during and after collisions (11, 12). The former program mainly provides information on particles' collisions with matter. The halting status and range are shown when ions of various energies react with matter. The software's latter program estimates the effect and energy changes after particles collide with the matter. During the simulation, types of particles and energy between 10 eV and 2 GeV can be chosen as input parameters. Materials, thickness, combinations of different materials and incident particle angles can be selected. Due to its ability to model ion distribution before conducting tests, the software is widely used in research and development. Monte Carlo simulation calculations are at the core of TRIM. When high-energy or charged particles reach the designated material, a computer simulation keeps track of the incident particles interacting with it. The motion analysis is based on the hypothesis of a continuous slowing process. The collision of incident particles and target materials is described by the binary collision approximation (BCA) method (13–15). The impact caused by neighboring atoms can be ignored. The collisions occur twice, so electronic interactions between the particles and the target material cause energy loss and slow down. Random sampling determines additional factors, such as the distance between two collisions. Factors such as trajectory, secondary particles, final position, and energy loss are recorded during the process. The corresponding expectation error values are obtained by statistical calculation. The penetration depth and the particles that collide are presented in a 3-D distribution. Due to energy loss, the particles gradually slow to a stop. The amount of kinetic energy created during the activity and the target's damage, phonon creation, sputtering, defect state, and ionization are also shown.

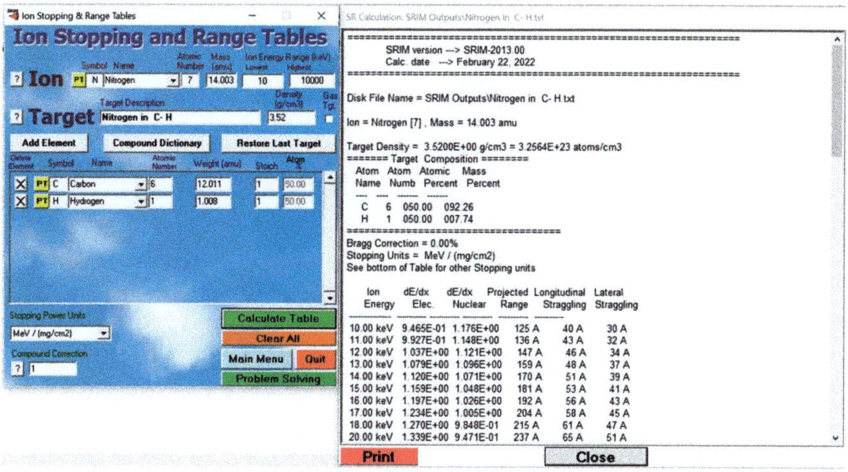

Figure 3.1 The interface of ion stopping and range tables.

Figure 3.1 shows the interface of ion stopping and range tables. The interface displays both the incident particle (ION) and the target. Here, different materials can be chosen for the simulation. The ion section allows for different types of elements, atomic mass, and energy. The element type, density, and percentages for various components can be chosen at the target selection screen. Then, the data from these input parameters can be interpreted as the energy loss, range of the incident particles, and target. This allows for a good understanding of situations with different incident particles of different energies.

Figure 3.2 shows the TRIM interface. The main parts are ION DATA, TARGET DATA, and Type of TRIM Calculation. In ION DATA, we can input the types and conditions of the incident particles, such as element name, atomic number, mass, incident angle, and energy. TARGET DATA allows for the entry of the target layers, thickness, density, and material. In addition, DAMAGE information, such as Disp, Latt, and Surf, which represent the displacement energy (E_{disp}), the lattice binding energy (E_{latt}), and the surface binding energy (E_{surf}), respectively, can be entered here. The TRIM Calculation allows for selecting the calculation method and information display method. The calculation methods include Quick and Full. The quick calculates the fixed energy of particles striking the target, the distribution, penetration depth of the particles in the target, the number of backscatter particles, and the number of penetrated particles. Ironically, the total calculation is relatively time-consuming, but outputs include much post-collision information, such as damage, sputtered particles, and recoil atoms.

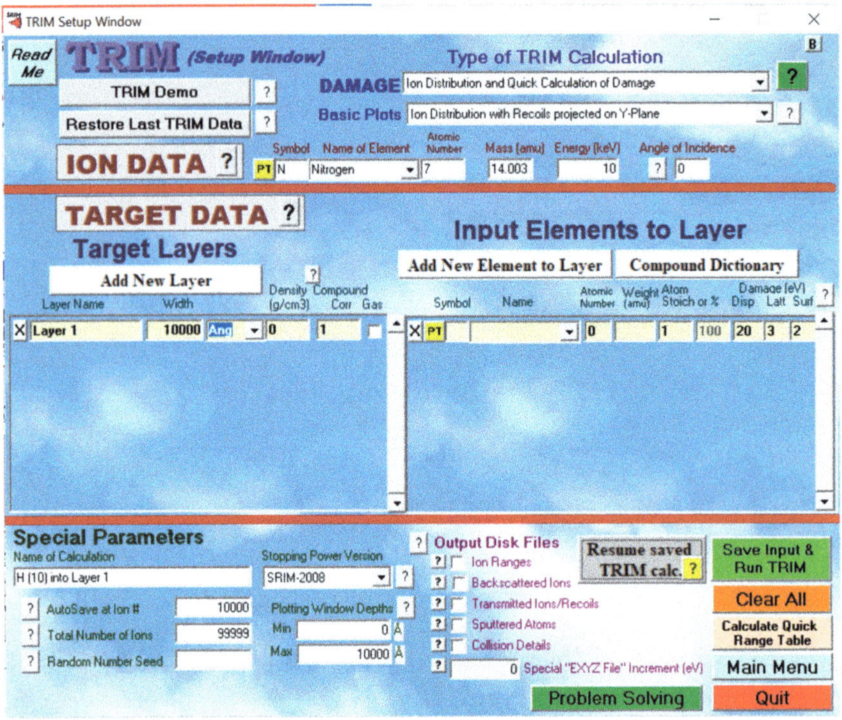

Figure 3.2 TRIM interface.

Varying the amount of incident particle energy creates different situations. For example, if a particle with energy E, which equals E1+E2, collides with the target atom, E1 is left in the collision particle, and E2 is transferred to the target atom:

- If E2 is less than E_{disp}, the target atom will return to its original place as a vibration. No displacement will occur, and the energy in the form of phonons will be released into the crystal. If E1 is less than E_{disp}, the particle will stop in the crystal lattice, becoming an interstitial atom, and the energy will be released as a phonon.
- If E2 is more significant than E_{disp}, the target atom will hit the original lattice position and cause a displacement. If the target atom moves, the energy minus the lattice energy will still be more significant than E_{disp}. The target atom will move on, striking other targets and creating more vacancies.
- When E2 is more significant than E_{disp}, E1 will also be greater than E_{disp} leading to the formation of vacancies. However, when E1 is less than E_{disp}, if the particles and the target are of the same element, the incident particles will replace the target atoms and stay in the vacancies. Then, the energy will be released in phonons, the so-called replacement collision. However, if the incident particles and the target atoms are of different elements, the incident particle will stop

inside the lattice as an interstitial atom. Then, the output interface will show the data obtained after the simulation, such as the particles' distribution and the damage level.

Before preparing an FND, the ion source will be chosen for a simulation. In this case, helium ions and protons were selected. First, the incident particle hydrogen atom was selected. The parameter values are built into the software. The simulation began with low collision particle energy and gradually increased it. The subject target we chose was the carbon atom because the diamond is a well-known allotrope of carbon. Thus, a diamond density of 3.52 g/cm^3 must be entered in this section. For the damage, three previously mentioned energies, E_{disp}, E_{latt}, and E_{surf}, were inputted as 37.5, 3, and 7.4 eV, respectively. The target thickness was set as 100 microns. The estimation time and accuracy were based on the calculation. The number of samples was 20 000. After completing the proton simulation, we switched the incident ion to helium. The remaining parameters do not need to be changed for the next simulation. This program can choose two ion sources simultaneously and make any change for simulation based on the user's needs.

In preparing FNDs, researchers typically use 40 keV of protons and 3 MeV of helium ions (16). According to the SRIM results, we can infer that a 3-MeV proton will cause a penetration depth of about 50 μm. When the energy is reduced to 40 keV, the penetration depth will be reduced to 230 nm. When a 40-keV helium ion source is used, the penetration depth is about 200 nm. Hence, different energies and ion sources result in different penetration depths. However, the question then arises as to how vacancies form. From the simulation results with 3-MeV protons, we can expect 13 vacancies, and when particle energy is reduced to 40 keV, the number of vacancies drops to four. When using 40-keV helium ions, we find 40 vacancies in FND. Given this, even though protons have lighter masses and deeper penetration, they cannot create additional vacancies for particles of the same energy. Heavier helium ions cannot penetrate as deeply but can create more vacancies. Therefore, when the target is an ND, a helium ion source effectively creates defects and reduces the needed ion dose and reaction time.

3.3 Production and Characterization of Fluorescent Nanodiamonds

After the previously mentioned simulation process, the FND is ready for hands-on preparation. Figure 3.3 shows a diagram of the related experimental setup (17), where He^+ ions are produced through a discharge of pure He^+ in a radio frequency (RF) positive ion source and then accelerated to 40 keV via a high-voltage acceleration tube. A summarized process representing the production of FNDs is shown in Figure 3.4. Typically, the nitrogen-containing ND is irradiated with a

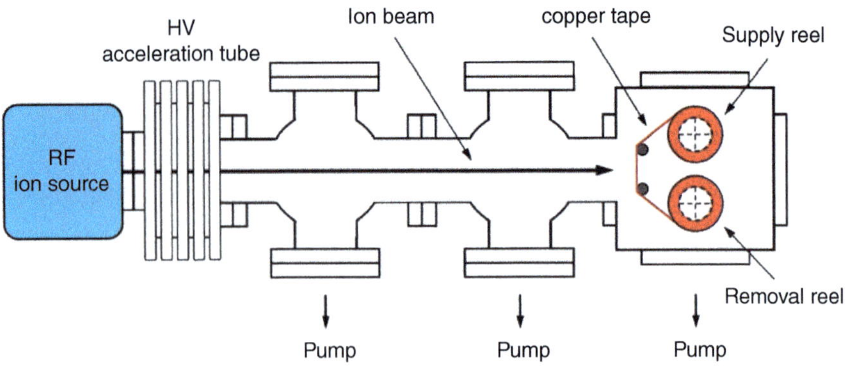

Figure 3.3 Schematic of the experimental setup for the production of FNDs with a medium-energy ion beam. *Source:* This figure is adapted from ref. (17).

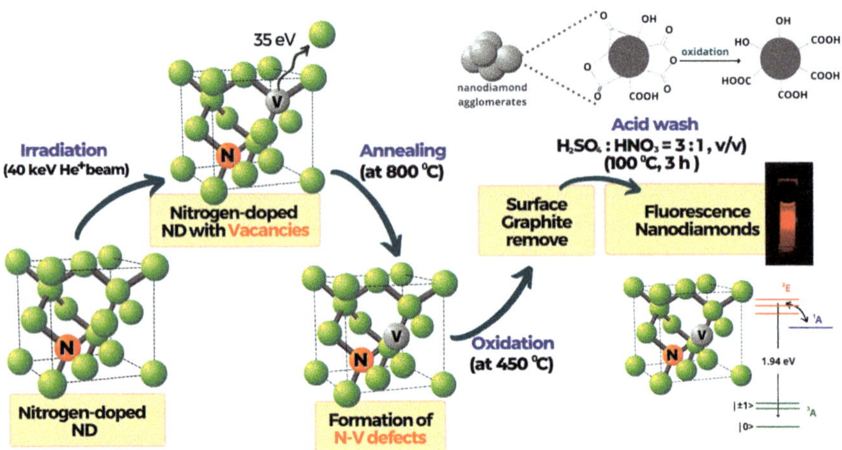

Figure 3.4 A schematic diagram representing the production of FNDs.

~40-keV He^+ beam, followed by annealing (800 °C) in a vacuum to form the NV^- centers, resulting in the creation of FND. Before fluorescence measurements, the freshly prepared FNDs were oxidized in air at 450 °C for 1 h (17, 18), followed by surface derivatization with carboxyl groups in strong oxidative acids to minimize fluorescence quenching graphitic surface structures possibly formed during ion bombardment and thermal annealing (16, 18, 19). Finally, FND was washed in a microwave reactor in a concentrated H_2SO_4:HNO_3 (3:1, v/v) solution at 100 °C.

According to the SRIM simulation, assuming an irradiation dose of 1×10^{16} ions cm^{-2} is employed, 1×10^4 vacancies can be produced in each 100-nm FND (equivalent to 100 ppm, correspondent with the density of nitrogen in each 100-nm ND). However, the density of NV^- centers in the final 100-nm FND product might be as much as 10 ppm (~10% efficiency as found in each bulk diamond) (20). Note that the NV^- centers in the FND lattice are distributed evenly. The distance between nitrogen and vacancy is estimated at approximately 10 nm.

Understanding and manipulating the behavior of fluorescent flaws in NDs remains challenging. For instance, the shape, size, and optical characteristics of individual FNDs in five distinct particle assemblages were studied with different preparation methods, including photoirradiation by a 15.5-MeV proton beam derived from the isochronous cyclotron U-120M for 70 minutes, electron irradiation by a 16.6-MeV electron beam extracted from the microtron MT25 for 21 hours, and He^{2+}+Li^+ irradiation (21, 22). In all samples except one (p^+), the most significant connection was seen between NV proportion and PL lifespan. The particles' fluorescence brightness is at best modest (He^{2+}) and frequently corresponds either faintly or not at all with their size. One sample's T1 loop relaxation time (p+) was substantially longer. In-solution spectroscopy revealed that all samples had a much greater NV proportion and a shorter PL lifetime than those produced on Si wafer substrates. This is due to the solution experiment's reduced excitation intensity and emphasizes the importance of the imaging state in general. Based on a knowledge of NV centers in ND, the He^{2+} sample was determined to have a substantially smaller NV proportion than the other samples, in contrast to single-particle analyses. This method can improve the quality of a single FND and gain better knowledge.

3.4 Latest Developments in High-Pressure, High-Temperature Fluorescent Nanodiamonds

In recent years, a novel type of FND material has been developed, manufactured from organic molecules utilizing the high-pressure, high-temperature (HPHT) process (23–25). Materials are composed under thermodynamically stable environments that are best for creating flawless diamond structures. They are also free of the metal contaminants present in typical HPHT synthesis using metallic media. When ND particles containing silicon-vacancy (SiV) and germanium-vacancy (GeV) centers were created by combining hydrocarbons with silicon and germanium dopants, narrow luminescence lines were obtained due to the diamond matrix's excellent structural quality (25, 26).

The creation of nano- and microdiamonds, including nitrogen-, silicon-, and germanium-related centers, demonstrates the feasibility of doping "organic" diamonds with optically active impurities (24, 27). Diamond crystallites with a single luminous core can be made by regulating the dopant concentration in organic precursors. In a combination of adamantane and $C_{24}H_{20}Ge$, for example, 50-nm diamonds with a single GeV core were created by lowering the concentration of Ge to 0.004 at% (27). Single-photon emission from this particle is confirmed by measuring the second-order autocorrelation function, g^2 (τ), for the GeV emission. A combination of adamantane and adamantane carbonitrile was used to produce larger crystallites. They ranged in size from 200 to 2 μm.

In the spectrum, three types of nitrogen-related faults with photoluminescence lines are seen: (i) negatively charged "nitrogen-vacancy" centers (NV^-); (ii) "double nitrogen-vacancy" (NVN or H3) centers; and (iii) neutrally charged "nitrogen-vacancy" centers (NV^0). Narrow zero-phonon lines (ZPLs) characterize the fluorescence of these centers at 638 nm (NV^-), 575 nm (NV^0), and 504 nm (H3), which are followed by a succession of broad lines (phonon replicas) at lesser wavelengths. Most of the smallest crystallites in these samples (200–300 nm) possessed one to three NV centers. By premixing the original organic molecules in a liquid solvent, good homogeneity in the distribution of nitrogen and NV centers in the sample volume is generated (28).

3.5 Production of Ultrasmall Fluorescent Nanodiamonds

The FND powder (median size of 50-nm FNDs provided by FND Biotech) usually has a broad size distribution. Approximately 10% of FND particles in the powder are smaller than 20 nm. Alternatively, we can use 100-nm FNDs (as a starting material) and then crush them with a 300-ton hydraulic press (30 times higher pressure than previous work (29)) to generate more sub-20-nm FNDs. After optimizing the experimental parameters in differential centrifugation, we can extract sub-20-nm FNDs from the 50-nm FNDs with multistep centrifugation procedures (30). The sizes of these FND particles can be characterized by dynamic light scattering (DLS), transmission electron microscopy (TEM), or atomic force microscopy (AFM).

For instance, FNDs were homogeneously dispersed on the surface of the crucible and put in a furnace at 500 °C for 1 h. FNDs would then be oxidized because the oxidation of FNDs makes the size smaller in the air (31). After the oxidation process, FNDs were then dispersed with distilled water. A different centrifugation process separated smaller-size FNDs from the suspension solution. Figure 3.5 shows the complete process of obtaining smaller FNDs. The parameters for this centrifugation process are 16 100 g and 95 min. This process led to sub-20-nm FNDs in the supernatant to the top of the tube. The solvent was then removed via the vacuum rotary evaporator, and ~17-nm FNDs could be obtained in the suspension solution. However, getting sub-10-nm FNDs requires 12 h at the speed of 16 100 g. It can also be obtained in a mere 20 min at the rate of 102 217 g. With these parameters, FNDs can be obtained with an average size of ~7 nm.

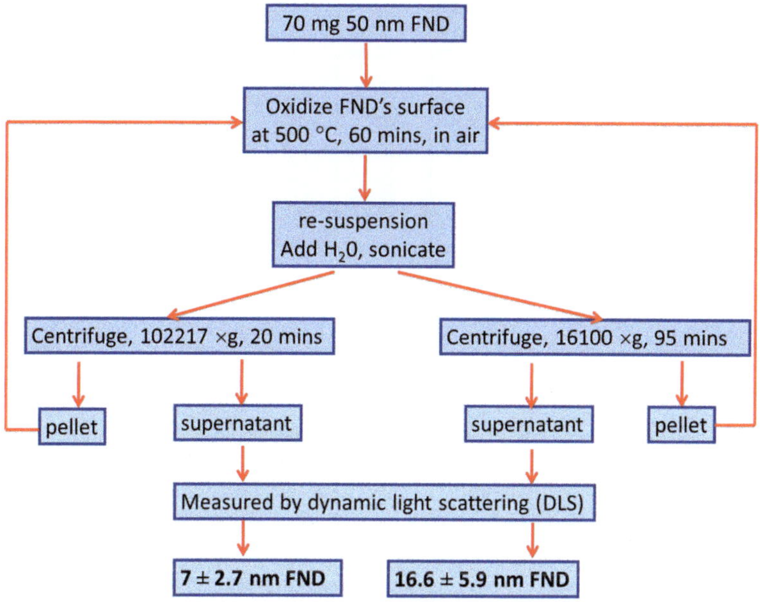

Figure 3.5 Process of obtaining smaller FNDs.

Unfortunately, technological limitations make it challenging to obtain 7-nm FNDs. Figure 3.6a shows the size distribution of FNDs, and Figure 3.6b shows the fluorescence spectra of different FND sizes. Note that the fluorescence intensity of 17-nm FNDs is two times dimmer than that of 35-nm FNDs. These sub-20-nm FNDs can potentially be used in developing a sensitive method to measure changes in intracellular transport.

The production of ultrasmall and bright FNDs is a highly challenging task. However, it is essential for providing specific intracellular transport readouts. Reduced FND size increases the fluorescent nanoprobe's mobility inside a cell, avoids changing the characteristics of targeted molecules, and improves its capacity to monitor intraneuronal transport. Moreover, sub-20-nm particles are more suitable than the submicrometer ones for biolabeling and intracellular transport studies (32).

To produce these ultrasmall nanotracers, high-quality FNDs in the 100–200 nm size range can be used as the starting material. Typically, these 100–200 nm FNDs are crushed down to be less than 20 nm using a hydraulic press (29). These sub-20-nm FNDs are also surface-derivatized with carboxylic groups by air oxidation and subsequent strong oxidative acid treatment (32). They can be covalently coupled with bioactive ligands and proteins by carbodiimide-based crosslinking and other methods through the surface-linked carboxylic group.

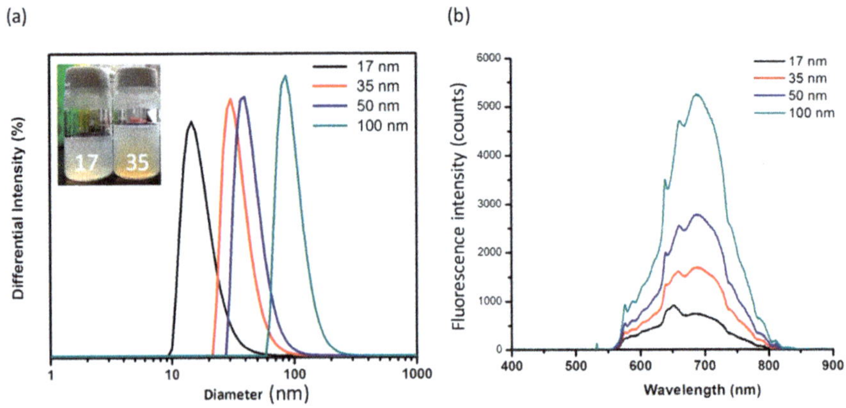

Figure 3.6 Size and fluorescence intensity of different fluorescent nanodiamond productions: (a) Size distribution. FND aqueous suspensions at 0.1-mg/ml concentration and (b) Fluorescence spectra of different FND sizes. Laser excitation at 532 nm in wavelength. *Source:* Image courtesy of Dr. Feng-Jen Hsieh.

The following list summarizes the methods used to make ultrasmall FNDs:

1) To manufacture FNDs, diamond powders with a nominal size of 100 nm were irradiated with a 40-keV helium ion beam, followed by annealing in a vacuum at 800 °C for 2 h and air oxidation at 450 °C for 2 h, as detailed in the previous study by Su *et al.* (29). 10-MeV electrons were employed to irradiate 100-nm FNDs on a larger scale (33, 34).
2) FND particles with sizes of 18 nm were created by crushing 100-nm FNDs into smaller pieces using a hydraulic oil press. FNDs with a size of around 100 nm were combined with NaF powders at a weight ratio of 1:2.5 and crushed into a pellet using a hydraulic oil press under a pressure of 20 tons. The pellet was then annealed in a vacuum at 800 °C for 2 hours before being dissolved in hot water to remove NaF and recover FNDs.

DLS and fluorescence spectroscopy can determine the size and intensity of these ultrasmall particles. Figure 3.7a demonstrates that the average diameter of the smallest FNDs extracted using a typical centrifuge is around 18 nm, with a size variation ranging from 10 to 30 nm. Figure 3.7b shows the results of the fluorescence measurements for these ultrasmall FND particles, demonstrating a twofold difference in fluorescence intensity before and after crushing the nanoparticles.

The methods of synthesizing FND from their molecular counterpart, adamantane, have also been established (24). According to Raman analysis, at a pressure of roughly 9 GPa and a temperature of 1250 °C, adamantane begins to convert into diamond in a titanium capsule. In the temperature range of 1250–1400 °C, a

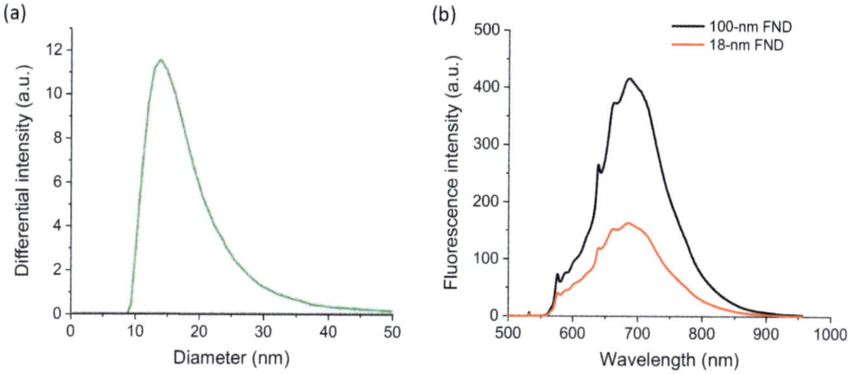

Figure 3.7 (a) Hydrodynamic diameter and (b) fluorescence intensity of sub-20-nm FNDs dispersed in water. *Source:* This figure is adapted from ref. (32).

nanocrystalline type of diamond predominates, but at greater temperatures, diamond microcrystals prevail (27). At a synthesis temperature of 1330 °C, diamond crystals as small as 3 nm were created.

3.6 Synthesis of Highly Dispersible Fluorescent Nanodiamonds

Various hydrophilizing strategies to improve FNDs' colloidal stability in biological applications have been identified to prevent nonspecific adsorption and aggregation of biomolecules in challenging biomedical conditions. Noncovalent adsorption of proteins on the FND surface, such as bovine serum albumin (BSA) and human serum albumin (HSA), to reduce nonspecific molecule adsorption and FND aggregation (35–37), is a commonly utilized method to address this constraint. An alternative approach is to cover FNDs with polyethylene glycol (PEG) or silica, although these coatings do not prevent aggregation or the adsorption of nonspecific biomaterials. Although the noncovalent manipulation of FND via electrostatic (37) and hydrophobic associations is easy, it restricts further chemical alteration of FND since the molecular connection on the nanoparticle surface is dependent on the solution environment. Nevertheless, biologists often cannot employ FNDs in therapeutic or other biological applications because these approaches require multiple complex syntheses and purification stages.

Even under physiological conditions, FNDs coated with hyperbranched polyglycerol (HPG) are harmless, have excellent solubility, and are biocompatible (38). HPG has many hydroxyl (OH) groups that are less hydrophobic than PEG chains'

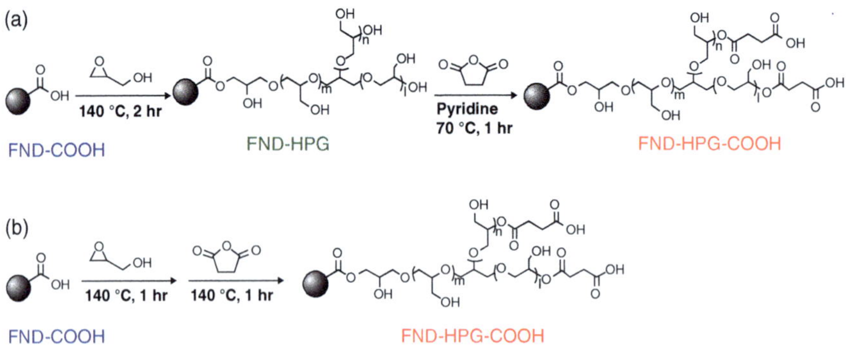

Figure 3.8 (a) Conventional and (b) one-pot synthesis of an HPG-coated fluorescent nanodiamond with terminal carboxyl groups (FND-HPG-COOH). *Source:* (40) Terada et al., 2018 / American Chemical Society.

ether groups. Furthermore, HPG's branched polyglycerol chains efficiently cover FNDs' hydrophobic interface. As a result, substituting FNDs with HPG increases their hydrophilicity and significantly reduces the nonspecific adsorption of biomolecules like proteins (39). As shown in Figure 3.8, this easy, one-pot synthetic process for generating FND-HPG-COOH allows the production of hydrophilic particulates that can conduct bioconjugation processes more rapidly and readily than the traditional method for FND-HPGs. The carboxylation in the conventional method requires additional reactions, like using organic base catalysts such as pyridine, and further purification step (40). Many other approaches to modifying the surface of ND to enhance its colloidal stability are available and will be discussed in the next chapter (Chapter 4).

3.7 Brightly Synthesized Fluorescent Nanodiamonds Using Solar Energy

This section outlines a simple, environmental, and low-cost production technique for FND. Low-grade coal is often used as a carbon precursor, and sunlight is the only energy source in NDs' production. High-powered magnifying glasses focused the sunlight (41). According to comprehensive characterization, the FND created has identical physicochemical and photo properties to NDs manufactured using other techniques. Due to oxygen-containing functional groups, the NDs produced from this method have high colloidal stability (such as hydroxyl and carboxyl). When treated with a near-infrared (NIR) laser, the NDs show significant fluorescence at 630 nm (excitation $\lambda = 590$ nm) and photothermal reaction. The FND's biocompatibility was demonstrated *in vitro*. The endocytosed

NDs fluoresced brightly, indicating they were within the cells' cytoplasm. When the NDs in cells were bombarded with a NIR laser, subsequent photothermal research demonstrated an effective dosage-dependent response. Ramana *et al.* established a simple, inexpensive, safe, and environmentally friendly synthetic technique, which can help extend ND research in various domains, including biomedical engineering (41).

As a carbon source, lignite was employed. Using a mortar and pestle, coal (1 g) was ground into a fine powder and dispersed in MilliQ water (100 ml) using sonication for 10 minutes. Impurities were removed from the dispersion by centrifugation at 10 000 rpm. The pelleted coal was dried at 300 °C for 5 hours to eliminate water and other contaminants. The powdered, dry coal pellets were redispersed in 450 ml of water, augmented with 45 ml of 30% hydrogen peroxide, and placed in a ceramic crucible with a crucible cover with an aperture in the center. To synthesize NDs, the dispersion was irradiated with intense sunlight focused using six high-power magnifying lenses (magnification of 30×) for 2 hours (41).

As illustrated in Figure 3.9, the magnifying glasses were oriented to direct the sun's rays through the aperture in the crucible lid toward the center of the crucible containing the coal dispersion. The reaction mixture's temperature was raised to around 250 °C, measured with a photothermal. The color of the reaction

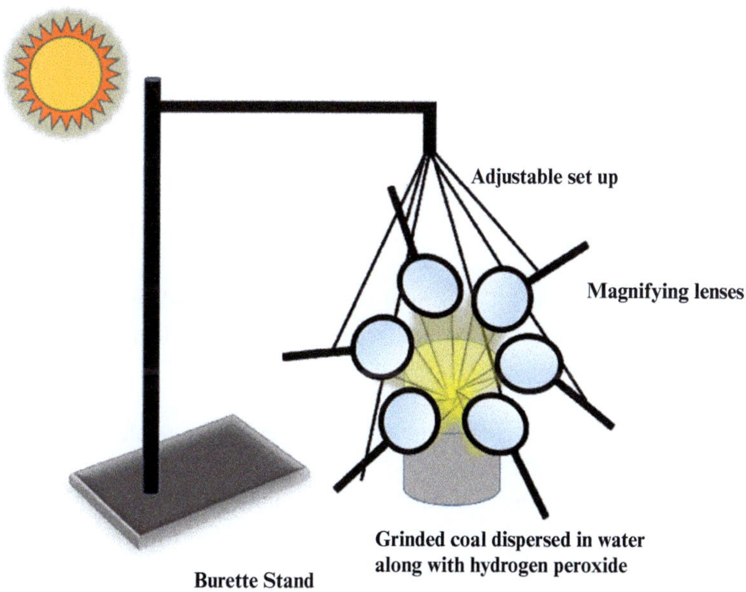

Figure 3.9 FND synthesis using sunlight. *Source:* (41) Ramana et al., 2021 / American Chemical Society.

mixture changed from gray to brown, indicating the creation of NDs. The reaction volume was decreased to around 40 ml after 2 hours of irradiation owing to water evaporation, which was centrifuged at 10 000 rpm to extract dust coal and generated NDs. The brown supernatant with NDs was collected and dialyzed for 5 hours against water. To get powdered NDs, the dialyzed sample was collected and dried in a 50 °C oven (41).

With the rapid development of nanoscience, many nanomaterials have been synthesized and discovered. FND is emerging as a type of nanomaterial that holds great promise for biological applications. Containing a high concentration of NV^- defect centers as a fluorophore, FND exhibits several remarkable features, including the emission of bright photoluminescence in the NIR window without photobleaching and photoblinking. In addition, many studies have shown that FND surfaces can be conveniently derivatized with functional groups for protein immobilization, realizing functional FNDs, which allows visualizing, tracking, and quantifying molecules as well as cells with a high spatial resolution both *in vitro* and *in vivo* for biological applications (35, 42). The fabrication method of FND outlined in this chapter not only benefits laboratories equipped with the infrastructure for producing FND but also has the potential to increase the yield of excellent FND by creating a higher number of vacancies.

Acknowledgments

This work was supported by the National Science and Technology Council (NSTC), Taiwan, under Grant No. 112-2221-E-011-031.

References

1 Hsiao WW-W, Hui YY, Tsai P-C, Chang H-C. Fluorescent nanodiamond: A versatile tool for long-term cell tracking, super-resolution imaging, and nanoscale temperature sensing. *Acc Chem Res*. 2016;49(3):400–7.
2 Balasubramanian G, Chan I, Kolesov R, Al-Hmoud M, Tisler J, Shin C, *et al.* Nanoscale imaging magnetometry with diamond spins under ambient conditions. *Nature*. 2008;455(7213):648–51.
3 Hsiao WW-W, Le T-N, Chang H-C. Applications of fluorescent nanodiamond in biology. In: Meyers, R.A. (Ed.) *Encyclopedia of Analytical Chemistry*. Hoboken, NJ, USA: John Wiley & Sons; 2022. pp. 1–43.
4 Gruber A, Dräbenstedt A, Tietz C, Fleury L, Wrachtrup J, Borczyskowski CV. Scanning confocal optical microscopy and magnetic resonance on single defect centers. *Science*. 1997;276(5321):2012.

5 Yu S-J, Kang M-W, Chang H-C, Chen K-M, Yu Y-C. Bright fluorescent nanodiamonds: No photobleaching and low cytotoxicity. *J Am Chem Soc.* 2005;127(50):17604–5.

6 Jelezko F, Wrachtrup J. Single defect centers in diamond: A review. *Phys Status Solidi A.* 2006;203(13):3207–25.

7 Davies G, Hamer MF, Price WC. Optical studies of the 1.945 eV vibronic band in diamond. *Proc R Soc A: Math Phys Eng Sci.* 1976;348(1653):285–98.

8 Mita Y. Change of absorption spectra in type-Ib diamond with heavy neutron irradiation. *Phys Rev B.* 1996;53(17):11360–4.

9 Fu CC, Lee HY, Chen K, Lim TS, Wu HY, Lin PK, *et al.* Characterization and application of single fluorescent nanodiamonds as cellular biomarkers. *Proc Natl Acad Sci USA.* 2007;104(3):727–32.

10 Acosta VM, Bauch E, Ledbetter MP, Santori C, Fu KMC, Barclay PE, *et al.* Diamonds with a high density of nitrogen-vacancy centers for magnetometry applications. *Phys Rev B.* 2009;80(11):115202.

11 Ziegler JF, Biersack JP. *The Stopping and Range of Ions in Matter. Treatise on Heavy-ion Science.* Springer; 1985. pp. 93–129.

12 Biersack JP, Haggmark LG. A Monte Carlo computer program for the transport of energetic ions in amorphous targets. *Nucl Instrum Methods.* 1980;174(1):257–69.

13 Robinson MT, Torrens IM. Computer simulation of atomic-displacement cascades in solids in the binary-collision approximation. *Phys Rev BB.* 1974;9(12):5008–24.

14 Smith R. *Atomic and Ion Collisions in Solids and at Surfaces.* Cambridge, United Kingdom: Cambridge University Press; 1997. January 01, 1997. p. 319.

15 Was GS. *Fundamentals of Radiation Materials Science: Metals and Alloys.* New York, NY: Springer; 2016.

16 Wee TL, Tzeng YK, Han CC, Chang HC, Fann W, Hsu JH, *et al.* Two-photon excited fluorescence of nitrogen-vacancy centers in proton-irradiated type ib diamond. *J Phys Chem A.* 2007;111(38):9379–86.

17 Chang Y-R, Lee H-Y, Chen K, Chang C-C, Tsai D-S, Fu C-C, *et al.* Mass production and dynamic imaging of fluorescent nanodiamonds. *Nat Nanotechnol.* 2008;3(5):284–8.

18 Huang LCL, Chang H-C. Adsorption and Immobilization of Cytochrome c on Nanodiamonds. *Langmuir.* 2004;20(14):5879–84.

19 Ushizawa K, Sato Y, Mitsumori T, Machinami T, Ueda T, Ando T. Covalent immobilization of DNA on the diamond and its verification by diffuse reflectance infrared spectroscopy. *Chem Phys Lett.* 2002;351(1):105–8.

20 Pezzagna S, Naydenov B, Jelezko F, Wrachtrup J, Meijer J. Creation efficiency of nitrogen-vacancy centres in diamond. *New J Phys.* 2010;12(6):065017.

21 Reineck P, Trindade LF, Havlik J, Stursa J, Heffernan A, Elbourne A, *et al.* Not all fluorescent nanodiamonds are created equal: A comparative study. *Part Part Syst Charact.* 2019;36(3):1900009.

22 Havlik J, Petrakova V, Rehor I, Petrak V, Gulka M, Stursa J, *et al.* Boosting nanodiamond fluorescence: Towards development of brighter probes. *Nanoscale.* 2013;5(8):3208–11.

23 Alkahtani MH, Alghannam F, Jiang L, Rampersaud AA, Brick R, Gomes CL, *et al.* Fluorescent nanodiamonds for luminescent thermometry in the biological transparency window. *Opt Lett.* 2018;43(14):3317–20.

24 Ekimov EA, Kudryavtsev OS, Mordvinova NE, Lebedev OI, Vlasov II. High-pressure synthesis of nanodiamonds from Adamantane: Myth or reality? *ChemNanoMat.* 2018;4(3):269–73.

25 Jantzen U, Kurz AB, Rudnicki DS, Schäfermeier C, Jahnke KD, Andersen UL, *et al.* Nanodiamonds carrying silicon-vacancy quantum emitters with almost lifetime-limited linewidths. *New J Phys.* 2016;18(7):073036.

26 Ekimov EA, Lyapin SGE, Boldyrev KN, Kondrin MV, Khmelnitskiy R, Gavva VA, *et al.* Germanium–vacancy color center in isotopically enriched diamonds synthesized at high pressures. *JETP Lett.* 2015;102(11):701–6.

27 Ekimov EA, Kondrin MV, Krivobok VS, Khomich AA, Vlasov II, Khmelnitskiy RA, *et al.* Effect of Si, Ge and Sn dopant elements on structure and photoluminescence of nano- and microdiamonds synthesized from organic compounds. *Diam Relat Mater.* 2019;93:75–83.

28 Shenderova OA, Shames AI, Nunn NA, Torelli MD, Vlasov I, Zaitsev A. Review article: Synthesis, properties, and applications of fluorescent diamond particles. *J Vac Sci Technol B.* 2019;37(3):030802.

29 Su L-J, Fang C-Y, Chang Y-T, Chen K-M, Yu Y-C, Hsu J-H, *et al.* Creation of high density ensembles of nitrogen-vacancy centers in nitrogen-rich type Ib nanodiamonds. *Nanotechnology.* 2013;24(31):315702.

30 Morita Y, Takimoto T, Yamanaka H, Kumekawa K, Morino S, Aonuma S, *et al.* A facile and scalable process for size-controllable separation of nanodiamond particles as small as 4 nm. *Small.* 2008;4(12):2154–7.

31 Gaebel T, Bradac C, Chen J, Say JM, Brown L, Hemmer P, *et al.* Size-reduction of nanodiamonds via air oxidation. *Diam Relat Mater.* 2012;21:28–32.

32 Mohan N, Tzeng YK, Yang L, Chen YY, Hui YY, Fang CY, *et al.* Sub-20-nm fluorescent nanodiamonds as photostable biolabels and fluorescence resonance energy transfer donors. *Adv Mater.* 2010;22(7):843–7.

33 Boudou J-P, Curmi PA, Jelezko F, Wrachtrup J, Aubert P, Sennour M, *et al.* High yield fabrication of fluorescent nanodiamonds. *Nanotechnology.* 2009;20(23):235602.

34 Lu HC, Peng YC, Chou SL, Lo JI, Cheng BM, Chang HC. Far-UV-excited luminescence of nitrogen-vacancy centers: Evidence for diamonds in space. *Angew Chem Int Ed.* 2017;129(46):14661–5.

35 Chang B-M, Lin H-H, Su L-J, Lin W-D, Lin R-J, Tzeng Y-K, *et al.* Highly fluorescent nanodiamonds protein-functionalized for cell labeling and targeting. *Adv Funct Mater.* 2013;23(46):5737–45.

36 Liu W, Yu F, Yang J, Xiang B, Xiao P, Wang L. 3D Single-molecule imaging of transmembrane signaling by targeting nanodiamonds. *Adv Funct Mater.* 2016;26(3):365–75.

37 Wu Y, Ermakova A, Liu W, Pramanik G, Vu TM, Kurz A, *et al.* Programmable biopolymers for advancing biomedical applications of fluorescent nanodiamonds. *Adv Funct Mater.* 2015;25(42):6576–85.

38 Zhao L, Takimoto T, Ito M, Kitagawa N, Kimura T, Komatsu N. Chromatographic separation of highly soluble diamond nanoparticles prepared by polyglycerol grafting. *Angew Chem Int Ed.* 2011;50(6):1388–92.

39 Sotoma S, Igarashi R, Iimura J, Kumiya Y, Tochio H, Harada Y, *et al.* Suppression of nonspecific protein–nanodiamond adsorption enabling specific targeting of nanodiamonds to biomolecules of interest. *Chem Lett.* 2014;44(3):354–6.

40 Terada D, Sotoma S, Harada Y, Igarashi R, Shirakawa M. One-pot synthesis of highly dispersible fluorescent nanodiamonds for bioconjugation. *Bioconjugate Chem.* 2018;29(8):2786–92.

41 Ramana LN, Agarwal V. Nanodiamonds synthesis using sustainable concentrated solar thermal energy: Applications in bioimaging and phototherapy. *Nanotechnology.* 2021;32(47):475602.

42 Wu T-J, Tzeng Y-K, Chang W-W, Cheng C-A, Kuo Y, Chien C-H, *et al.* Tracking the engraftment and regenerative capabilities of transplanted lung stem cells using fluorescent nanodiamonds. *Nat Nanotechnol.* 2013;8(9):682–9.

4

Surface Modification of Nanodiamonds

Stefanny Angela, Tzu-Chun You, Dinh Minh Pham, Trong-Nghia Le, and Wesley Wei-Wen Hsiao**

* *Corresponding authors*

Nanodiamonds (NDs) are biocompatible and noncytotoxic, composed entirely of carbon atoms. Researchers have been drawn to these unique characteristics and their future use in analytical chemistry and biology. However, three major limitations arising from the ND's intrinsic properties must be overcome for them to be successfully used in analytical and biological applications: (i) difficulty in directly modifying the surface of NDs; (ii) weak colloidal stability owing to ND particles' agglomeration under physiological conditions; and (iii) nonspecific protein adsorption to the surface of NDs. Typically, the colloidal stability of unmodified ND decreases with increasing ionic strengths in solution, as seen with numerous other nanoparticles. This can create significant difficulties in utilizing NDs in biological systems involving physiological buffers or high salt concentration culture mediums. Therefore, before any application, it is essential to comprehend the surface modification of NDs and the factors affecting the colloidal stability. This chapter describes three strategies for ND surface modification, including functionalization, encapsulation, and bioconjugation.

4.1 Surface Modification of Nanodiamonds

In the context of biological research, NDs hold significance only when they are present as a stable colloidal dispersion (1, 2). Dynamic light scattering (DLS) is a reliable technique for evaluating the properties and stability of

Nanodiamonds in Analytical and Biological Sciences: Principles and Applications, First Edition.
Edited by Wesley Wei-Wen Hsiao and Trong-Nghia Le.
© 2023 John Wiley & Sons Ltd. Published 2023 by John Wiley & Sons Ltd.

colloids. By analyzing the hydrodynamic particle sizes in the solution, DLS can determine the size values both before and after the surface modification. The zeta potential strongly influences the stability and susceptibility for agglomeration of a colloid. For instance, the colloid is stable if the zeta potential is more diminutive than −30 mV or higher than +30 mV. NDs are found to form colloids in the zeta potential range of −40 to +40 mV, depending on their surface modifications (3). Thus, particle agglomeration and nonspecific interaction impacts are expected to be prevented by appropriately modifying the surface.

4.2 Surface Functionalization of Nanodiamonds

NDs produced by any method, whether detonation, high-pressure, high-temperature (HPHT) or chemical vapor deposition, contain various moieties and generally retain sp^2 carbon atoms on the surface and impurities generated during synthesis. Additionally, due to their chemical inertness, NDs should undergo essential surface functionalization procedures before being used in practical applications. To efficiently eliminate these impurities and residual chemical components, researchers use powerful oxidative acids to treat as-formed ND materials (4, 5). Robust oxidizing solutions of H_2SO_4, concentrated hydrogen peroxide (H_2O_2), and supercritical water are typically used for NDs' purification (6). Air and ozone treatments at extreme temperatures have also been introduced as a more economical and ecologically friendly approach (7–10). As a result, a variety of oxidized function groups can be obtained, including carboxylic (COOH), carbonyl (C=O), different alcohol (tertiary, secondary, primary) and ether groups, etc. Specifically, approximately 7% of the total surface of HPHT ND after oxidative treatment to form COOH is sufficient for covalent and noncovalent bioconjugation (11). To analyze the composition and surface terminations of these NDs, Raman spectroscopy, Fourier transform infrared (FTIR) spectroscopy, thermal desorption mass spectrometry (TDMS), and X-ray photoelectron spectroscopy (XPS) are all routinely applied. Researchers can use FTIR to distinguish between many kinds of functional groups and adsorbed substances on NDs' surfaces. These techniques allow researchers to observe minute variations in the chemical composition of the surface before and after alteration. However, the superposition usually occurs in this range due to the spectra of carbonyl, aldehyde, carboxylic acid, ketone, ester, and other oxygen-containing groups. Thus, by borane reduction, Kruger *et al.* developed a homogeneous hydroxyl group layer on the NDs' surface to simplify the FTIR spectra analysis (12).

Another route to modify NDs' surface is gas treatments at high temperatures. By heating NDs in NH_3, a heterogeneity of nitrogen-containing functional groups such as NH_2, C≡N, and moieties containing C=N can be formed (13). When NDs are heated in Cl_2, they can form acyl chlorides; in F_2, they create C–F groups. The surface carboxylic or any oxidative function groups can be reduced to C–OH or even C–H groups. High-temperature treatment can also eliminate all hydroxyl groups by extended hydrogenation durations or hydrogen plasma treatment. Furthermore, using N_2, Ar, or vacuum conditions for heating at elevated temperatures can remove all functional groups and convert DNDs to graphitic carbon nano-onions. Krueger *et al.* have described all available strategies for chemically treating and functionalizing ND surfaces (13).

To conclude, either wet chemistry or high-temperature gas treatments have both made it possible to bind a wide array of functional groups to NDs' surfaces. These chemical modifications enable scientists to conjugate NDs with additional bioactive ligands or biomolecules for biomedical and biotechnology purposes. Figure 4.1 summarizes the typical techniques for ND surfaces' chemical modifications (14).

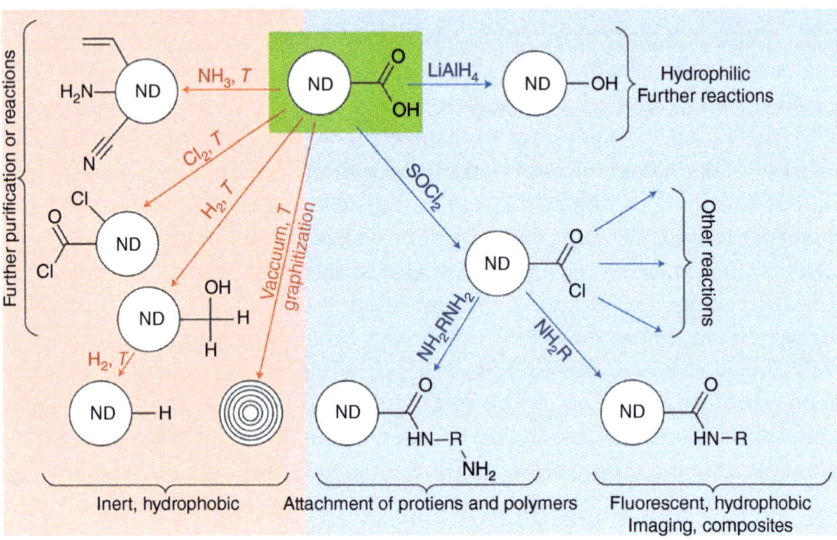

Figure 4.1 Summary of the most commonly applied approaches for ND surface modification. *Source:* (14) Mochalin et al., 2012 / Springer Nature.

4.3 Encapsulation of Nanodiamonds

Although direct surface modification is a straightforward way to modify ND characteristics, polymer grafting offers another potential approach. The polymer used to graft on the ND surface is obtained in two ways: the "grafting to" method and the "grafting from" method. The "grafting to" approach requires a chemical interaction between complementary functional groups and end-functionalized polymer chains on the surface of a solid substrate. In contrast, covalently attaching a polymer in situ to a solid substrate is the process known as "grafting from" (15).

One of the first polymers to be covalently grafted onto the carboxylated surface of NDs was poly-L-lysine by forming an amide link (16). Later, organosilanes or polyelectrolytes (such as polyallylamine) were used to functionalize NDs' surface with amino groups (17). NDs were also covered with polydopamine (PDA) shells. The abundance of catechol/quinine groups on the surface of the PDA shell might facilitate functionalization with thiol/amine-containing molecules through Michael addition or Schiff base reactions under oxidizing conditions. Therefore, NDs modified by thiol-terminated poly(ethylene glycol) (PEG-SH) molecules have improved colloidal stability and biocompatibility (18). Atom transfer radical polymerization was then discovered in 1995 and applied as a practical approach for adhering polymer brushes onto NDs' surfaces terminated with an initiator (19, 20). It was discovered in both water and buffers that are covering NDs with hyperbranched polyglycerol by ring-opening polymerization remarkably increased their colloidal stability. Recently, copolymers composed of poly[*N*-(2-hydroxypropyl)methacrylamide] have been developed as a nontoxic and protein-resistant covering. Under biological conditions, these hydrophilic and highly flexible polymers inhibit ND agglomeration and enable the bioorthogonal attachment of diverse molecules via "click chemistry" (21).

Many innovative scientists have developed silica-coated NDs (21–26). Due to its outstanding stability, flexibility of control over the coating process, controlled porosity, chemical inertness, optical clarity, processability, and biocompatibility, amorphous silica has been employed as a coating material. The silica-wrapped ND surface includes sustainable free silanol groups, which enables biomolecules to attach to the particles. The coating of NDs with methacrylamide copolymers produced from an ultrathin silica shell is presented in Figure 4.2. Through click chemistry, the copolymer can bear both fluorescent probes (e.g., Alexa Fluor 488) and targeting peptides (e.g., cRGD) (22). Chemical treatment enables the selective binding of molecules, which then significantly increases the colloidal stability of the particles. Furthermore, as shown in Figure 4.2c, the shell coating can correct the ND particle's irregular form (23).

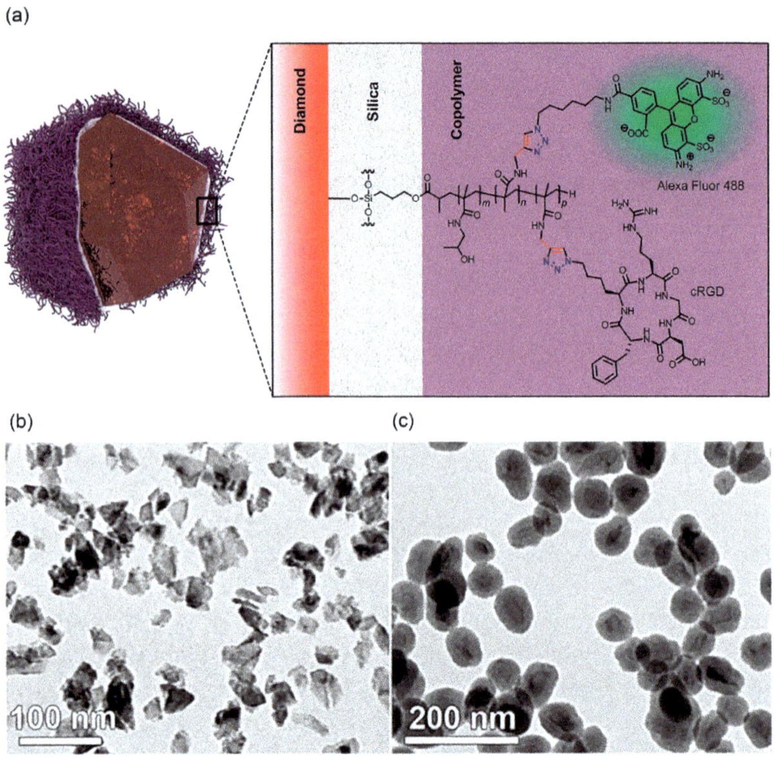

Figure 4.2 (a) Silica structure coated on ND particle for subsequent conjugation with polymeric molecules. *Source:* (22) Mochalin et al., 2012 / Royal Society of Chemistry. Transmission electron microscopy images of (b) before and (c) after coating silica ND particles. *Source:* Reproduced from ref. (23) / John Wiley & Sons.

NDs can combine with liposomes to generate complexes. Chang *et al.* wrapped HPHT NDs within cationic cholesterol-based lipids following surface reduction and silanization (Figure 4.3) (27). The encapsulated particles were quickly diffused, at the rate of more than one order of magnitude, into the cytoplasm of live cells. Ho *et al.* studied ND–lipid hybrid particles (NDLPs), which showed a robust interaction between the ND surface and small molecules and were self-assembled. Figure 4.4 depicts their proposed mechanism for the delivery of imaging or therapeutic payloads to specific cells (26, 28). Due to the simplicity of ND encapsulation in liposomes, particles can flexibly combine with bioactive proteins or ligands for targeted applications on the lipid layer.

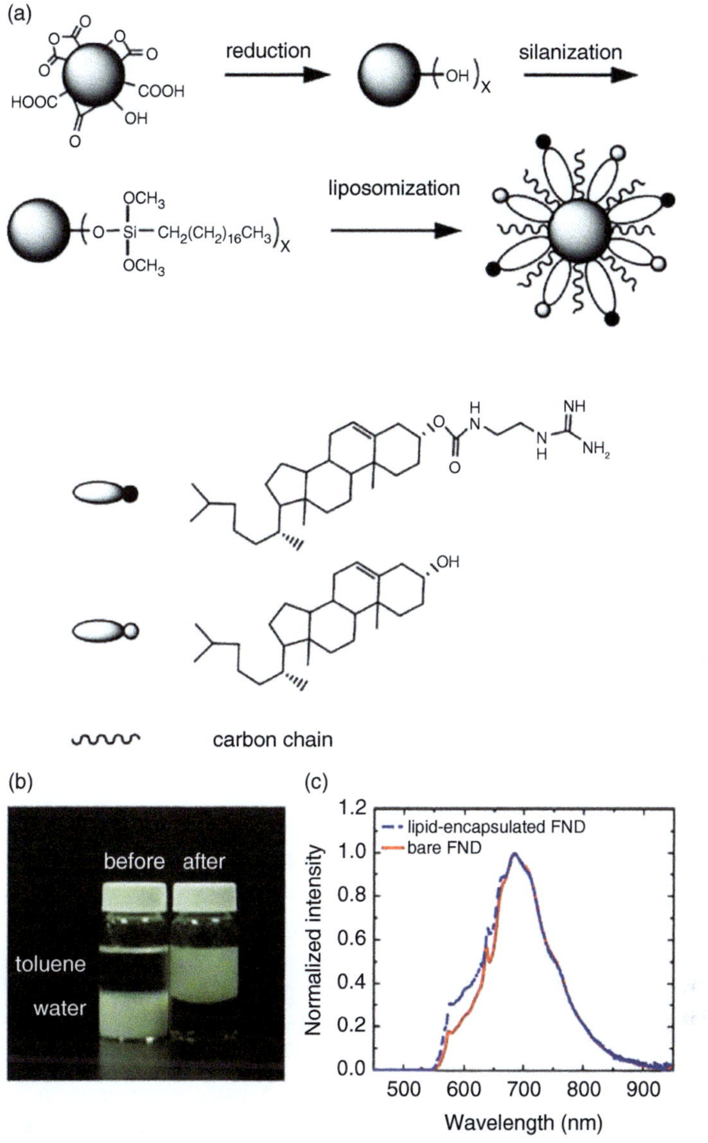

Figure 4.3 (a) Synthesis of lipid-encapsulated fluorescent nanodiamonds (FNDs). *Source:* (27) Hui et al., 2010 / Optica Publishing Group (b) 40-nm FNDs suspended in toluene/water before and after the reduction and silanization treatments. (c) Fluorescence spectra of bare and lipid-encapsulated 40-nm FNDs suspended in water. *Source:* Reproduced from ref. (27) / Optica Publishing Group.

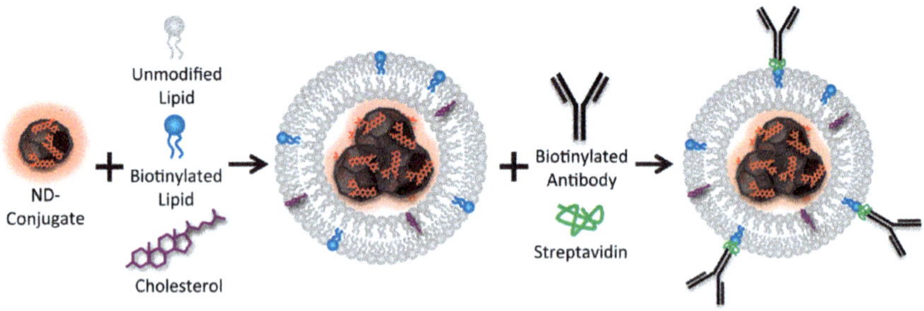

Figure 4.4 Wrap NDs in liposome by rehydrating lipid thin films comprising cholesterol and biotinylated lipids in concentrated ND solution. *Source:* This figure is adapted from refs. (26, 28).

4.4 Bioconjugation of Nanodiamonds

The NDs' surface can be highly functionalized in many ways, depending on the manufacturing technique and subsequent chemical treatment. Functionalized NDs can exhibit a wide range of properties, from inertness to sensitivity, and can even cause harm to cells and animals. In addition, biomolecules can be attached to NDs via noncovalent interactions or covalent bonding. For example, noncovalent adsorption of proteins, such as serum albumin, on NDs has been reported to stabilize nanoparticle colloids in phosphate-buffered saline (PBS) over seven days (29, 30).

Figure 4.5 summarizes some organic and biological molecules attached to NDs via covalent or noncovalent interaction. There are chemotherapeutic drugs (31, 32), carbohydrates (13, 33), peptides (17, 34), proteins (32), small interfering RNA (35), and DNA (36–38). When attaching the biomolecules in the form of proteins to the surface of NDs, it is essential to preserve their functionality. An excellent illustration of this is in the Nguyen *et al.* project, which immobilized lysozymes on HPHT NDs (11). These researchers observed the enzymes' hydrolytic activity upon adsorption and discovered that it reduced with decreasing surface coverage. The ND surface was blocked with extra proteins such as cytochrome C to overcome this problem, resulting in more "crowded" conditions. As a result of this strategy, the activity of the ND-bound lysozymes was boosted from 60% to 70%.

The technique of inserting spacers between biomolecules and NDs was introduced to overcome the limitation and retain the adsorbates' activity more

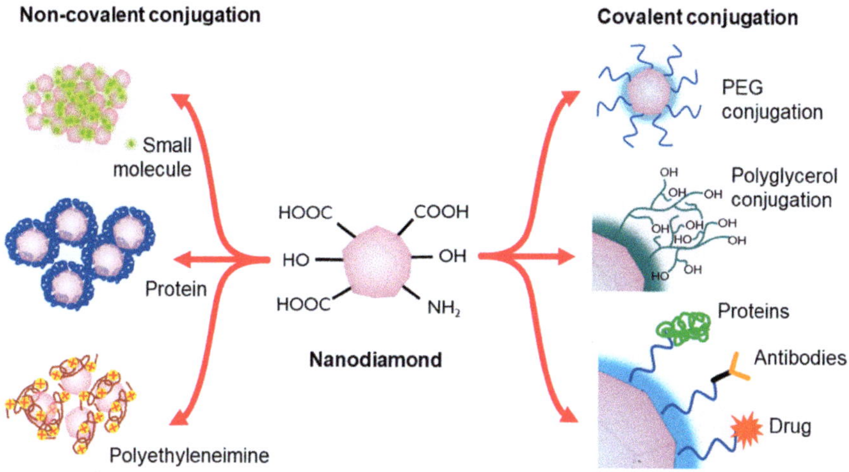

Figure 4.5 A scheme illustrating the bioconjugation of NDs with various types of biomolecules.

efficiently. This step is crucial for containing active site enzymes that may be sterically hampered after attaching to the ND's surface. Spacers also suppress nonspecific interactions and protein conformational changes generated by intense biomolecule–particle interactions. Polyethylene glycol is generally used as a spacer (19) due to its low toxicity and lack of particular interactions with biomolecules. In addition, it can make the covalent conjugation with the surface carboxylate group of ND due to its amino groups on termini or other conjugation processes (Figure 4.6). There are many types of spacers classed by discrete lengths. They contain carboxyl groups on the other termini to facilitate crosslinking carbodiimide with bioactive ligands (39) or proteins (39) that aim to label cells or target applications.

The distribution of many potent medications, including those used to treat cancer, is problematic. For instance, certain drugs are soluble in hazardous polar aprotic solvents but insoluble in polar protic solvents, such as water. To address this issue, a novel delivery method was created by combining surface modification with drug loading on NDs. Using NDs can achieve a synergy benefit of compatibility and high drug-loading capacity. In recent years, researchers have used ND surfaces to form chemical bonds with drugs (40, 41). However, most of this research has focused on adsorption (40, 41).

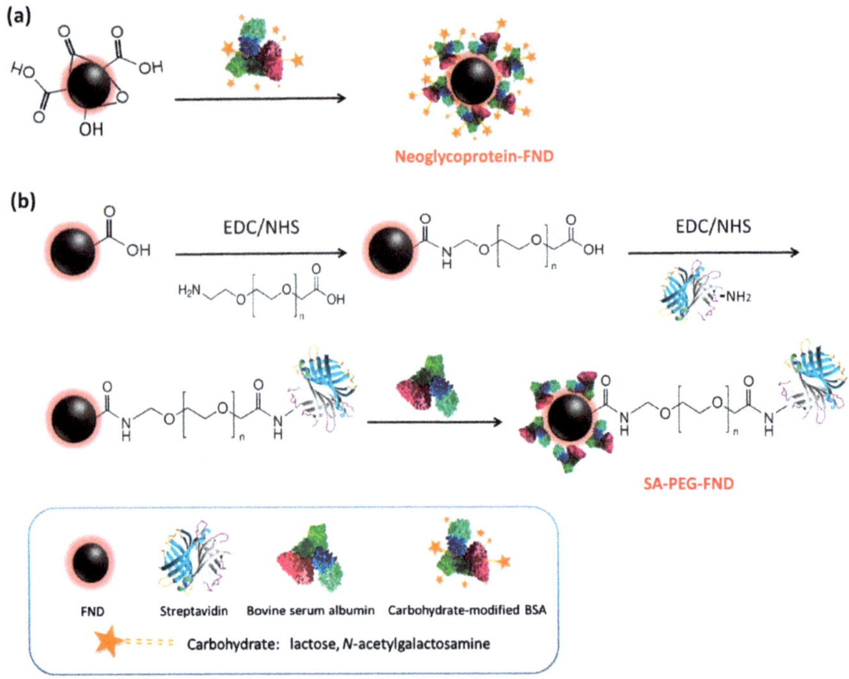

Figure 4.6 Processes of HPHT ND bioconjugation via (a) physical adsorption and (b) covalent bonding through activating the surface carboxyl groups. Abbreviations: FND: fluorescent nanodiamond; EDC: *N*-(3-dimethyl aminopropyl)-*N*′-ethyl carbodiimide hydrochloride; NHS: *N*-hydroxysuccinimide; PEG: polyethylene glycol. *Source:* (40) Chang et al., 2013 / John Wiley & Sons.

4.5 Application of Surface-Modified Nanodiamonds

Surface-modified NDs could assist in the adsorption or conjugation of various biomolecules, resulting in diverse applications in biotechnology and biomedicine. Some of the most attractive applications relate to the use of NDs as bactericidal agents, biomolecule transporters, enzyme immobilization substrates, and supports for solid-phase extraction (SPE). However, the current focus of modern science is mainly on the delivery of chemotherapeutic agents, with the long-term goal being applying surface-modified NDs to clinical processing to aid in the detection and medicaments of numerous illnesses.

4.5.1 Biotechnological Applications

a. Solid-Phase Extraction

In bioanalytical chemistry, the selective isolation of biomolecule-like proteins from complex mixtures is crucial. Several strategies have been introduced to overcome this challenge. One uses affinity chromatography columns containing micrometer-sized agarose beads to catch the target molecules from the raw sample solutions. After that, those beads are retrieved and examined using mass spectrometry or gel electrophoresis. Due to the interference from the micrometer-sized beads in ion production and extraction, direct analysis of these surface-bound proteins frequently experiences low mass resolution and accuracy.

As a result of their tiny size, superior chemical stability, and simplicity of purification, NDs could serve as an SPE device in bioanalytical applications (42). Chang *et al.* developed an efficacy strategy for protein analysis using acid-treated 100-nm HPHT NDs (11, 43). These particles were said to have a high protein loading capacity of 100 mg g^{-1} and a remarkable affinity for polypeptides and proteins. Proteins could be efficiently extracted from complex mediums or highly diluted solutions within minutes. The resonance of many interactions, such as ionic interactions, hydrogen bonding, and hydrophobic interactions, can result in the high affinity of NDs. Therefore, protein–ND complexes can withstand repeated exposure to deionized water with little loss.

Protein–ND complexes can also be quickly analyzed via gel electrophoresis or mass spectrometry following separation by either centrifugation or filtration. A platform for proteomic analysis named "SPEED (solid-phase extraction and elution on diamond)" has been developed (44). As there is no requirement to remove the ND adsorbent, this platform has an opportunity for the purification, concentration, and enzymatic digestion of intact proteins for subsequent sodium dodecyl sulfate-polyacrylamide gel electrophoresis (SDS-PAGE) or analysis using MALDI-MS, or matrix-assisted laser desorption/ionization technology. Additionally, the particles may be directly subjected to a one-pot process that includes disulfide bond reduction, free cysteine residue protection, and proteolytic digestion of the adsorbed proteins.

The affinity of NDs can be adjusted via chemical modification with many types of biomolecules, such as polyarginine (PA), and polylysine (PL), or to form a positive charge layer on the ND surface. These particles show a great attraction to negative charge species such as DNA (45), phosphopeptides (46), and sulfated glycosaminoglycans (47). Consequently, this development enables a rapid quantitative study of these biomolecules using high-affinity purification and MALDI-MS. Similarly, boronic acid-modified NDs exhibit glycopeptide selectivity (48). Aside from their inertness, NDs offer a distinct benefit in biomolecule selective extractors since they prohibit reactions with adsorbed analytes prior to PAGE or MS analysis.

b. Enzyme Immobilization

Enzymes' shortcomings, including the lack of long-term stability and difficulties in recovery and recycling, restrict their use in commercial applications. By immobilizing the molecules on HPHT ND or DND surface, some of the drawbacks of enzymes, such as their transient stability and difficulties in recovery and recycling, can be lessened or enhanced. The abundance of carboxylic groups on HPHT ND or DND surfaces can effectively undergo bioconjugation with enzymes owing to the coupling agents of 1-ethyl-3-(3-dimethylaminopropyl)-carbodiimide hydrochloride (EDC) and *N*-hydroxysuccinimide (NHS). This immobilization offers numerous advantages in terms of high stability, reusability, simple separation from the reaction mixture, and improved catalytic activities (49). One advantage of adopting NDs over other nanoparticles is their biocompatibility. In addition, there were no substantial changes to immobilized trypsin on DND surfaces (43).

In comparison with free trypsin, the Michaelis constant value and the maximum effective velocity of immobilizations were of the same order of magnitude. Notably, the immobilized trypsin showed significantly effective catalysis and did not suffer autolysis compared with trypsin alone. One investigation found similar outcomes for the cleavage of myoglobin and bovine serum albumin when utilizing immobilized trypsin for 5 minutes and free trypsin for 12 hours.

Alshawafi *et al.* demonstrated that, when combined with NDs, immobilized horseradish peroxidase (HRP) on polymethyl methacrylate nanofibers had better stability than the pure enzyme (50). Furthermore, 60% of the initial enzyme activity was maintained after 10 reuses. This study was notable in that the immobilized peroxidase withstood denaturation owing to increased temperature. Immobilized HRP shows greater thermal stability than soluble HRP, giving it a competitive edge in industrial applications requiring high temperatures to accelerate reaction rates. Additionally, the immobilized HRP was significantly more stable in organic solvents than soluble enzymes, which is advantageous in practical applications.

c. Antibacterial Activity

NDs serving as bactericidal agents are generally made up of NDs and ND–protein complexes. Early research in 2004 from Huang *et al.* found that NDs noncovalently modified by particular immunoglobulin (IgG) antibodies interacted with *Salmonella typhimurium* and *Staphylococcus aureus* in solution (51). A later study in 2007 from Perevedentseva *et al.* showed HPHT NDs conjugated with lysozymes could form bonds with *Escherichia coli*, while nonmodified NDs were unable to bond with this bacterium (52). The high antibacterial properties of the ND–lysozyme complex were illustrated via bacterial survival tests. These findings emphasize the critical role of ND–protein combinations in detecting and treating bacteria or other pathogens (even viruses) in solution.

DNDs also have the ability to kill both Gram-positive and Gram-negative bacteria, even though they are compatible with eukaryotic cells. Six distinct types of

DNDs containing various oxygen surfaces were obtained by annealing in H_2 or air at 450–500 °C to examine the relationship between their surface compositions and bactericidal activity (53). Researchers found that NDs' antibacterial properties mainly stemmed from their oxidized and negatively charged surfaces. The bactericidal potential of these NDs can be inhibited by proteins such as fetal bovine serum. Thus, the scope and duration of NDs' antibacterial activity can be regulated and modified for specific purposes.

Other studies using milled HPHT NDs have shed light on the effect of nanoparticle concentration and size (54). The antibacterial activity of NDs was not observed to be monotonous in this investigation. Researchers estimated the average particle mass and surface to reveal the relationship between the size and concentration of particles. This resulted in the inference of a link between colony-forming ability (CFA) decrease and total NDs surface area. The increased cumulative surface area led to an increase in NDs' antibacterial activity, indicating the critical role of NDs' surface chemistry in their interaction with bacteria.

Similar investigations have advanced our understanding of the antibacterial activity of NDs with various sizes, surface chemistry, mass concentrations, bacteria strains, and culture conditions (55). In this research, detonation nanodiamonds (DND), carboxylated DND (cDND), fluorescent nanodiamonds (FNDs), and carboxylated FNDs (cFNDs) were prepared to compare the influence of NDs with different surface chemistry on bacteria degradation. Four different sizes and concentrations of each ND and seven different types of cell culture medium were employed to treat Gram-positive and Gram-negative bacteria. NDs/bacteria aggregation and its effect on the bacteria's colony-forming abilities were also evaluated using sonication treatment interference. Notably, bacterial clumping and particle aggregation behavior are taken into consideration. However, there was no antibacterial treatment technique in conjunction with a specific material attribute. This led to the development of a model that blames the accumulation of bacteria and NDs in different suspension media for the decline in colony-forming units.

4.5.2 Biomedical Applications

a. Protein Delivery

NDs can serve as a delivery vehicle for biomolecules due to the following properties:

- A sufficient loading capacity.
- Strong binding of bioactive ligands or biomolecules to its surface.
- A functional mechanism for targeted release.

As stated in previous sections, ND particles (size of 100 nm) have enormous specific surface areas, are highly biocompatible, and are readily surface-modifiable with various functional groups. Its high loading capacity allows NDs to deliver a high load concentration while using smaller quantities of the delivery agent. Due to their delayed release mechanism, innovative drug delivery strategies, such as controlled and sustained release, may be developed using ND. Notably, NDs can protect and sustain the intrinsic therapeutic properties of attached molecules. Therefore, they are an ideal choice for a drug delivery system.

Lin *et al.* attached three types of proteins (myoglobin, bovine serum albumin, and insulin) with varying sizes, molecular masses, conformations, and biological functions to two kinds of NDs: positively charged j-DND and negatively charged h-HPHT ND (56). Myoglobin and bovine serum albumin, in particular, exhibit Langmuir-type adsorption, creating a 1:1 complex at saturation, whereas insulin folds into a firmly bound multimer prior to adsorption. The notable differences in adsorption patterns are attributed to the affinity of the individual proteins for NDs without the effects of protein concentration.

NDs have been used to release insulin and transform growth factor-beta (TGF-β) (57, 58). These ND–insulin models helped promote wound healing and vascularization for burn victims. Furthermore, ND–insulin complexes can be separated by pH, allowing this strategy to target bacterial infections alongside serious wounds. NDs were also applied to deliver the potential anti-scarring agent TGF-β. Interestingly, TGF-β–ND complexes can be kept without risk in PBS since incubation in serum-containing media triggers their release. Enzyme-linked immunosorbent assay (ELISA) results show that the structural integrity of the adsorbed protein molecules was firmly maintained (57).

Moore and coworkers examined the ability of DNDs to deliver bone morphogenetic proteins (BMP-2) and essential fibroblast growth factor (bFGF). Injection of BMP-2 and bFGF can stimulate differentiation and proliferation in osteoblast progenitor cells. Both proteins can be easily adsorbed on the surface of NDs and produce a stable colloidal solution. Notably, the attached proteins can be released in slightly acidic conditions. Therefore, NDs are expected to be an effective injectable alternative for delivering BMP-2 and bFGF to promote bone formation.

Numerous studies have also been published on the topic of ND immobilization. NDs' colloidal stability and enterosorbent potential in both aqueous and nonaqueous conditions enable them to be employed efficiently for targeted drug administration. NDs are easily complex with any agent, including antigens, antibodies, and immunoglobulins. The antigen carrier is critical in managing the antigen's structure and exposing or covering up essential functions; it is hypothesized that NDs' immobilization leads to less protein conformational distortion, allowing for improved antibody binding and, hence, a more excellent immune response. By these findings, Kossovsky *et al.* immobilized an antigen using NDs coated with a disaccharide and then administered it directly to rabbits to stimulate an immune response (59). Huang *et al.* revealed how antibodies and microorganisms are immobilized on NDs (51). Kong *et al.* used this approach to capture proteins for MS analysis (45, 60). Additionally, research has revealed that NDs' seeded electrodes are either biofunctionalized and employed for pathogen detection by UV-alkene surface chemistry on diamond films. This involves nanostructuring biosensing electrodes with NDs to facilitate antibody immobilization (61). Thus, surface-modified diamond nanoparticles as antigen-delivery vehicles represent an interesting conjugation of materials science, chemistry, and biology.

Taken together, those findings show the prospective applications of NDs in cancer drug delivery, wound healing, and anti-scarring. Due to the regulated growth factor release provided by ND-enabled technology, several regenerative medicine techniques, such as bone, cardiovascular, neuron tissue production, and numerous other agents, also profit from this technology. The most essential characteristics of ND-based delivery methods are reducing the burst release and flexibility in attaching or depositing different polymeric matrices, planar, or biomolecules.

b. Gene Delivery

A biological technology procedure called "gene therapy" involves inserting foreign genetic material into host cells to address genetic diseases, prevent diseases, or endow them with new biological functions. The combination of gene delivery with nanomedicine recently gained favor as a result of its potential combined efficacy and efficiency (37, 41). However, minimal progress has been accomplished toward combining these two areas into a single platform, primarily due to the lack of safe, generally applicable delivery techniques. On their own, naked nucleic acids are insufficient for disease treatment due to their inefficient cellular transport. The main reason for this stems from their poor physicochemical and pharmacological properties, which includes instability, innate immune responses, and anionic and hydrophilic features. Small interfering RNA (siRNA), short double-stranded RNA molecules used to suppress genes by specifically targeting complementary mRNA strands and then triggering mRNA degradation, also suffer from these obstacles and are easily excreted through glomerular filtration (62–64). Many strategies, such as lipid-based systems, polymer-based systems, aptamer-based systems, and inorganic systems, have been investigated in an attempt to overcome these issues. However, the poor penetrance and low efficiency of disease treatments limit their clinical applications.

Recently, researchers have pointed to the potential advantages of NDs, exceptionally functional NDs, in genomics, including DNA and RNA. ND was functionalized with amine groups via surface adsorption of 800 Da polyethyleneimine

(PEI800) for plasmid DNA delivery. This model exhibited outstanding properties compared with other materials due to its minimal cytotoxicity and high transfection efficiency. Accordingly, ND-PEI800s promote a 400- and 800-fold improvement in transfection effectiveness compared with amine- and carboxyl-terminated NDs (37). The most promising aspect of this technique is scalability, allowing this material to be used for various gene therapy applications. This polymer has effectively silenced the expression of green fluorescent protein via siRNA delivery and has been shown to be more efficient and less harmful for cells than Lipofectamine (35). Some of the following studies use ND-PEI hybrids in treating certain diseases, such as Ewing's sarcoma, a kind of pediatric bone cancer (65). The administration of siRNA can prevent EWS-Fli1 from expression, reducing the capacity for cancer cell proliferation.

In summary, NDs have the potential to heighten the transfection ability of polymers with an excellent biocompatibility with cell lines. As a result, this approach could provide insight into the potential uses of these adaptable functionalized NDs in nanomedicine.

c. Preclinical Validation of Cancer Therapy

The spectrum of major advancements in tumor therapy effectiveness and safety has been made possible by combining NDs with doxorubicin (DOX) (31, 40, 66). These complexes of NDs–DOX significantly diminish brain tumors by convection-enhanced intracranial delivery, suggesting that NDs have the potential to be used in the treatment of other nervous system-related diseases and injuries. Thanks to carboxylic groups on the surface, ND suspensions remain stable in water and can form complexes with water-insoluble drugs like purvalanol A (a drug used for liver cancer treatment), 4-hydroxytamoxifen (a breast cancer treatment drug), and dexamethasone (an anti-inflammatory agent). In another study, the carboxylate group of NDs prevented the epithelial–mesenchymal transition process via the TGF-β signaling pathway, thereby reducing the migration of cancer cells *in vitro*. Carboxylated NDs significantly reduced the metastasis of murine B16 melanoma cells in the *in vivo* lung metastasis model. As a result, carboxylated NDs could potentially be used for specialized cancer therapy (67). They are increasing dispersibility in water results from NDs' ability to adsorb medicines on surfaces while maintaining their therapeutic efficacy. NDs drastically reduced the particle size of these drugs in the water while improving the zeta potential, enhancing their dispersibility and cellular uptake capacity. These findings suggest that NDs may have a role to play in developing injectable formulations for medications that are not water-soluble.

However, DOX is highly toxic and may result in side effects like cardiotoxicity and myelosuppression. While there is no doubt about the performance of this drug, the concept of drug delivery should be further investigated to prevent it from auto-releasing on the delivery route. It has been demonstrated that using NDs as drug carriers has an advantage in the amount of potent DOX on the particle surface, delaying release and activity. Quantitative real-time polymerase chain reaction experiments and numerous cell lines have been used to assess the NDs–drug complex toxicity. There has been no evidence of a rise in inflammatory cytokines expressions, such as IL-6, tumor necrosis factor-alpha, or inducible nitric oxide synthase (31). As a second benefit, compared with bare DOX, the ND–DOX hybrid had more significant toxicity toward DOX-resistant mouse LT2-Myc liver and 4T1 mammary tumor models (40). It also fared better than bare DOX in preventing tumor growth over long-term treatment. NDs prevented DOX from being effluxed prematurely from tumor cells and mitigated the harmful effects of naked DOX by significantly lowering myelosuppression and early death (40). The circulation half-time of DOX NDs increased from 0.83 to 8.43 hours, confirming the sequestering action of NDs predicted in previous investigations (31).

To sum up, raising loading capacity, improving aqueous dispersibility, maintaining customizable release, and improving retention in chemoresistant cells are all benefits of developing chemotherapeutic drugs using NDs. These characteristics of the NDs are made possible by their hydrophilic functional group-enriched surface, significant surface-to-volume ratio, capacity to form loose clusters, enhanced cellular transport, and superior biocompatibility. According to research, the ND provides an ideal platform for constructing chemotherapeutic drugs that overcome many of medicine's significant obstacles. A road map to clinical translation includes the discovery of ND-therapeutic molecules that outperform clinical standards in terms of effectiveness, safety, and cost.

Due to their numerous applications in biological research, surface-modified NDs garner increasing interest in both *in vitro* and *in vivo* systems. It has been shown that carbon-based nanoparticles have higher biocompatibility levels and lower toxicity for both cells and organisms compared with other kinds of material (68). However, carbon nanoparticles exhibit varying toxicity *in vitro* and *in vivo*. Among carbon black, carbon nanotubes, graphene, and NDs, NDs offer the best biocompatibility and the lowest cytotoxicity for treated cells (69). However, further evaluation of the relative effects of these nanomaterials on animal models is desirable.

Acknowledgments

This work was supported by the National Science and Technology Council (NSTC), Taiwan, under Grant No. 112-2221-E-011-031 and the National Foundation for Science and Technology Development (NAFOSTED), Vietnam, under Grant No. 106.02-2018.38.

References

1 Xing Y, Dai LM. Nanodiamonds for nanomedicine. *Nanomedicine-Uk*. 2009;4(2):207–18.

2 Hsiao WW-W, Le T-N, Chang H-C. Applications of fluorescent nanodiamond in biology. In: Meyers, R.A. (Ed.) *Encyclopedia of Analytical Chemistry*. Hoboken, NJ, USA: John Wiley & Sons; 2022. pp. 1–43.

3 Slegerova J, Rehor I, Havlik J, Raabova H, Muchova E, Cigler P. *Nanodiamonds as Intracellular Probes for Imaging in Biology and Medicine*. Netherlands: Dordrecht: Springer; 2014.

4 Huang LCL, Chang H-C. Adsorption and immobilization of cytochrome c on nanodiamonds. *Langmuir*. 2004;20(14):5879–84.

5 Basso L, Cazzanelli M, Orlandi M, Miotello A. Nanodiamonds: Synthesis and application in sensing, catalysis, and the possible connection with some processes occurring in space. *Appl Sci*. 2020;10(12):4094.

6 Anikeev V, Zaikovskii V. Treatment of detonation carbon in supercritical water. *Russ J Appl Chem*. 2010;83(7):1202–8.

7 Osswald S, Yushin G, Mochalin V, Kucheyev SO, Gogotsi Y. Control of sp2/sp3 carbon ratio and surface chemistry of nanodiamond powders by selective oxidation in air. *J Am Chem Soc*. 2006;128(35):11635–42.

8 Bradac C, Osswald S. Effect of structure and composition of nanodiamond powders on thermal stability and oxidation kinetics. *Carbon*. 2018;132:616–22.

9 Shenderova O, Koscheev A, Zaripov N, Petrov I, Skryabin Y, Detkov P, *et al*. Surface chemistry and properties of ozone-purified detonation nanodiamonds. *J Phys Chem C*. 2011;115(20):9827–37.

10 Ackermann J, Krueger A. Efficient surface functionalization of detonation nanodiamond using ozone under ambient conditions. *Nanoscale*. 2019;11(16):8012–9.

11 Nguyen TTB, Chang HC, Wu VWK. Adsorption and hydrolytic activity of lysozyme on diamond nanocrystallites. *Diam Relat Mater*. 2007;16(4–7):872–6.

12 Kruger A, Liang Y, Jarre G, Stegk J. Surface functionalisation of detonation diamond suitable for biological applications. *J Mater Chem*. 2006;16(24):2322–8.

13 Hartmann M, Betz P, Sun Y, Gorb SN, Lindhorst TK, Krueger A. Saccharide-modified nanodiamond conjugates for the efficient detection and removal of pathogenic bacteria. *Chem Eur J*. 2012;18(21):6485–92.

14 Mochalin VN, Shenderova O, Ho D, Gogotsi Y. The properties and applications of nanodiamonds. *Nat Nanotechnol*. 2012;7(1):11–23.

15 Minko S. *Grafting on Solid Surfaces: "Grafting to" and "Grafting from" Methods*. Stamm ME, editor. Heidelberg: Springer; 2008.

16 Fu CC, Lee HY, Chen K, Lim TS, Wu HY, Lin PK, *et al.* Characterization and application of single fluorescent nanodiamonds as cellular biomarkers. *Proc Natl Acad Sci USA*. 2007;104(3):727–32.

17 Vial S, Mansuy C, Sagan S, Irinopoulou T, Burlina F, Boudou JP, *et al.* Peptide-grafted nanodiamonds: Preparation, cytotoxicity and uptake in cells. *Chembiochem*. 2008;9(13):2113–9.

18 Jung HS, Cho KJ, Seol Y, Takagi Y, Dittmore A, Roche PA, *et al.* Polydopamine encapsulation of fluorescent nanodiamonds for biomedical applications. *Adv Funct Mater*. 2018;28(33):1801252.

19 Zhang XY, Fu CK, Feng L, Ji Y, Tao L, Huang Q, *et al.* PEGylation and polyPEGylation of nanodiamond. *Polymer*. 2012;53(15):3178–84.

20 Dahoumane SA, Nguyen MN, Thorel A, Boudou JP, Chehimi MM, Mangeney C. Protein-functionalized hairy diamond nanoparticles. *Langmuir*. 2009;25(17): 9633–8.

21 Rehor I, Lee KL, Chen K, Hajek M, Havlik J, Lokajova J, *et al.* Plasmonic nanodiamonds: Targeted core-shell type nanoparticles for cancer cell thermoablation. *Adv Healthcare Mater*. 2014;4:460–8.

22 Slegerova J, Hajek M, Rehor I, Sedlak F, Stursa J, Hruby M, *et al.* Designing the nanobiointerface of fluorescent nanodiamonds: Highly selective targeting of glioma cancer cells. *Nanoscale*. 2015;7:415–20.

23 Rehor I, Slegerova J, Kucka J, Proks V, Petrakova V, Adam MP, *et al.* Fluorescent nanodiamonds embedded in biocompatible translucent shells. *Small*. 2014;10(6):1106–15.

24 Prabhakar N, Näreoja T, von Haartman E, Karaman DŞ, Jiang H, Koho S, *et al.* Core–shell designs of photoluminescent nanodiamonds with porous silica coatings for bioimaging and drug delivery II: Application. *Nanoscale*. 2013;5(9):3713–22.

25 von Haartman E, Jiang H, Khomich AA, Zhang J, Burikov SA, Dolenko TA, *et al.* Core-shell designs of photoluminescent nanodiamonds with porous silica coatings for bioimaging and drug delivery I: Fabrication. *J Mater Chem B*. 2013;1(18):2358–66.

26 Bumb A, Sarkar SK, Billington N, Brechbiel MW, Neuman KC. Silica encapsulation of fluorescent nanodiamonds for colloidal stability and facile surface functionalization. *J Am Chem Soc*. 2013;135(21):7815–8.

27 Hui YY, Zhang B, Chang Y-C, Chang C-C, Chang H-C, Hsu J-H, *et al.* Two-photon fluorescence correlation spectroscopy of lipid-encapsulated fluorescent nanodiamonds in living cells. *Opt Express*. 2010;18(6):5896–905.

28 Moore L, Chow EK-H, Osawa E, Bishop JM, Ho D. Diamond-lipid hybrids enhance chemotherapeutic tolerance and mediate tumor regression. *Adv Mater*. 2013;25(26):3532–41.

29 Tzeng YK, Faklaris O, Chang BM, Kuo Y, Hsu JH, Chang HC. Superresolution imaging of albumin-conjugated fluorescent nanodiamonds in cells by stimulated emission depletion. *Angew Chem Int Ed*. 2011;50(10):2262–5.

30 Lee JW, Lee S, Jang S, Han KY, Kim Y, Hyun J, *et al*. Preparation of non-aggregated fluorescent nanodiamonds (FNDs) by non-covalent coating with a block copolymer and proteins for enhancement of intracellular uptake. *Mol Biosyst*. 2013;9(5):1004–11.

31 Huang H, Pierstorff E, Osawa E, Ho D. Active nanodiamond hydrogels for chemotherapeutic delivery. *Nano Lett*. 2007;7(11):3305–14.

32 Li J, Zhu Y, Li W, Zhang X, Peng Y, Huang Q. Nanodiamonds as intracellular transporters of chemotherapeutic drug. *Biomaterials*. 2010;31(32):8410–8.

33 Barras A, Martin FA, Bande O, Baumann J-S, Ghigo J-M, Boukherroub R, *et al*. Glycan-functionalized diamond nanoparticles as potent E. coli anti-adhesives. *Nanoscale*. 2013;5(6):2307–16.

34 Genjo T, Sotoma S, Tanabe R, Igarashi R, Shirakawa M. A Nanodiamond-peptide bioconjugate for fluorescence and ODMR microscopy of a single actin filament. *Anal Sci*. 2016;32(11):1165–70.

35 Chen M, Zhang XQ, Man HB, Lam R, Chow EK, Ho DA. Nanodiamond vectors functionalized with polyethylenimine for siRNA delivery. *J Phys Chem Lett*. 2010;1(21):3167–71.

36 Kaur R, Chitanda JM, Michel D, Maley J, Borondics F, Yang P, *et al*. Lysine-functionalized nanodiamonds: Synthesis, physiochemical characterization, and nucleic acid binding studies. *Int J Nanomedicine*. 2012;7:3851–66.

37 Zhang XQ, Chen M, Lam R, Xu XY, Osawa E, Ho D. Polymer-functionalized nanodiamond platforms as vehicles for gene delivery. *Acs Nano*. 2009;3(9):2609–16.

38 Edgington R, Spillane KM., Papageorgiou G, Wray W, Ishiwata H, Labarca M, *et al*. Functionalisation of detonation nanodiamond for monodispersed, soluble DNA-nanodiamond conjugates using mixed silane bead-assisted sonication disintegration. *Sci Rep*. 2018;8(1):1–11.

39 Chang B-M, Lin -H-H, Su L-J, Lin W-D, Lin R-J, Tzeng Y-K, *et al*. Highly fluorescent nanodiamonds protein-functionalized for cell labeling and targeting. *Adv Funct Mater*. 2013;23(46):5737–45.

40 Chow Edward K, Zhang X-Q, Chen M, Lam R, Robinson E, Huang H, *et al*. Nanodiamond therapeutic delivery agents mediate enhanced chemoresistant tumor treatment. *Sci Transl Med*. 2011;3(73):73ra21.

41 Chen M, Pierstorff ED, Lam R, Li SY, Huang H, Osawa E, *et al*. Nanodiamond-mediated delivery of water-insoluble therapeutics. *Acs Nano*. 2009;3(7):2016–22.

42 Wu C-C, Han C-C, Chang H-C. Applications of surface-functionalized diamond nanoparticles for mass-spectrometry-based proteomics. *J Chin Chem Soc*. 2010;57(3B):583–94.

43 Wei LM, Zhang W, Lu HJ, Yang PY. Immobilization of enzyme on detonation nanodiamond for highly efficient proteolysis. *Talanta*. 2010;80(3):1298–304.

44 Chen WH, Lee SC, Sabu S, Fang HC, Chung SC, Han CC, *et al*. Solid-phase extraction and elution on diamond (SPEED): A fast and general platform for proteome analysis with mass spectrometry. *Anal Chem*. 2006;78(12):4228–34.

45 Kong XL, Huang LCL, Liau SCV, Han CC, Chang HC. Polylysine-coated diamond nanocrystals for MALDI-TOF mass analysis of DNA oligonucleotides. *Anal Chem*. 2005;77(13):4273–7.

46 Chang C-K, Wu C-C, Wang Y-S, Chang H-C. Selective extraction and enrichment of multiphosphorylated peptides using polyarginine-coated diamond nanoparticles. *Anal Chem*. 2008;80(10):3791–7.

47 Hsieh C-C, Guo JY, Hung S-U, Chen R, Nie Z, Chang H-C, *et al*. Quantitative analysis of oligosaccharides derived from sulfated glycosaminoglycans by nanodiamond-based affinity purification and matrix-assisted laser desorption/ionization mass spectrometry. *Anal Chem*. 2013;85(9):4342–9.

48 Xu GB, Zhang W, Wei LM, Lu HJ, Yang PY. Boronic acid-functionalized detonation nanodiamond for specific enrichment of glycopeptides in glycoproteome analysis. *Analyst*. 2013;138(6):1876–85.

49 Bornscheuer UT. Immobilizing enzymes: How to create more suitable biocatalysts. *Angew Chem Int Ed*. 2003;42(29):3336–7.

50 Alshawafi WM, Aldhahri M, Almulaiky YQ, Salah N, Moselhy SS, Ibrahim IH, *et al*. Immobilization of horseradish peroxidase on PMMA nanofibers incorporated with nanodiamond. *Artif Cells Nanomed Biotechnol*. 2018;46(sup3):S973–81.

51 Huang TS, Tzeng Y, Liu YK, Chen YK, Walker KR, Guntupalli R, *et al*. Immobilization of antibodies and bacterial binding on nanodiamond and carbon nanotubes for biosensor applications. *Diam Relat Mater*. 2004;13(4–8):1098–102.

52 Perevedentseva E, Cheng CY, Chung PH, Tu JS, Hsieh YH, Cheng CL. The interaction of the protein lysozyme with bacteria E-coli observed using nanodiamond labelling. *Nanotechnology*. 2007;18(31):315102.

53 Wehling J, Dringen R, Zare RN, Maas M, Rezwan K. Bactericidal activity of partially oxidized nanodiamonds. *Acs Nano*. 2014;8(6):6475–83.

54 Ong S, Van Harmelen R, Norouzi N, Offens F, Venema I, Najafi MH, *et al*. Interaction of nanodiamonds with bacteria. *Nanoscale*. 2018;10(36):17117–24.

55 Norouzi N, Ong Y, Damle VG, Najafi MBH, Schirhagl R. Effect of medium and aggregation on antibacterial activity of nanodiamonds. *Mater Sci Eng C*. 2020;112:110930.

56 Lin C-L, Lin C-H, Chang H-C, Su M-C. Protein attachment on nanodiamonds. *J Phys Chem A*. 2015;119(28):7704–11.

57 Smith AH, Robinson EM, Zhang XQ, Chow EK, Lin Y, Osawa E, *et al*. Triggered release of therapeutic antibodies from nanodiamond complexes. *Nanoscale*. 2011;3(7):2844–8.

58 Shimkunas RA, Robinson E, Lam R, Lu S, Xu XY, Zhang XQ, *et al.* Nanodiamond-insulin complexes as pH-dependent protein delivery vehicles. *Biomaterials.* 2009;30(29):5720–8.

59 Kossovsky N, Gelman A, Hnatyszyn HJ, Rajguru S, Garrell RL, Torbati S, *et al.* Surface-modified diamond nanoparticles as antigen delivery vehicles. *Bioconjugate Chem.* 1995;6(5):507–11.

60 Kong X, Huang L, Hsu C-M, Chen W-H, Han C-C, Chang H-C. High-affinity capture of proteins by diamond nanoparticles for mass spectrometric analysis. *Anal Chem.* 2005;77(1):259–65.

61 Zhang W, Patel K, Schexnider A, Banu S, Radadia AD. Nanostructuring of biosensing electrodes with nanodiamonds for antibody immobilization. *Acs Nano.* 2014;8(2):1419–28.

62 Whitehead KA, Langer R, Anderson DG. Knocking down barriers: Advances in siRNA delivery. *Nat Rev Drug Discov.* 2009;8(2):129–38.

63 Gary DJ, Puri N, Won YY. Polymer-based siRNA delivery: Perspectives on the fundamental and phenomenological distinctions from polymer-based DNA delivery. *J Control Release.* 2007;121(1–2):64–73.

64 Takahashi Y, Nishikawa M, Takakura Y. In vivo siRNA delivery to tumor cells and its application to cancer gene therapy. *Yakugaku Zasshi.* 2007;127(10):1525–31.

65 Alhaddad A, Adam MP, Botsoa J, Dantelle G, Perruchas S, Gacoin T, *et al.* Nanodiamond as a vector for siRNA delivery to Ewing sarcoma cells. *Small.* 2011;7(21):3087–95.

66 Vaijayanthimala V, Lee DK, Kim SV, Yen A, Tsai N, Ho D, *et al.* Nanodiamond-mediated drug delivery and imaging: Challenges and opportunities. *Expert Opin Drug Del.* 2015;12(5):735–49.

67 Guo Q, Li L, Gao G, Liu R, Einaga Y, Zhi J. Nanodiamonds inhibit cancer cell migration by strengthening cell adhesion: Implications for cancer treatment. *ACS Appl Mater Interfaces.* 2021;13(8):9620–9.

68 Yong K-T, Law W-C, Hu R, Ye L, Liu L, Swihart MT, *et al.* Nanotoxicity assessment of quantum dots: From cellular to primate studies. *Chem Soc Rev.* 2013;42(3):1236–50.

69 Vaijayanthimala V, Cheng P-Y, Yeh S-H, Liu K-K, Hsiao C-H, Chao J-I, *et al.* The long-term stability and biocompatibility of fluorescent nanodiamond as an in vivo contrast agent. *Biomaterials.* 2012;33(31):7794–802.

5

Toxicity Assessments of Nanodiamonds

Duc-Thang Vo, Tzu-Syuan You, Yu-Teng Lin, Stefanny Angela, Trong-Nghia Le, and Wesley Wei-Wen Hsiao**

* *Corresponding authors*

Toxicity assessments have been evaluated both *in vitro* and *in vivo*. Carbon nanoparticles are more likely to be biocompatible and less hazardous to cells and animals than metal- or semiconductor-based nanomaterials. Nanodiamonds (NDs), in particular, are more biocompatible and have lower cytotoxicity than other carbon-based nanomaterials (e.g., carbon nanotubes, graphene, and carbon black). The cytotoxicity of NDs depends on their properties and the exposure time. NDs must enter the cell in most cellular applications; therefore, toxicity assessments are required to validate the uses of NDs.

5.1 Cellular Uptake of Fluorescent Nanodiamonds

The biocompatibility of NDs with cells or living organisms is essential for their use in biological applications (1–3). Compared with other nanoparticles, NDs are more likely to be biocompatible and less hazardous to cells and animals (3–5). Various reports have shown that some mammalian cells in cultures, such as HeLa cervical cancer (6–8), macrophages, A549 lung cancer, HFL-1 fibroblast-like human fetal lung (9), Beas-2b human bronchial epithelial, 3T3-L1 preadipocytes, and 489-2.1 multipotent stromal cells, readily ingest unfunctionalized FNDs, even without surface modification, specific targeting moiety, or chemical stimulus (10–12). For instance, FNDs with an average size of 46 nm enter HeLa cells mostly by clathrin-mediated endocytosis after 2 hours of incubation (13). FNDs of 100-nm diameter can be internalized by cancer and stem cells after 4 hours of incubation via macropinocytosis (14). The intracellular localization of FNDs implies that these nanoparticles can be used for the intracellular delivery of therapeutic molecules. However, not all types of cells exhibit this behavior. For cells that do not readily take up particles, electroporation and chemical transformation protocols have been designed to facilitate FND internalization into nonmammalian cells. For

Nanodiamonds in Analytical and Biological Sciences: Principles and Applications, First Edition.
Edited by Wesley Wei-Wen Hsiao and Trong-Nghia Le.
© 2023 John Wiley & Sons Ltd. Published 2023 by John Wiley & Sons Ltd.

instance, this technique works on yeast cells, in particular *Saccharomyces cerevisiae*. Indeed, it can be applied to all varieties of cells with a cell wall, while preserving cell viability (15).

It is well known that the cellular uptake of FNDs by cells happens mainly through energy-dependent and clathrin-mediated endocytosis, but not by caveolae-mediated pathways (16). Once internalized by cells, FND particles are retained as distinct populations within endosomes and are quickly released into the cell cytoplasm (17, 18). As a result, the morphology (such as size and shape) and surface identity of FND particles affect their cellular uptake pathway as well as their translocation at biological barriers. For instance, sharp-shaped FNDs are more easily taken up by cells and move swiftly from endosomes to cytoplasm (19). The use of electropositive coatings (16, 20) or a pH-sensitive dextran-coated layer (8) can enhance FND particles' internalization by cells. By the uncertainty principle, the nanoparticles do not accumulate within the nucleus (11). Earlier findings reported that nuclear targeting of nanoparticles has multiple subcellular barriers such as cell membrane selectivity, endosomal trapping, lysosomal degradation, cytoplasmic crowding, and, most significantly, nuclear entry (21). FNDs possess a sharp-edged shape, which disrupts the endosomal membrane (22). This allows FND particles to escape the endosome, thus reducing the degradation of chemical molecules like antibodies bound on the FND's surface. While most prior research only showed the perinuclear localization of FND particles with no translocation to the nucleus (13, 14), recent studies have shown that Fenton-treated NDs can successfully enter the nuclei of HeLa cells (23). The internalization of FNDs in cells is likely affected by their surface properties, incubation duration, and other physiochemical factors such as size, shape, and aggregation. The functionalization of FND-based platforms for selective nuclear targeting or gene silencing in cancer cells (24) as well as FNDs linked to antibodies for targeting the nucleus of yeast cells like *S. cerevisiae*, as a model organism for aging (25), has recently been developed. In addition, a facile and robust approach that simultaneously detects signals from nitrogen-vacancy (NV) centers and spectroscopic Raman cells has been developed to visualize spectral colocalization of unmodified high-pressure high-temperature (HPHT) FNDs within the cell nucleus (26). The success of FND particles' specific accumulation and penetration into the nucleus makes them good candidates for a number of applications, such as delivery agents for nucleic acids.

Labeling ND particles inside living cells is a valuable feature for detecting individual cells and tracing cellular progeny. For instance, a single-particle FND-labeled HeLa cell is tracked continuously for eight days by flow cytometry (18), and the albumin-conjugated FNDs for single-cell imaging and quantitative detection of human mesenchymal stem or progenitor cells in the lungs of the miniature pigs after intravenous administration (27). Due to FNDs' long-term stability and biocompatibility at the cellular level (28), other comparable applications can be developed. For instance, FND carrying either NV (red) or N–V–N (green) defect centers can be used as long-term cell labels and trackers (29). Such a strategy could be used to regulate a tumor's response to therapeutics after a certain time or to track the progression and spread of metastases.

5.2 *In Vitro* Assessments

In vitro cytotoxicity tests are an important part of assessing NDs' safety because they are simple, relevant, repeatable, less expensive, and easy to control (30). NDs exhibit high biocompatibility and low cytotoxicity in *in vitro* assessments. Cytotoxicity effects can be evaluated by detecting metabolite, enzymes, nucleotide uptake activity, cell proliferation, differentiation, differential gene, protein expression, adenosine tripolyphosphate (ATP) generation, and oxidative stress level (5, 30).

Several cellular properties of numerous cell lines have been studied to determine the direct effects of NDs on cell viability (5). In one research, low cytotoxic effects were detected by measuring metabolic activity directly with a 3-(4,5-dimethylthiazolyl-2)-2, 5-diphenyltetrazolium bromide (MTT) assay, a 3-(4,5-dimethylthiazol-2-yl)-5-(3-carboxymethoxyphenyl)-2-(4-sulfophenyl)-2H-tetrazolium (MTS) assay, or a 2,3-bis-(2-methoxy-4-nitro-5-sulfophenyl)-2H-tetrazolium-5-carboxanilide (XTT) assay (5). However, nanoparticles could interfere with various colorimetric cytotoxicity assays (MTT, XTT, etc.), probably due to light scattering and/or direct nanoparticle interaction with the chemical components of the assays. As a result, researchers suggested an additional centrifugation step, which removed NDs from the plate before reading the assays (31). After incubation (with or without NDs), similar morphological and structural cells were observed by a fluorescent microscope, a transmission electron microscope, and a scanning electron microscope. Atomic force microscopy (AFM) is another method of assessing ND toxicity in *in vitro* models. AFM offers information on topographical, mechanical, and chemical characteristics by scanning a cantilever with an extremely sharp tip at a precise accuracy (below 10 nm) to measure interactions between a specimen and a scanning probe point (32). Because AFM does not require a vacuum, it is ideal for imaging soft/biological materials such as live cells and biomolecules in their natural settings. A low oxidation stress level was shown by measuring the generation of reactive oxygen species (ROS) via the fluorescence intensity of dichlorofluorescein (DCF) (31). Similar ATP production was shown after incubation with or without NDs and was characterized by luminescence (31).

In vitro biocompatibility of detonation nanodiamonds (DNDs), HPHT NDs, and their modified derivatives have been studied using the aforementioned assessments. Many studies have found that DNDs are intrinsically biocompatible (31, 33, 34). Some researchers, however, have challenged the idea of inherent biocompatibility. They claim that DNDs might cause hazardous reactions under specific circumstances and that their toxicity varies according to their origin, surface chemistry (35, 36), the kind of cell lines utilized for the evaluation (35), the composition of the culture medium (1), and their concentration (5).

According to Schrand *et al.*, the morphology of neuroblastoma cells incubated with DNDs (2–10 nm, 100 μg ml^{-1}, 24 hours) was similar to control cells. They showed intact mitochondrial membranes with an increase in neurite outgrowth and branching. The DNDs were found to be difficult to eliminate from the cell after internalization (31). In another study, Thomas *et al.* reported that the morphology of Raw 264.7 macrophages has no apparent differences in the incubation without or with the addition of DNDs (50 μg ml^{-1} at the various sizes, 6–500 nm). The DNDs were also found at cell borders and within the cells (37).

After removing the DNDs by centrifugation, the metabolic activity of neuroblastoma cells and macrophages was investigated by the reduction of MTT. There was no significant difference in viability compared with controls at concentrations up to 100 μg ml^{-1} for the neuroblastoma cells. Still, there was a slight effect on macrophages, depending on the concentration of DNDs (31). Thomas *et al.* also investigated the effect of DND sizes and concentrations on Raw 264.7 macrophages (37). In this study, cytotoxicity was evaluated by WST-1 assay (metabolic activity) and BrdU immunoassay (cell proliferation) (37). Similar results to Schrand *et al.* for neuroblastoma cells and macrophages were obtained. There were no significant variations in the incubation of the macrophages with DNDs (6 nm, up to 100 μg ml^{-1}) (31, 37). However, there was a reduction in both cell proliferation and metabolic activity at a twofold increase in DNDs concentration (6 nm, 200 μg ml^{-1}) (37). Cell proliferation (BrdU assay) but not metabolic activity (WST-1 assay) decreased in wells with smaller particles (6, 60, and 100 nm) in comparison with 250- and 500-nm ND particles (37). This can be explained by the increase of particle number and surface area of smaller size DND when they are at the same concentration (37). On the other hand, Li *et al.* found that the viability of HeLa cells, A549 cells, and K562 cells also depended on the presence of serum in the culture medium (1, 5). The viabilities of cells in complete medium remained around 100%, even at DND concentrations (2–10 nm) up to 100 μg ml^{-1} (1). However, in medium without serum, the viability of cells decreased rapidly as the concentration of DNDs as well as incubation time increased (1). Almost all of the cells in the serum-free medium were dead at concentrations of 100 μg ml^{-1} after 6 hours of incubation (1). A high adsorption capacity of serum protein in DNDs was observed (48 wt% at 37°C) (1). Serum proteins coating the DNDs serve as a protective layer for mammalian cells (1). In contrast to Li *et al.*, Thomas *et al.* did not find significant variations in either serum-free medium or complete medium (1, 5, 37). However, they investigated different medium compositions and cells. In addition, they observed a significant reduction in viable cells, an increase in late apoptotic cells due to necrosis, and a reduction in gene expression of TNF-α, IL-1β, chemokine Ccl2, and platelet-derived growth factor (PDGF) irrespective of DND particle size at concentrations of 50 μg ml^{-1} (37). In another study, Moore *et al.* found a modest increase in cell death and apoptosis of HepG2s (hepatocellular carcinoma) without diminished cellular metabolic activity at concentrations of 1 mg DNDs ml^{-1} (38). In contrast to HepG2s, HeLa cells showed a modest reduction in cellular metabolic activity without increased cell death or apoptosis at 1 mg DNDs ml^{-1} (38). However, consistent with the investigation of Li *et al.*, no effects on cellular metabolism, cell death, or caspase were found in either HepG2s or HeLa cells at more reasonable concentrations (< 500 μg ml^{-1}). The transcriptional response to the response of HepG2s and HeLa cells to DNDs (25 μg ml^{-1}, 24 hours) was examined using reverse transcription quantitative polymerase chain reaction (38). It displayed a lack of apoptosis response or induction. Furthermore, a relatively low level of ROS produced in cells incubated with DNDs was also obtained (31).

The potential cytotoxic effect of ND particles in human SaOS-2 osteoblast-like cells and U937 monoblastoid cell lines, as well as their impact on SaOS-2 function as bone depositing cells, was studied by Ibrahim *et al.* (39). They discovered that ND particles increased osteogenicity in osteoblast-like cells while causing negligible cytotoxicity. However, in reacting to these particles, the monoblastoid cell line U937 showed concentration-dependent cytotoxic and inflammatory responses, likely via Toll-like receptor 4 and activation of the NF-κB signaling pathway. It is worth noting that the maximum concentrations employed in this study were somewhat greater than what would be predicted in the case of total ND particle leakage from bone implants (39).

Longer-term functional effects of NDs have also been studied (5). According to Mytych *et al.* and Dworak *et al.*, the cell viability depends on the kind of cell lines and concentration of NDs after three days of incubation (40, 41). Mytych *et al.* did not find any significant decreases in viability in human hepatocyte (liver cells) at 10 μg ml^{-1} of NDs, while Dworak *et al.* found cytotoxic effects in human lymphocytes (white blood cells) at concentrations of NDs up to 50 μg ml^{-1}. Researchers also discovered that while ROS generation was not found in human hepatocytes, significant induction of ROS in human lymphocytes did occur. Changes in chromatin stability were also detected in human lymphocytes, even at low concentrations of NDs (1 μg ml^{-1}) (40). Some genotoxic effects were reported through an increase in DNA repair proteins (embryonic stem cells) or in inflammatory gene expression (human lung cells) (5). In addition, NDs after uptake into cell accumulate in the cytoplasm over time and generate some morphological changes concerning the size and spreading of the cell (5).

The effects of NDs on cytotoxicity were studied in two-dimensional (2-D) and three-dimensional (3-D) models by Khanal *et al.* These studies found that NDs' toxicity varied depending on the model employed (42). In 2-D models, after four days of therapy, NDs were found to lower the metabolic activity of cells. This decrease might be linked to the genotoxic impact of oxidized NDs. This is time-dependent since cell metabolic activity fell after four days of ND exposure. Cells demonstrated equal metabolic activity regardless of concentration beyond a threshold of roughly 25 $g\,ml^{-1}$. The ingestion of NDs was concentration-dependent. When cells reached the saturation threshold (25 $g\,ml^{-1}$), they formed colonies and became resistant to additional nanoparticle absorption.

A scaffold-free 3-D model was created using magnetic levitation. The rate of ring closure, which correlates to the ability of cells to migrate and proliferate (viability and functioning) in the 3-D model, was shown to be comparable across all treatment groups and the control group. Furthermore, cell viability was equivalent in both the control and treatment groups in the Live/Dead test. It suggests that when cells were grown as 3D compact structures, NDs had no significant negative influence. Differences in ND toxicity in 2-D and 3-D models might be explained by changes in nanoparticle intake diffusional pathways, the presence of extracellular matrix in 3-D models, and the total number of cells instantly exposed to a given dose of NDs (42).

Khanal and colleagues, for example, studied the impact of NDs on external and intracellular proteins (fibronectin-FN and proteins inside liver cells, respectively) using infrared spectroscopy AFM (AFM-IR) (43). They discovered that peptides, both intrinsic and extrinsic, exhibited conformational changes at the ND–protein interface, including protein aggregation and the development of β-sheets, intermolecular β-sheets, and antiparallel β-turns, for both intracellular and extracellular proteins, implying that individual NDs have a negative effect on protein structure. Although AFM-IR does not allow researchers to quantify the concentration at which macroscopic cytotoxicity occurs, the high-resolution data on nanoscale chemistry provides us with a precise knowledge of nanoparticle–protein/macromolecule interactions. This information has significant implications for our understanding of nanoparticle toxicity in the biological environment. As mentioned previously, nanoparticle toxicity is dependent on the particles' size, shape, and surface chemistry and varies from cell to cell (healthy versus malignant). The AFM-IR approach is critical in determining ideal nanoparticle features, such as surface conditions, which can help inform numerous biological applications such as drug administration and bioimaging (43).

By adjusting the sp^3/sp^2 carbon ratio without other physicochemical characteristics such as shape, size, or metal impurities, Lee *et al.* sought to understand the function of the structural parameters of sp^3/sp^2 carbon ratios on the toxicity of DND (44). A 2′7′-dichlorofluorescein diacetate (DCFH-DA) assay was used to estimate the intrinsic ROS generation potential of DNDs. Results showed a strong negative correlation with sp^3/sp^2 carbon ratios, implying that ROS generation increased as sp^3/sp^2 carbon ratios decreased. These findings imply that the sp^3/sp^2 carbon ratio is a critical predictor of DNDs' ROS generation capacity, and that the produced ROS are the primary cause of cytotoxicity and proinflammatory cytokines *in vitro* (IL-6, L-1β, MIP-1β, and TNF-α) (44). Figure 5.1 shows that the levels of ROS generation potential had strong correlation coefficients with concentrations of several proinflammatory cytokines. While the findings of this study do not show that DND is a flammogenic or poisonous substance, they do suggest that sp^2 carbon contamination in DND might hinder the particles' use in biological applications, which could be useful information for developing NDs that are safer by design (44). These results are in line with research conducted *in vivo*, which is discussed in the next section.

NDs are chemically inert and can be surface functionalized easily with carboxyl groups and their derivatives (45). Additional investigations have revealed that surface modification using biomolecules such as peptides (46) do not affect HPHT ND biocompatibility. HPHT NDs purified by air oxidation and strong

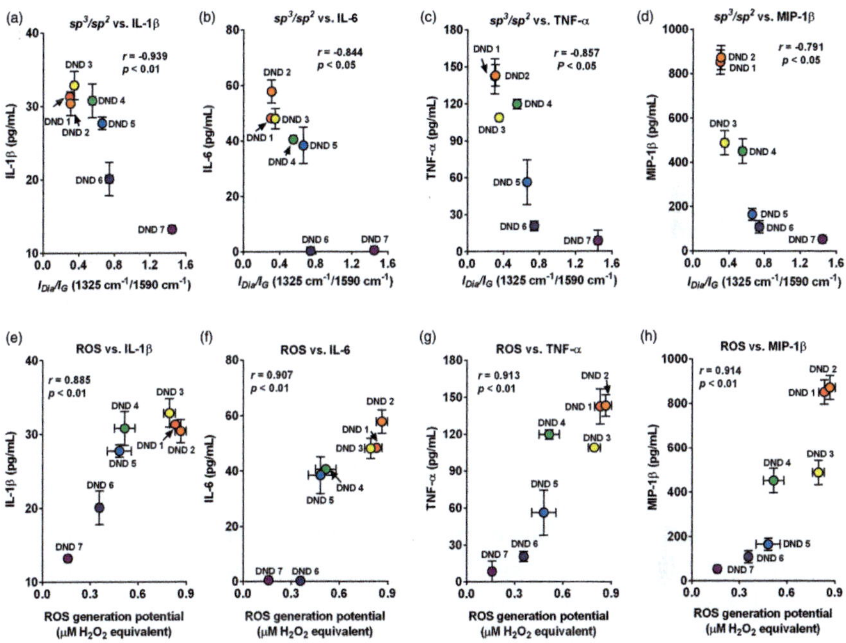

Figure 5.1 In *in vitro* investigations, the intensity ratio (I_{Dia}/I_G) or the levels of ROS production potential were plotted against levels of proinflammatory cytokines. The doses used in ROS production and *in vitro* toxicity studies were 272 and 170 $cm^2\ ml^{-1}$, respectively. The intensity ratio (I_{Dia}/I_G) levels were shown to have high association correlations with various proinflammatory cytokine data, including IL-1β (a), IL-6 (b), TNF-α (c), and MIP-1β (d). Similarly, the DCFH-DA test revealed strong correlation coefficients between levels of ROS generation potential of DND samples and different proinflammatory cytokine data, including IL-1β (e), IL-6 (f), TNF-α (g), and MIP-1β (h). The correlation coefficient and p value for each figure was calculated using a two-tailed Pearson correlation test. *Source:* (45) Lee et al., 2020 / Taylor & Francis.

oxidative treatments showed good hemocompatibility in human blood, with minor hemolytic and thrombogenic activities, according to further study by the same research group. Oxidized HPHT NDs were used to treat human primary endothelial cells with diameters of 35–500 nm and showed no cytotoxicity (47). Recent research found that HPHT NDs may successfully penetrate human cells. However, exposure levels up to 250 μg ml^{-1} showed no significant cytotoxic or genotoxic effects on six cell lines (48).

FNDs hold great promise in many biological applications as cell tracking and localization markers. They show no sign of photobleaching and can be taken up by mammalian cells with minimal cytotoxicity (45). According to Yu *et al.*, FNDs (100 nm) translocated through the cell's membrane and formed aggregates within the cell. But they did not significantly diminish cell reduction activity on 293T human kidney cells at concentrations up to 400 μg ml^{-1} (45).

To summarize, NDs exhibit low cytotoxicity in *in vitro* tests, depending on many factors. Cytotoxicity can occur at high concentrations of NDs or during long incubation periods at certain concentrations. Therefore, *in vitro* tests should be studied to evaluate NDs' effect on the cells. Other assessments using *in vitro* studies are summarized in Table 5.1.

5.3 *In Vivo* Assessments

As opposed to research using living cells, *in vitro* and *in vivo* biocompatibility analyses use animal models to investigate the possible effects of NDs on human health. The toxicity of NDs *in vivo* has been studied using a range of animal models, including *Xenopus* embryos (36), *Tetrahymena thermophile* and *Paramecium caudatum* (56), *Drosophila melanogaster* embryos (57), zebrafish embryos (54, 58), mice, rats, and rabbits, as can be seen in Figure 5.2 (30).

Caenorhabditis elegans is an excellent model organism for determining nanoparticle toxicity. The Annelida is tiny (1 mm in length) and see-through to the naked eye. Because of its limited life span, simple operation, and high sensitivity to a wide range of stresses, *C. elegans* have been used as a benchmark system for ecotoxicological investigations. Researchers have fed or microinjected surface-oxidized HPHT NDs into nematodes and discovered that worms' growth and development were normal (59). According to toxicity tests that used lifespan and reproductive capacity as physiological indications, NDs are nontoxic and do not cause any observable stress to the worms.

Rabbits were one of the first animals employed in this research (60). Researchers gave the rabbits a large dose (125 mg) of DNDs by intravenous injection, yet no fatalities occurred. The red blood cell count and hemoglobin level of the rabbits remained steady for 15 minutes after the NDs were administered. The quantities

Table 5.1 *In vitro* and *in vivo* assessments to evaluate the cytotoxicity of NDs.

ND type	Biological model	ND concentration	Parameters evaluated	Conclusion	Reference
5 nm DND	Human lymphocytes	1–100 $\mu g\ ml^{-1}$	Membrane damage, cell death, chromosome damage, oxidative stress	Evident cytotoxicity at 50 $\mu g\ ml^{-1}$. At 1–10 $g\ ml^{-1}$, aneugenicity and genotoxicity are seen. Oxidative stress is dose-dependent	(40)
50 and 100 nm F-HPHT ND	Human and rat red blood cells (RBC)	20 and 1000 $\mu g\ ml^{-1}$	Cytokine production, oxygenation/deoxygenation dynamic of RBC	Immune response is not stimulated	(49)
4.4 nm DND	*Corbicula fluminea* Asian clam	0.1–10 $\mu g\ ml^{-1}$	Histopathology, oxidative stress	Stress-induced oxidative stress, persistent toxicity, and changes in the cells of the digestive system	(50)
4–5 nm DND	HepG2, HeLa cells	1–500 $\mu g\ ml^{-1}$	Cell death, metabolic activity, gene expression, apoptosis	No toxicity	(38)
ND	Murine		Metabolism biocompatibility, oxidative stress markers	Glutathione S-transferase activity increased after 7 days and catalase activities were observed on 14 days, respectively	(51)
270 nm HPHT ND	Mesenchymal stem cells	1.5 $\mu g\ ml^{-1}$	Immunotoxicity, differentiation of cells, and morphology	No signs of inflammation or toxicity	(52)
4–5 nm DND	Male CD-1:IGS rats, nonhuman cynomolgus monkeys	5 $mg\ ml^{-1}$ (rats) and 15–25 $mg\ kg^{-1}$ (primates)	Body weight, repeated dosing biocompatibility, serum chemistry, complete blood count, urinalysis	For primates, the treatment dose is 6.75–13.5 $mg\ kg^{-1}$; no organ dysfunctions in either models, and no deleterious effects were observed during a 6-month period	(53)
100 nm F-HPHT ND	Zebra fish (*Dani rerio*) embryo	1000–5000 $\mu g\ ml^{-1}$	Biocompatibility, embryogenesis, ND movement, and velocity	For high concentrations of ND tested, dose-dependent toxicity has been found	(49, 54)
3, 5, and 7 nm DND	BEAS-2B cells	25–400 $\mu g\ ml^{-1}$	Membrane disruption mechanisms	Membrane disruption fades in the order 7 nm > 5 nm > 3 nm DND	(55)

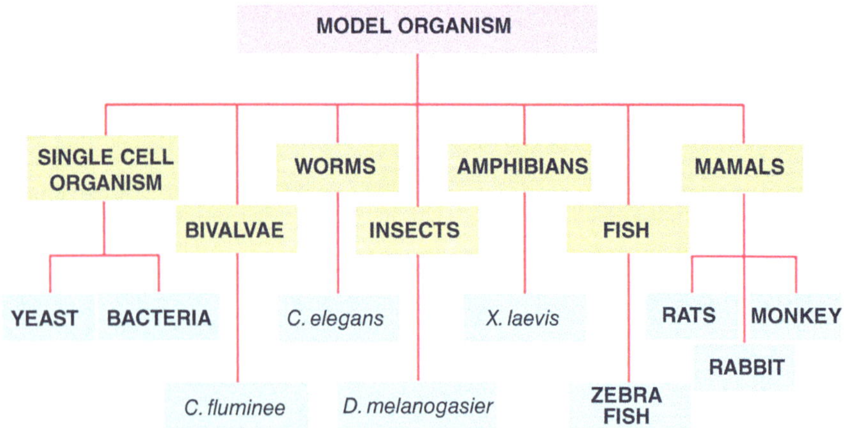

Figure 5.2 Variety of animal models for *in vivo* studies of ND biocompatibility.

of biological molecules such as total bilirubin, low-density lipoprotein, and triglyceride were altered significantly over a longer period (48 hours after the injection). However, the animals showed no signs of irritation no inflammation three months later, indicating that the NDs exhibit long-term biocompatibility.

The intratracheal instillation of DNDs has been reported to have deleterious effects on the lungs, kidneys, liver, and hematological systems of mice in investigations employing mouse models (61). There were no changes in body weight or appearance of abnormal diseases. Due to the rapid absorption and long retention time of the NDs in lung tissue, the lungs sustained the most severe toxicological repercussions, becoming irritated and suffering tissue damage. In contrast, Yuan *et al.* found no pulmonary damage in mice following intratracheal instillations of DNDs. The discrepancies in these results might be linked to DND sources and how their surfaces are changed, as seen in cellular studies. After microinjecting HPHT NDs into rats over five months, researchers found no *in vivo* toxicity or observable adverse effects (e.g., stress reactions). There were no significant changes between control groups and ND-treated organisms. Interleukin-1b (IL-1b) and interleukin-6 (IL-6) inflammatory cytokine levels were not significantly elevated in mice following the intravenous injection of HPHT ND particles (62).

Moore *et al.* have looked at DND biocompatibility in nonhuman primates (Cynomolgus monkeys) and rats (53). The rat study was conducted as a two-week-long multiple-dosage subacute inquiry that included serum, histology, and urine analyses. The nonhuman primate study was conducted as a dual-gender, multiple-dose, long-term inquiry in both standard/clinically relevant and increased dosing cohorts, with thorough urine, serum, body weight data, and histology. In addition to controls ($n = 3$), this study tested both male and female monkeys with two DND dosage levels, a regular dose of 15 mg kg^{-1} ($n = 5$) and an escalated dose of

25 mg/kg ($n = 4$). These findings point to a downstream maximum tolerated dosage and a level of DND administration with no obvious harmful effects. The lack of organ impairment seen in this investigation at both 15 and 25 mg kg^{-1} lays the groundwork for the clinical translation of DND-based therapeutic and contrast agents. According to the findings of these investigations, NDs are well tolerated at clinically relevant dosages (53).

NDs' biocompatibility has been demonstrated with HPHT ND (100-nm) particles. However, researchers are now starting to turn their attention to the safety of FND particles. The pharmacokinetics, safety, and biodistribution of NDs delivered to rats by intravenous infusion in a single high dosage have been investigated using FND particles (NV)-Z800. These particles have a much larger average size of 800 nm and contain NV fluorescence centers (63). This research indicated that the systemic administration of a single massive dose (60 mg kg^{-1}) of FNDs to mice over two weeks is both safe and well tolerated. Nanomaterials were mostly found in the spleen and liver, which matches the findings of many other investigations (53, 64–66). There were no fatalities in any of the animals exposed to high doses of NDs, and the animals maintained their body and organ weights well. All motor/sensory function tests, as well as blood cell counts and liver and kidney function, were normal.

Doxorubicin (DOX) is one of the most commonly used chemotherapeutic medications for cancer treatment; however, it must be used with caution owing to dose-related toxicity (67). The 4T1 mouse mammary cancer cell line has been used to test the doxorubicin–polyglycerol–nanodiamond conjugate (ND–DOX) as a cytostatic agent. In this study, ND–DOX or pure DOX was injected into the tail vein of mice. Despite having a lower apparent therapeutic efficacy than DOX, the tumor-bearing mice tolerated ND–DOX well (68, 69). DOX treatment caused a continual decline in animal body weight, indicating severe systemic toxicity. However, ND–DOX-treated animals maintained a constant body weight (69). In both the systemic and tumor contexts, this study found that ND can reduce chemoresistance and restore cancer-induced immunosuppression. Ryu *et al.* observed that folic acid (FA) and doxorubicin modified (FA/DOX–ND) clusters are highly concentrated in tumor tissues and that FA/DOX–ND clusters significantly reduced tumor volume after intravenous injection into tumor-bearing animals (70). These findings suggest that NDs can perhaps be used to facilitate the development of new cancer therapy options.

In mouse cancer models, an ND drug delivery system coupled with DOX (NDX) has been shown to improve therapeutic effectiveness and reduce toxicity. Biocompatibility tests also show that NDs appear to be well tolerated in nonhuman primates. While large-animal studies of nanomedicine drug safety are uncommon, Wang *et al.* offer a significant step forward in bridging prior achievements with clinical evaluation (Figure 5.3) (67). The safety of NDs as a medicine delivery platform was tested in Naïve Beagle dogs. The researchers used intravenous (IV)

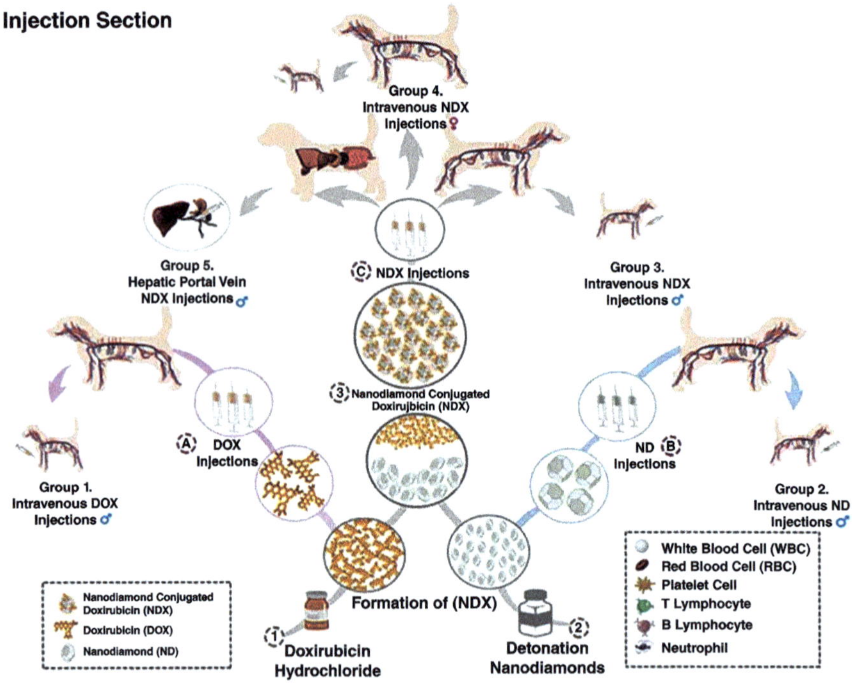

Figure 5.3 ND toxicity assessment in Naïve Beagle dogs. *Source:* (68) Wang et al., 2021 / Springer Nature.

and hepatic portal vein injection techniques to provide DOX, ND, and NDX to a dual-gender animal model. Canines were tracked for five days to see if their health phenotype changed. In addition, blood and urine samples were collected and used in clinical pathology studies. Results showed that the ND drug delivery platform considerably reduced DOX toxicity in the Naïve Beagle dog model (67).

Ion implantation was used to incorporate the therapeutically authorized radionuclide ^{32}P into the lattice of ND particles (71). *In vivo* study on the chicken egg model was conducted to investigate the characteristics, uptake, safety, and biodistribution. Employing autoradiographic techniques and analytical methodologies, ^{32}P tagging has provided a reliable, safe, and efficient means of localizing and measuring NDs. Results of this research have revealed that the distribution of ^{32}P-ND in growing chicken embryos can be observed after several days. The biocompatibility of the tagged NDs was investigated and confirmed by observing their effects on the vitality and metabolic activity of human bone-derived mesenchymal stromal cells. ^{32}P-ND may be utilized for imaging at low concentrations. It can also be employed to trigger cell death in a dose-dependent manner at higher doses (71).

Lee *et al.* suggest that the sp^3/sp^2 carbon ratio is a key predictor of DND's ROS generation capability and that ROS is the major cause of acute inflammation *in vivo* (44). In this study, the intratracheal instillation technique was used to treat six-week-old specific-pathogen-free female Wistar rats with DND samples and reference materials, carbon black, and nickel oxide (NiO). Based on the lung sensitivity model for inflammation and the bolus injection DND generating a transient inflammation, ND should be classified as a low-toxicity, low-solubility nanomaterial. This study does not prove that DND is dangerous. Still, it does imply that sp^2 carbon contamination in DND should be taken into consideration in the development of biological DND applications. Figure 5.4 shows the intensity ratio (I_{Dia}/I_G) or ROS production potential plotted against *in vivo* acute inflammatory endpoints (44).

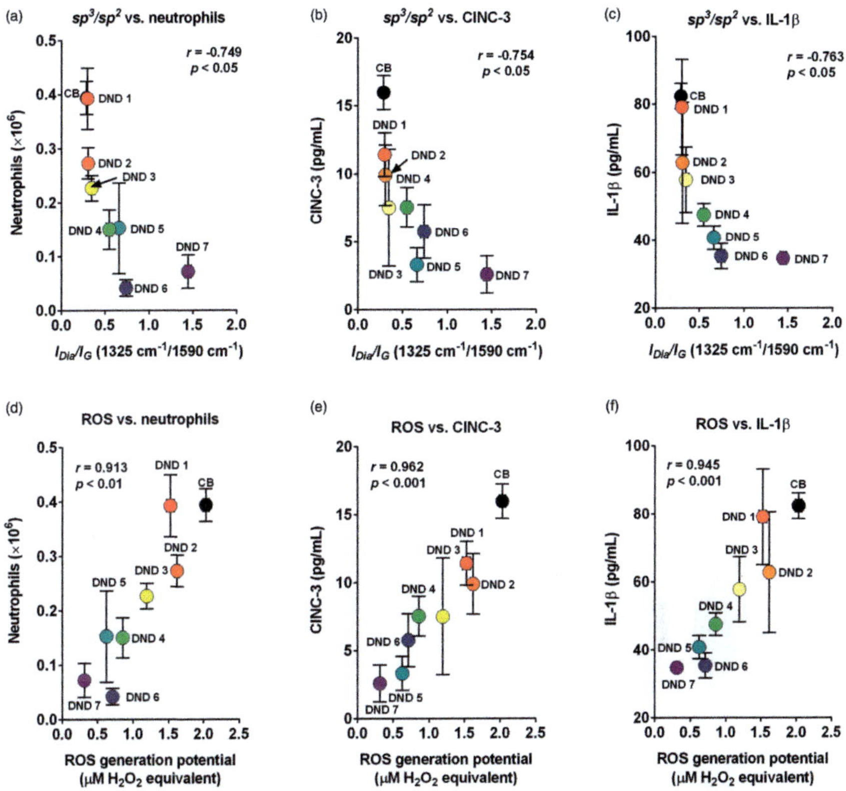

Figure 5.4 Intensity ratio (I_{Dia}/I_G) or ROS generation potential plotted against *in vivo* acute inflammatory endpoints. The number of neutrophils (a), CINC-3 (b), and IL-1β (c) were all highly correlated when the I_{Dia}/I_G values (1325 cm^{-1}/1590 cm^{-1}) were shown. Likewise, the number of neutrophils (d), CINC-3 (e), and IL-1β (f) were all highly correlated when levels of ROS production potential of the DNDs were plotted at 272 cm^2 ml^{-1}. The correlation coefficient and p value for each figure were calculated using a two-tailed Pearson correlation test. *Source:* (45) Lee et al., 2020 / Taylor & Francis.

Recent research has shown that NDs are considered antioxidants because they can scavenge ROS. However, research on the antioxidant activities of NDs is limited, and their impact on oxidative stress caused by hydrogen peroxide (H_2O_2) is unclear. Furthermore, owing to the rising use of NDs in nanomedical, nanocosmetics, and biomedical goods, toxicity studies are required. Kaluç *et al.* looked at the effects of a carboxylated ND (cND) on H_2O_2-induced oxidative stress and its 24-hour toxicity in *S. cerevisiae*, a unicellular eukaryotic model (72). They discovered that by lowering H_2O_2 levels in the medium, cND dramatically reduces cell mortality and ROS generation in response to H_2O_2-induced oxidative stress ($p < 0.05$). Although higher dosages (1000 mg ml^{-1} and above, $p < 0.05$) inhibited colony development, even 10.000 mg ml^{-1} cND administration for 24 hours could not entirely prevent colony formation. cND also lowered the amount of ROS produced during normal metabolism ($p < 0.05$) without causing cell death. The experiment results can be seen in Figure 5.5. In *S. cerevisiae*, these findings imply that cND possesses ROS-scavenging functions but presents no toxicity (72).

To conclude, existing research in the area has demonstrated that NDs are highly biocompatible in various cell lines and animal models, with minimum or no cytotoxicity. As a result, NDs offer a lot of promise in biological applications. However, immunogenicity, diffusivity, and the metabolism of NDs, as well as nonspecific particle accumulation in reticuloendothelial systems when administered *in vivo*, are all issues that still need to be addressed (73). More assessments of *in vivo* studies about ND can be found in Table 5.1.

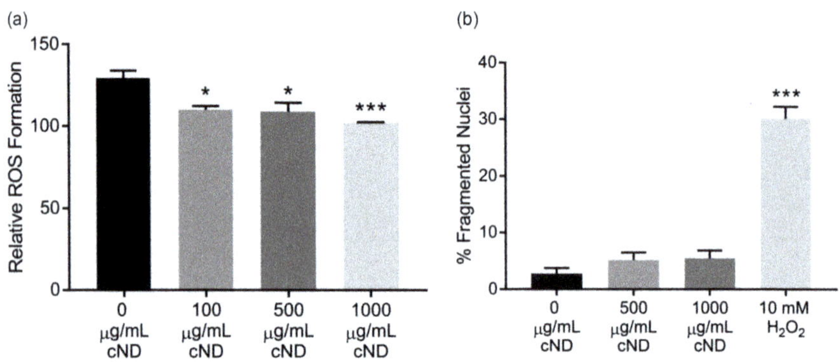

Figure 5.5 ROS formation (a) and nuclear fragmentation (b) in response to cND treatment: (a) relative ROS formation in untreated, 100-, 500-, and 1000-mg ml^{-1} cND treated cultures for 24 hours (*$p < 0.05$, ***$p < 0.001$, compared with untreated 24-hour, one-way ANOVA, $n = 3$); (b) percentage of fragmented nuclei in response to 0-, 500-, and 1000-mg ml^{-1} cND treatment for 24 hours; 10-mM H_2O_2 treatment for 1 hour was used as the positive control (***$p < 0.001$, $n = 3$, one-way ANOVA). *Source:* (73) Kaluç et al., 2021 / Taylor & Francis.

5.4 Nanodiamonds' Phytotoxic Effect

The reaction of *Chlorella pyrenoidosa* to five different doses of NDs was studied in depth using phenotypic and transcriptional analyses by Zhang *et al.* (67). The toxicity of NDs seems to be dose-dependent. After 48 hours of exposure, greater concentrations of NDs (50 mg l^{-1}) inhibited growth by 75.4%, increased oxidative stress, and caused deformed morphology in microalgae. High concentrations of NDs incited higher ROS in microalgal cells, which was equivalent to oxidative damage, as seen in Figure 5.6a–c. Unlike the evenly dispersed cells

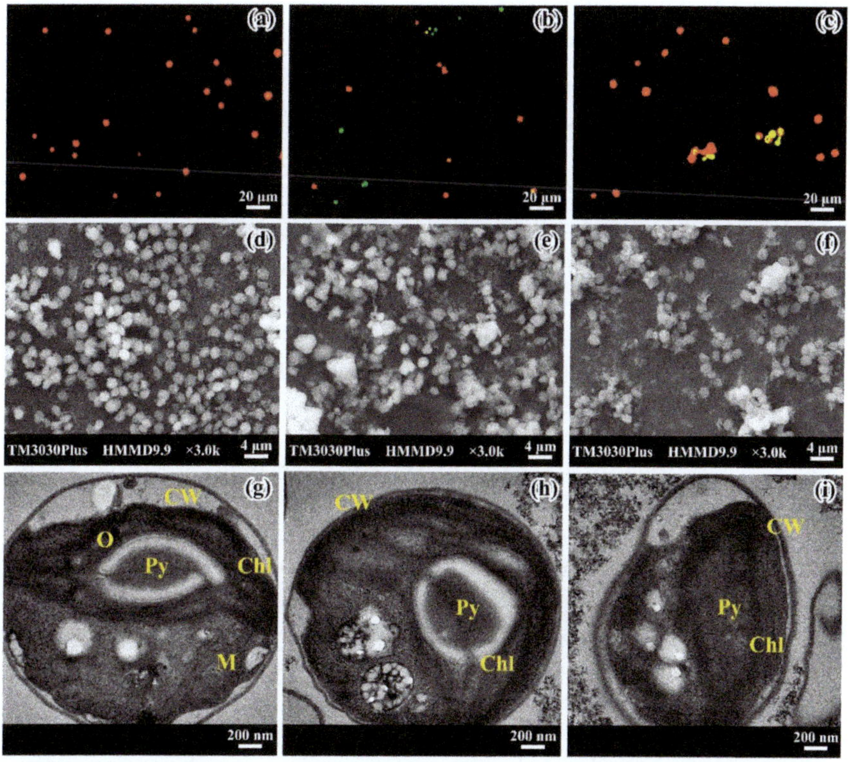

Figure 5.6 ROS expression in the control (a), 20 mg l^{-1} NDs (b), and 50 mg l^{-1} NDs (c) following 48 hours of exposure, fluorescence microscopy procedures. Morphology of cells in the control (d), 20-mg l^{-1} NDs treatment (e), and 50-mg l^{-1} NDs treatment (f) after 48 hours of exposure by scanning electron microscopy. Submicroscopic structure of the microalgae in control (g), 20-mg l^{-1} NDs treatment (h), and 50-mg l^{-1} NDs treatment (i) after 48 hours of exposure by transmission electron microscopy. (Note: Chl = microalgal chloroplast; CW = microalgal cell wall; M = microalgal mitochondria; Py = microalgal pyrenoid; O = microalgal osmiophilic globule). *Source:* Reproduced from ref. (68) / Springer Nature.

without ND treatment (Figure 5.6a and d), the flocs generated by microalgal cells seemed to be agglomerated following exposure to NDs (Figure 5.6b, c, e, and f). Meanwhile, under 50-mg l^{-1} NDs' activation, the aggregated microalgae produced numerous flocs, as seen in Figure 5.6c and f. Aggregated microalgal flocs were generated, perhaps preventing further NDs from entering the microalgal cells. When exposed to NDs, chlorophyll content changed. Although NDs' dramatically improved photosynthesis, excessive electron production during this process might have been a major cause of ROS creation. Finally, in response to increasing ND concentrations, cellular regulating mechanisms were comprehensively exposed, offering important insights into aquatic biological dangers of NDs. This study reminds us that the often-overlooked issues of the aquatic ecosystem should be given more attention before NDs are widely used in these contexts (67).

5.5 Role of Oxidative Stress in Nanodiamonds' Toxicity

As mentioned previously, insoluble nanoparticles, such as NDs, result in intracellular ROS production (33, 44, 67, 74). In nanotoxicology, the generation of oxidative stress is crucial. Nano-object ingestion by phagocytosis and intracellular metal ion release are two major processes in producing oxidative stress by nano-objects. According to *in vitro* studies, nanoparticles can cause intracellular ROS. Cumulative oxidative stress induces cell death and apoptosis in cells. *In vivo*, on either hand, initial oxidative stress increases ROS production and phagocyte recruitment, resulting in secondary oxidative stress. In the living body, primary and secondary oxidative stresses are indistinguishable. Nanoparticle-induced oxidative stress is triggered by intracellular ROS production, according to *in vitro* studies. The release of intracellular metal ions from nanoparticles is a key element in the formation of intracellular ROS. The kind of metal determines the amount of oxidative stress (75). As a result, CNTs may be hazardous due to the production of ROS by catalytic metals, notably iron. However, recent research has indicated that CNT toxicity is generated by a mechanism unrelated to oxidative stress. Furthermore, calcium-related nanoparticles have biological effects that are independent of oxidative stress. Nanoparticles' capacity to cause oxidative stress is determined by physical and chemical features such as crystalline phase, protein adsorption ability, and zeta potential. NDs induce intracellular ROS based on its zeta potential (76). The process of intracellular ROS production is still not fully understood. Furthermore, understanding the relationship between oxidative stress and *in vivo* toxicity, as well as the mechanism of intracellular ROS formation by insoluble nanoparticles, is critical (74).

To conclude, the present body of published work indicates that NDs are highly biocompatible in a variety of cell lines and animal models with low or no cytotoxicity. As a result, NDs offer a lot of promise in biological applications even though other parameters, including particle purity, surface chemistry, and size, may all contribute to the existence of cytotoxicity. Further attention should be given to the immunogenicity, diffusivity, and metabolism of NDs, including the particles' nonspecific accumulation in reticuloendothelial systems following *in vivo* administration. NDs are expected to be applied in nanomedicine technology. Therefore, toxicity assessments are critical, as it is important to demonstrate the safety of NDs in animal models before progressing to human clinical trials.

References

1 Li J, Zhu Y, Li W, Zhang X, Peng Y, Huang Q. Nanodiamonds as intracellular transporters of chemotherapeutic drug. *Biomaterials*. 2010;31(32):8410–8.

2 Hsiao WW-W, Le T-N, Chang H-C. Applications of Fluorescent Nanodiamond in Biology. In: Meyers, RA (Ed.), *Encyclopedia of Analytical Chemistry*. Hoboken, NJ, USA: John Wiley & Sons; 2022. pp. 1–43.

3 Chipaux M, van der Laan KJ, Hemelaar SR, Hasani M, Zheng T, Schirhagl R. Nanodiamonds and their applications in cells. *Small*. 2018;14(24):1704263.

4 Vaijayanthimala V, Cheng P-Y, Yeh S-H, Liu K-K, Hsiao C-H, Chao J-I, *et al*. The long-term stability and biocompatibility of fluorescent nanodiamond as an in vivo contrast agent. *Biomaterials*. 2012;33(31):7794–802.

5 Zhu Y, Li J, Li WX, Zhang Y, Yang XF, Chen N, *et al*. The biocompatibility of nanodiamonds and their application in drug delivery systems. *Theranostics*. 2012;2(3):302–12.

6 Hemelaar SR, Saspaanithy B, L'Hommelet SR, Perona Martinez FP, Van der Laan KJ, Schirhagl R. The response of HeLa cells to fluorescent nanodiamond uptake. *Sensors*. 2018;18(2):355.

7 McGuinness LP, Yan Y, Stacey A, Simpson DA, Hall LT, Maclaurin D, *et al*. Quantum measurement and orientation tracking of fluorescent nanodiamonds inside living cells. *Nat Nanotechnol*. 2011;6(6):358–63.

8 Nie L, Zhang Y, Li L, van Rijn P, Schirhagl R. pH sensitive dextran coated fluorescent nanodiamonds as a biomarker for HeLa cells endocytic pathway and increased cellular uptake. *Nanomaterials*. 2021;11(7):1837.

9 Perevedentseva E, Hong S-F, Huang K-J, Chiang I-T, Lee C-Y, Tseng Y-T, *et al*. Nanodiamond internalization in cells and the cell uptake mechanism. *J Nanoparticle Res*. 2013;15(8):1834.

10 Chang Y-R, Lee H-Y, Chen K, Chang C-C, Tsai D-S, Fu C-C, *et al*. Mass production and dynamic imaging of fluorescent nanodiamonds. *Nat Nanotechnol*. 2008;3(5):284–8.

11 Faklaris O, Garrot D, Joshi V, Druon F, Boudou JP, Sauvage T, *et al.* Detection of single photoluminescent diamond nanoparticles in cells and study of the internalization pathway. *Small.* 2008;4(12):2236–9.

12 Hemelaar SR, de Boer P, Chipaux M, Zuidema W, Hamoh T, Martinez FP, *et al.* Nanodiamonds as multi-purpose labels for microscopy. *Sci Rep.* 2017;7(1):720-.

13 Faklaris O, Joshi V, Irinopoulou T, Tauc P, Sennour M, Girard H, *et al.* Photoluminescent diamond nanoparticles for cell labeling: Study of the uptake mechanism in mammalian cells. *Acs Nano.* 2009;3(12):3955–62.

14 Liu K-K, Wang C-C, Cheng C-L, Chao J-I. Endocytic carboxylated nanodiamond for the labeling and tracking of cell division and differentiation in cancer and stem cells. *Biomaterials.* 2009;30(26):4249–59.

15 Hemelaar SR, van der Laan KJ, Hinterding SR, Koot MV, Ellermann E, Perona-Martinez FP, *et al.* Generally applicable transformation protocols for fluorescent nanodiamond internalization into cells. *Sci Rep.* 2017;7(1):5862.

16 Vaijayanthimala V, Chang HC. Functionalized fluorescent nanodiamonds for biomedical applications. *Nanomedicine-Uk.* 2009;4(1):47–55.

17 Prabhakar N, Khan MH, Peurla M, Chang H-C, Hänninen PE, Rosenholm JM. Intracellular trafficking of fluorescent nanodiamonds and regulation of their cellular toxicity. *ACS Omega.* 2017;2(6):2689–93.

18 Fang C-Y, Vaijayanthimala V, Cheng C-A, Yeh S-H, Chang C-F, Li C-L, *et al.* The exocytosis of fluorescent nanodiamond and its use as a long-term cell tracker. *Small.* 2011;7(23):3363–70.

19 Chu Z, Zhang S, Zhang B, Zhang C, Fang C-Y, Rehor I, *et al.* Unambiguous observation of shape effects on cellular fate of nanoparticles. *Sci Rep.* 2014;4:4495.

20 Zheng T, Perona Martínez F, Storm IM, Rombouts W, Sprakel J, Schirhagl R, *et al.* Recombinant protein polymers for colloidal stabilization and improvement of cellular uptake of diamond nanosensors. *Anal Chem.* 2017;89(23):12812–20.

21 Tammam SN, Azzazy HM, Lamprecht A. How successful is nuclear targeting by nanocarriers? *J Control Release.* 2016;229:140–53.

22 Chu Z, Miu K, Lung P, Zhang S, Zhao S, Chang H-C, *et al.* Rapid endosomal escape of prickly nanodiamonds: Implications for gene delivery. *Sci Rep.* 2015;5:11661.

23 Martín R, Álvaro M, Herance JR, García H. Fenton-treated functionalized diamond nanoparticles as gene delivery system. *Acs Nano.* 2010;4(1):65–74.

24 Leung HM, Chan MS, Liu LS, Wong SW, Lo TW, Lau C-H, *et al.* Dual-function, cationic, peptide-coated nanodiamond systems: Facilitating nuclear-targeting delivery for enhanced gene therapy applications. *ACS Sustain Chem Eng.* 2018;6(8):9671–81.

25 Morita A, Hamoh T, Sigaeva A, Norouzi N, Nagl A, van der Laan KJ, *et al.* Targeting nanodiamonds to the nucleus in yeast cells. *Nanomaterials.* 2020;10(10):1962.

26 Gulka M, Salehi H, Varga B, Middendorp E, Pall O, Raabova H, *et al.* Simultaneous label-free live imaging of cell nucleus and luminescent nanodiamonds. *Sci Rep.* 2020;10(1):9791.

27 Su L-J, Wu M-S, Hui YY, Chang B-M, Pan L, Hsu P-C, *et al.* Fluorescent nanodiamonds enable quantitative tracking of human mesenchymal stem cells in miniature pigs. *Sci Rep.* 2017;7(1):45607.

28 Gerstenhaber JA, Marcinkiewicz C, Barone FC, Sternberg M, D'Andrea MR, Lelkes PI, *et al.* Biocompatibility studies of fluorescent diamond particles-(NV)~800nm (part V): In vitro kinetics and in vivo localization in rat liver following long-term exposure. *Int J Nanomedicine.* 2019;14:6451–64.

29 Prabhakar N, Peurla M, Shenderova O, Rosenholm JM. Fluorescent and electron-dense green color emitting nanodiamonds for single-cell correlative microscopy. *Molecules.* 2020;25(24):5897.

30 Chauhan S, Jain N, Nagaich U. Nanodiamonds with powerful ability for drug delivery and biomedical applications: Recent updates on in vivo study and patents. *J Pharm Anal.* 2020;10(1):1–12.

31 Schrand AM, Huang HJ, Carlson C, Schlager JJ, Osawa E, Hussain SM, *et al.* Are diamond nanoparticles cytotoxic? *J Phys Chem B.* 2007;111(1):2–7.

32 Brown CP. Advancing musculoskeletal research with nanoscience. *Nat Rev Rheumatol.* 2013;9(10):614–23.

33 Schrand AM, Dai L, Schlager JJ, Hussain SM, Osawa E. Differential biocompatibility of carbon nanotubes and nanodiamonds. *Diam Relat Mater.* 2007;16(12):2118–23.

34 Schrand AM, Hens SAC, Shenderova OA. Nanodiamond particles: Properties and perspectives for bioapplications. *Crit Rev Solid State Mater Sci.* 2009;34(1–2):18–74.

35 Xing Y, Xiong W, Zhu L, Osawa E, Hussin S, Dai L. DNA damage in embryonic stem cells caused by nanodiamonds. *Acs Nano.* 2011;5(3):2376–84.

36 Marcon L, Riquet F, Vicogne D, Szunerits S, Bodart JF, Boukherroub R. Cellular and in vivo toxicity of functionalized nanodiamond in Xenopus embryos. *J Mater Chem.* 2010;20(37):8064–9.

37 Thomas V, Halloran BA, Ambalavanan N, Catledge SA, Vohra YK. In vitro studies on the effect of particle size on macrophage responses to nanodiamond wear debris. *Acta Biomater.* 2012;8(5):1939–47.

38 Moore L, Grobárová V, Shen H, Man HB, Míčová J, Ledvina M, *et al.* Comprehensive interrogation of the cellular response to fluorescent, detonation and functionalized nanodiamonds. *Nanoscale.* 2014;6(20):11712–21.

39 Ibrahim M, Xue Y, Ostermann M, Sauter A, Steinmueller-Nethl D, Schweeberg S, *et al.* In vitro cytotoxicity assessment of nanodiamond particles and their osteogenic potential. *J Biomed Mater Res A.* 2018;106(6):1697–707.

40 Dworak N, Wnuk M, Zebrowski J, Bartosz G, Lewinska A. Genotoxic and mutagenic activity of diamond nanoparticles in human peripheral lymphocytes in vitro. *Carbon*. 2014;68:763–76.

41 Mytych J, Lewinska A, Zebrowski J, Wnuk M. Nanodiamond-induced increase in ROS and RNS levels activates NF-κB and augments thiol pools in human hepatocytes. *Diam Relat Mater*. 2015;55:95–101.

42 Khanal D, Hau H, Kondyurin A, Fu D, Ramzan I, Chrzanowski W. Nanotoxicity of nanodiamond in two and three dimensional liver models. *Int J Nanotechnol*. 2017;14(1–6):133–54.

43 Khanal D, Kondyurin A, Hau H, Knowles JC, Levinson O, Ramzan I, *et al*. Biospectroscopy of nanodiamond-induced alterations in conformation of intra- and extracellular proteins: A nanoscale IR study. *Anal Chem*. 2016;88(15):7530–8.

44 Lee D-K, Ha S, Jeon S, Jeong J, Kim D-J, Lee SW, *et al*. The sp3/sp2 carbon ratio as a modulator of in vivo and in vitro toxicity of the chemically purified detonation-synthesized nanodiamond via the reactive oxygen species generation. *Nanotoxicology*. 2020;14(9):1213–26.

45 Yu S-J, Kang M-W, Chang H-C, Chen K-M, Yu Y-C. Bright fluorescent nanodiamonds: No photobleaching and low cytotoxicity. *J Am Chem Soc*. 2005;127(50):17604–5.

46 Vial S, Mansuy C, Sagan S, Irinopoulou T, Burlina F, Boudou JP, *et al*. Peptide-grafted nanodiamonds: Preparation, cytotoxicity and uptake in cells. *Chembiochem*. 2008;9(13):2113–9.

47 Lee JW, Lee S, Jang S, Han KY, Kim Y, Hyun J, *et al*. Preparation of non-aggregated fluorescent nanodiamonds (FNDs) by non-covalent coating with a block copolymer and proteins for enhancement of intracellular uptake. *Mol Biosyst*. 2013;9(5):1004–11.

48 Moche H, Paget V, Chevalier D, Lorge E, Claude N, Girard HA, *et al*. Carboxylated nanodiamonds can be used as negative reference in in vitro nanogenotoxicity studies. *J Appl Toxicol*. 2017;37(8):954–61.

49 Tsai L-W, Lin Y-C, Perevedentseva E, Lugovtsov A, Priezzhev A, Cheng C-L. Nanodiamonds for medical applications: Interaction with blood in vitro and in vivo. *Int J Mol Sci*. 2016;17(7):1111.

50 Cid A, Picado A, Correia JB, Chaves R, Silva H, Caldeira J, *et al*. Oxidative stress and histological changes following exposure to diamond nanoparticles in the freshwater Asian clam *Corbicula fluminea* (Müller, 1774). *J Hazard Mater*. 2015;284:27–34.

51 Ali AA-M, Mohamed HRH. Genotoxicity and oxidative stress induced by the orally administered nanosized nickel and cobalt oxides in male albino rats. *J Basic Appl Zool*. 2019;80(1):2.

52 Blaber SP, Hill CJ, Webster RA, Say JM, Brown LJ, Wang S-C, *et al.* Effect of labeling with iron oxide particles or nanodiamonds on the functionality of adipose-derived mesenchymal stem cells. *PLOS ONE*. 2013;8(1):e52997.

53 Moore L, Yang J, Lan TTH, Osawa E, Lee D-K, Johnson WD, *et al.* Biocompatibility assessment of detonation nanodiamond in non-human primates and rats using histological, hematologic, and urine analysis. *Acs Nano*. 2016;10(8):7385–400.

54 Chang C-C, Zhang B, Li C-Y, Hsieh C-C, Duclos G, Treussart F, *et al.*, editors. Exploring cytoplasmic dynamics in zebrafish yolk cells by single particle tracking of fluorescent nanodiamonds. *Advances in Photonics of Quantum Computing, Memory, and Communication V*. San Francisco, California: SPIE; 2012. February 13.

55 Eldawud R, Reitzig M, Opitz J, Rojansakul Y, Jiang W, Nangia S, *et al.* Combinatorial approaches to evaluate nanodiamond uptake and induced cellular fate. *Nanotechnology*. 2016;27(8):085107-.

56 Lin YC, Perevedentseva E, Tsai LW, Wu KT, Cheng CL. Nanodiamond for intracellular imaging in the microorganisms in vivo. *J Biophotonics*. 2012; 5(11–12):838–47.

57 Simpson DA, Thompson AJ, Kowarsky M, Zeeshan NF, Barson MSJ, Hall LT, *et al.* In vivo imaging and tracking of individual nanodiamonds in *Drosophila melanogaster* embryos. *Biomed Opt Express*. 2014;5(4):1250–61.

58 Mohan N, Zhang B, Chang C-C, Yang L, Chen C-S, Fang C-Y, *et al.* Fluorescent nanodiamond – A novel nanomaterial for in vivo applications. *Mater Res Soc Symp Proc*. 2011;1362:null–null.

59 Mohan N, Chen C-S, Hsieh H-H, Wu Y-C, Chang H-C. In vivo imaging and toxicity assessments of fluorescent nanodiamonds in *Caenorhabditis elegans*. *Nano Lett*. 2010;10(9):3692–9.

60 Puzyr AP, Baron AV, Purtov KV, Bortnikov EV, Skobelev NN, Moginaya OA, *et al.* Nanodiamonds with novel properties: A biological study. *Diam Relat Mater*. 2007;16(12):2124–8.

61 Zhang X, Yin J, Kang C, Li J, Zhu Y, Li W, *et al.* Biodistribution and toxicity of nanodiamonds in mice after intratracheal instillation. *Toxicol Lett*. 2010;198(2):237–43.

62 Li H-C, Hsieh F-J, Chen C-P, Chang M-Y, Hsieh PCH, Chen C-C, *et al.* The hemocompatibility of oxidized diamond nanocrystals for biomedical applications. *Sci Rep*. 2013;3(1):3044.

63 Barone FC, Marcinkiewicz C, Li J, Sternberg M, Lelkes PI, Dikin DA, *et al.* Pilot study on biocompatibility of fluorescent nanodiamond-(NV)-Z~800 particles in rats: Safety, pharmacokinetics, and bio-distribution (part III). *Int J Nanomedicine*. 2018;13:5449–68.

64 Moore LK, Caldwell MA, Townsend TR, MacRenaris KW, Moyle-Heyrman G, Rammohan N, *et al.* Water-soluble nanoconjugate for enhanced cellular delivery of receptor-targeted magnetic resonance contrast agents. *Bioconjugate Chem.* 2019;30(11):2947–57.

65 Miller MR, Raftis JB, Langrish JP, McLean SG, Samutrtai P, Connell SP, *et al.* Inhaled nanoparticles accumulate at sites of vascular disease. *Acs Nano.* 2017;11(5):4542–52.

66 Purtov K, Petunin A, Inzhevatkin E, Burov A, Ronzhin N, Puzyr A, *et al.* Biodistribution of different sized nanodiamonds in mice. *J Nanosci Nanotechno.* 2015;15(2):1070–5.

67 Wang L, Su W, Ahmad KZ, Wang X, Zhang T, Yu Y, *et al.* Safety evaluation of nanodiamond-doxorubicin complexes in a Naïve Beagle canine model using hematologic, histological, and urine analysis. *Nano Res.* 2021;15.

68 Li T-F, Xu Y-H, Li K, Wang C, Liu X, Yue Y, *et al.* Doxorubicin-polyglycerol-nanodiamond composites stimulate glioblastoma cell immunogenicity through activation of autophagy. *Acta Biomater.* 2019;86:381–94.

69 Yuan S-J, Xu Y-H, Wang C, An H-C, Xu H-Z, Li K, *et al.* Doxorubicin-polyglycerol-nanodiamond conjugate is a cytostatic agent that evades chemoresistance and reverses cancer-induced immunosuppression in triple-negative breast cancer. *J Nanobiotechnology.* 2019;17(1):110.

70 Ryu T-K, Baek SW, Lee G-J, Rhee C-K, Choi S-W. Targeted tumor therapy based on nanodiamonds decorated with doxorubicin and folic acid. *Macromol Biosci.* 2017;17(2):1600180.

71 Happel P, Waag T, Schimke M, Schweeberg S, Muzha A, Fortak K, *et al.* Intrinsically 32P-labeled diamond nanoparticles for in vivo imaging and quantification of their biodistribution in chicken embryos. *Adv Funct Mater.* 2018;28(36):1802873.

72 Kaluç N, Thomas PB. A carboxylated nanodiamond reduces oxidative stress and shows no sign of toxicity in yeast. *Fuller Nanotub Carbon Nanostructures.* 2021;30: 1–8.

73 Perevedentseva E, Lin Y-C, Cheng C-L. A review of recent advances in nanodiamond-mediated drug delivery in cancer. *Expert Opin Drug Del.* 2021; 18(3):369–82.

74 Horie M, Tabei Y. Role of oxidative stress in nanoparticles toxicity. *Free Radic Res.* 2021;55(4):331–42.

75 Horie M, Fujita K, Kato H, Endoh S, Nishio K, Komaba LK, *et al.* Association of the physical and chemical properties and the cytotoxicity of metal oxide nanoparticles: Metal ion release, adsorption ability and specific surface area†. *Metallomics.* 2012;4(4):350–60.

76 Horie M, Komaba LK, Kato H, Nakamura A, Yamamoto K, Endoh S, *et al.* Evaluation of cellular influences induced by stable nanodiamond dispersion; the cellular influences of nanodiamond are small. *Diam Relat Mater.* 2012;24:15–24.

6

In vitro Bioimaging of Fluorescent Nanodiamonds

Trong-Nghia Le, Yen-Tse Chiang, Yuen Yung Hui, Thi-Hong-Hanh Le, Yan-Kai Tzeng, Neha Sharma, Wei-Hung Chiang, and Wesley Wei-Wen Hsiao**

* *Corresponding authors*

Negative nitrogen-vacancy (NV^-) centers are lodged deeply in the diamond nanoparticles, and their fluorescence is not influenced by any surface modification and bioconjugation. Together with the remarkable biocompatibility of carbon-based nanomaterials, these unique properties make fluorescent nanodiamond (FND) especially well-suited for long-term cell tracking applications, most notably in stem cell studies. This chapter summarizes current advancements in FND-based technology, particularly emphasizing super-resolution imaging and long-term cell tracking. These modern and multifunctional technologies are in tune with existing contemporary imaging modalities and offer a wide range of applications.

6.1 Bioimaging Applications

Bioimaging is an important facet of current life science research. It allows researchers not only to observe biological processes in cells and organisms but also to track molecules or cells in real time and in three dimensions for disease diagnosis and therapy. Being noninvasive, sensitive, and quantitative are three unique characteristics of bioimaging probes. For example, as the result of many fluorescence imaging studies, we can now connect advanced metabolic processes with the activities of proteins in living cells.

In the twenty-first century, scholars developed various kinds of fluorescent nanoparticles for bioimaging, including dye-doped nanospheres (1, 2), quantum dots (QDs) (3), metallic nanoclusters (4, 5), carbon-based nanoparticles (6), and polymeric nanoparticles (7). Biological macromolecules such as peptides, nucleic acids, antibodies, or ligands can be conjugated onto those fluorescent nanoparticles, allowing them to function as fluorescent markers. However, there are certain challenges in the field of fluorescent bioimaging. For instance, the color of fluorescent markers can change, biological samples can undergo photodamage, and strong light scattering from tissue, cell autofluorescence can

Nanodiamonds in Analytical and Biological Sciences: Principles and Applications, First Edition.
Edited by Wesley Wei-Wen Hsiao and Trong-Nghia Le.
© 2023 John Wiley & Sons Ltd. Published 2023 by John Wiley & Sons Ltd.

occur. QDs, which are semiconductor nanoparticles typically made of cadmium or selenium, hold the most promise. Compared with organic fluorophores, QDs have greater absorption coefficients and notably higher levels of brightness and photostability. However, their inherent toxicity has hampered their use as biolabels for living cells and organisms (8). Although upconversion nanoparticles are attractive alternatives (9), these inorganic nanocrystallites are doped with transition metals, lanthanide, or actinide ions, and as a result, their biocompatibility remains a concern.

With the recent invention of FND, a new age of nanoparticle-based cell labeling, imaging, and tracking has begun (10, 11). This is due to FND's exceptionally high biocompatibility and unique optical properties (10). Recent research on surface-functionalized FNDs has recently been assessed for use in quantum sensing, bioimaging, and drug or gene delivery. Results have shown that FND surfaces can be easily functionalized with protein and nucleic acid (10). FND with NV color centers can also generate brilliant nonphotobleaching, nonphotoblinking, and tissue-penetrating red photons when exposed to green–yellow light. FND was first shown to be useful in cellular fluorescence imaging in 2005 (12) and for *in vivo* stem cell tracking in 2013 (13). As both technological and biomedical demands increase, surface-modified FNDs are predicted to be among the most beneficial optical nanoprobes for bioimaging and diagnostics. FNDs enable researchers to observe, track, and measure molecules or cells at a high spatial resolution – a capability required to properly comprehend complex biological systems both *in vitro* and *in vivo*.

Advancements in biological imaging technology have helped shed light on life's most crucial processes. FND-based technology is in synchronicity with existing modern imaging modalities. This chapter will discuss a broad range of FND-based bioimaging technologies, including multiphoton excitation, super-resolution, cathodoluminescence (CL), and magnetic resonance imaging. Before going into the details of this topic, the reader may choose to review several fundamental concepts, such as important spectral aspects of diamond color centers and optical properties of NV^0, NV^-, and H3.

6.2 Epifluorescence and Confocal Fluorescence Imaging

FNDs' excellent chemical stability, biocompatibility, and photostability make them an interesting alternative for tracking temporal and spatial activities in living cells. The fluorescence bank of the NV^- center has a peak of around 700 nm, which is well separated from cellular autofluorescence caused by endogenous fluorophores (14, 15). This allows the nanoparticle to play the role of a long-term, three-dimensional (3-D) labeling and tracking device in cells. Neugart *et al.* studied FNDs diffusion in HeLa cells and found that particles coated with sodium dodecyl sulfate were readily taken up by the cells, with most FNDs being immobilized in the cytosol (16). Colocalization analysis by Faklaris *et al.* revealed that most ingested FND particles become trapped in endocytic vesicles, with some smaller particles making their way to the cytosol (17, 18). Since FNDs as delivery vehicles have myriad applications in the transportation of therapeutic agents into cells, these single-particle tracking (SPT) experiments bolster the potential of the nanoparticles in this realm.

The surfaces of NDs are versatile platforms for conjugating biomolecules through chemical modification (Chapter 4). Because their fluorescent centers are deeply buried in the diamond lattice, all of the strategies discovered for NDs may be fruitfully applied to FNDs without appreciably affecting their optical characteristics. Consequently, FNDs have been used in specific targeting, acting as probes to track the interaction of transferrin (Tf) with its receptors on HeLa cells (19). Confocal fluorescence imaging confirmed the receptor-mediated uptake of the FND–Tf bioconjugates formed by amide linkage between carboxylated FND and Tf (Figure 6.1). Zhang *et al.* have similarly illustrated how FNDs can be applied to the receptor-mediated targeting of cancer cells with folic acid (20). The authors covalently coupled FNDs with folic acid using biocompatible polyethylene glycol (PEG) as a cross-linked buffer layer. Ensemble average measurements with flow cytometry showed that the folate-conjugated FND particles could be taken up by HeLa cells through receptor-mediated endocytosis. The majority of FND particles were accumulated in intracellular endocytic vesicles in the perinuclear range.

Another noteworthy development in the field is the synthesis of fluorescence-enhanced and protein-functionalized FNDs for the targeted tagging of mammalian cells (21). Researchers have identified two types of cellular targeting strategies. In the first method, strong oxidative acid is used to treat FND to give it a noncovalent coating with glycoproteins or neoglycoproteins (i.e., proteins chemically modified with sugar residue). This allows the particles to target hepatocytes owing to

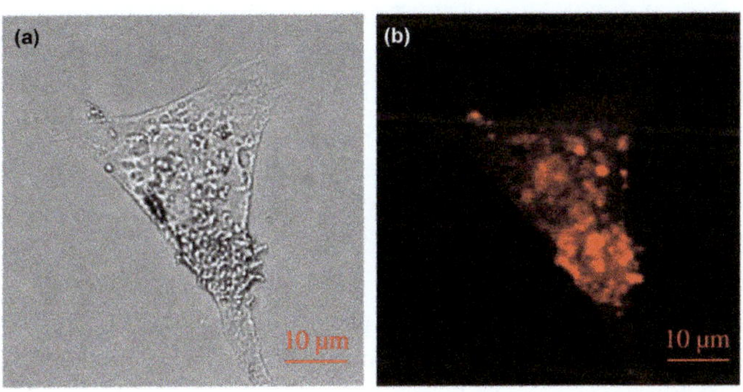

Figure 6.1 (a) Bright-field image and (b) confocal scanning image of FND–Tf uptaken by HeLa cells. *Source:* Reproduced from ref. (21) / ELSEVIER.

carbohydrate receptors. In another method, FND is PEGylated and then covalently conjugated with streptavidin. As a result, biotin-labeled antibody linking was formed. Different human cell strains (such as hepatoma, HepG2, and breast cancer cells, ASB145-1R, MCF-7, and MDA-MB-231) were used to demonstrate a high targeting specificity for bioconjugated FNDs. Such cell-specific targeting and tagging strategies can be applied in various situations.

Cigler *et al.* recently enclosed FNDs in translucent silica shells, which were then coated with a biocompatible copolymer shell to target cancer cells (22, 23). The copolymers can bear both fluorescent probes (e.g., Alexa Fluor 488) and targeting peptides (e.g., cyclic RGD). These RGD-attached nanoparticles allowed for the specific targeting of integrin $\alpha v\beta 3$ receptors on glioblastoma cells, with high internalization effectiveness.

Other notable advances in the field include the study of the exocytosis of FNDs in cells (24), the use of FND as a probe for intracellular trafficking in primary neurons (25), the enhanced cellular uptake of FNDs by coating with block copolymers (26), the shape effect on the cellular fate of FNDs (27), the labeling and imaging of neuron cells with FNDs (28), and the effect of FNDs on neuronal survival and morphogenesis (29). Although FNDs had no cytotoxicity in primary neurons in either the central or peripheral nervous systems, they did shorten neurites due to spatial impediment on the advancing axonal development cone.

6.3 Single-Particle Tracking

A notable use of FND is SPT, which is used to examine the cytoplasmic movements of live cells. The use of SPT allows for the direct detection of molecular or vesicle mobility inside biological matrices in conjunction with other single-molecule spectroscopies, including fluorescence correlation spectroscopy (FCS), fluorescence recovery after photobleaching (FRAP), and Förster resonance energy transfer (FRET) (30–32). SPT can measure trajectories to derive quantitative data on the forces driving a particle's motion if there is enough spatial and temporal resolution. Since each particle can be observed separately, SPT may be used to separate populations of particles with different motility characteristics. Additionally, seeing individual FND-labeled cells can shed light on intracellular processes, including drug transport and viral infection.

Fu *et al.* introduced FND to HeLa cells and monitored the particles' two-dimensional (2-D) movements inside the cell to provide the first report of FND for SPT in cells (14). Later, Chang *et al.* improved the SPT method by observing the desired FND in three dimensions (3-D) while vertically mounting an objective microscope lens on a piezoelectric translational stage (33). Figure 6.2 depicts

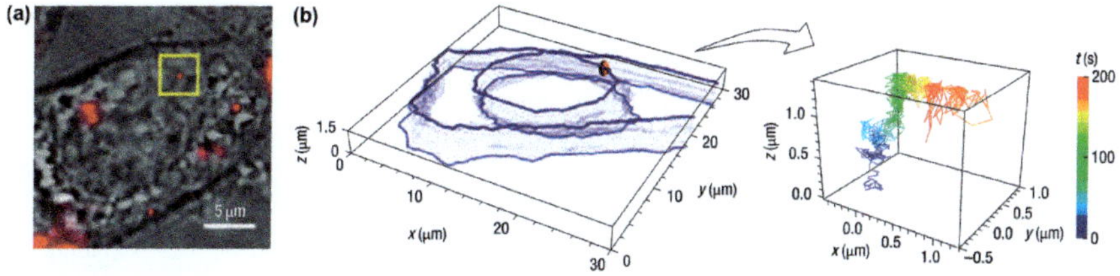

Figure 6.2 Single-particle tracking of 35-nm FND in a living HeLa cell. (a) Bright-field/fluorescent merged image. *Source:* Reproduced from ref. (34) / Springer Nature (b) 3-D movement path of a single FND inside the cell over 200 s. *Source:* (34) Chang et al., 2008 / Springer Nature.

the 3-D tracking of a single 35-nm FND in a living HeLa cell. Axial sectioning of the cell confirmed the internalization of the FND particle. With a servo system that controlled the position of the microscope's objective lens, the motion of a single FND particle inside the cell was tracked over a 200-s period. A mean squared displacement analysis for the 3-D trajectory determined the diffusion coefficient of the internalized FND as under 0.01 $\mu m^2/s$. FND uptake has a diffusion coefficient equivalent to QDs enclosed within an endosome. Additionally, McGuinness *et al.* revealed that the optically detected magnetic resonance (ODMR) method might be utilized to continuously monitor the dynamics of FND trapped in the endosomal or lysosomal compartments of HeLa cells for more than 16 hours (34). These experiments show that FND works well as a fluorescent marker for intracellular monitoring.

In another study, by facilitating protein transport via membrane tunneling nanotubes (TNTs) formed between cells, such as human embryonic kidney cells, Epperla *et al.* used the FND-based SPT approach to observe intercellular dynamics (35). TNTs are cytoplasmic transport networks that facilitate the movement of proteins and other cytoplasmic elements. In this study, FNDs coated with green fluorescent protein (EGFP) and bovine serum albumin (BSA) were able to pass through the TNT. Then, fluorescence microscopy was used to track a single particle in the transient membrane connections utilizing these FND bioconjugates. According to the time-lapse series analysis depicted in Figure 6.3, BSA-coated EGFP-PA-FND particles are ingested by endocytosis and end up within payloads that migrate through TNT from the cytoplasm of one cell to another. Notably, stop-and-go and to-and-fro movements driven by molecular motors were found to be involved in the active transport of FND bioconjugates trapped in endosomal vehicles. These findings show that FND can aid in the transport of biomolecules across cells. Additionally, these particles may be used to constantly follow the entire delivery process over an extended period.

An FND-based SPT technique was created by Haziza *et al.* to investigate intraneuronal transport abnormalities brought on by genetic risk factors for brain

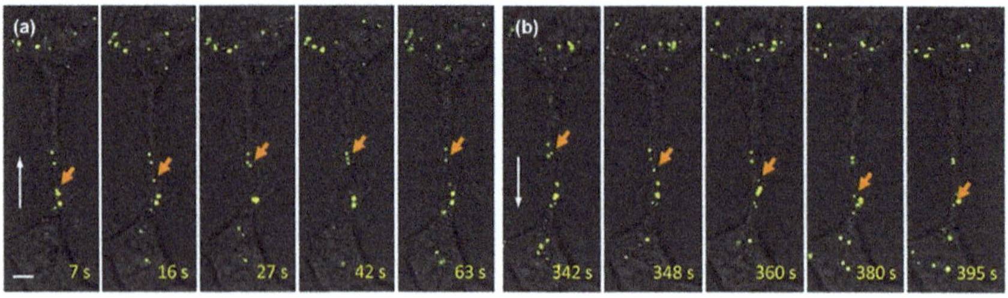

Figure 6.3 (a) Forward and (b) backward movement tracking of two different BSA-coated EGFP-PA-FND particles. Orange arrows indicate the particle of interest. Scale bar: 5 μm. *Source:* Reproduced from ref. (36) / John Wiley & Sons.

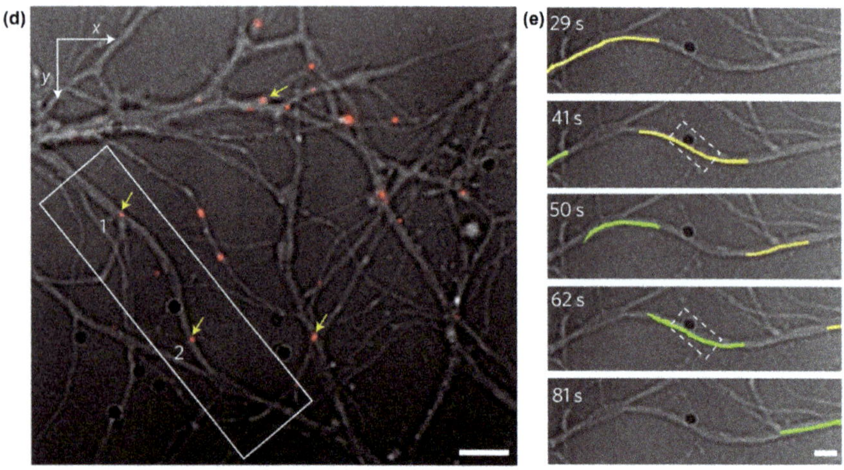

Figure 6.4 FND tracking for the observation of intraneuronal transport. (Left) Four FNDs can be seen traveling inside dendrites of the neuronal branches. Scale bar: 5 mm. (Right) Location of two FND particles (1 in yellow, 2 in green), as estimated by particle tracking with a 10-second persistence. Scale bar: 1 mm. *Source:* Reproduced from ref. (15) / Springer Nature.

disease (15). They leveraged FND's unique optical properties to perform high-throughput, long-term tracking with 12-nm spatial resolution inside the branches of dissociated neurons for a period of 50 ms (Figure 6.4). Then, two transgenic mouse lines that mimicked the slight variations in protein concentration (30%) observed in patient brains underwent the FND tracking experiment. The FND-enabled approach was sensitive enough in both instances to pick up on changes in intraneuronal trafficking.

In addition, by analyzing the connection between fluctuations in fluorescence intensity within a tightly confined zone, FCS is another excellent method of analyzing FND with high contrast (16). FCS is a single-molecule-based technique that measures fluorescence intensity correlations from individual fluorescent molecules. FCS gives information about the fluorescence fluctuations of a population of single molecules over time. Kinetic and thermodynamic properties are also provided by FCS for fluorescent molecules (36). Average sizes and intensities of FNDs can be measured via fluorescence intensity fluctuation of FND particles diffusing through the excitation focal volume (37, 38). In Figure 6.5, a monomeric red fluorescent protein called DsRed-Monomer (4 nm in size) is compared with the 11-nm FND stimulated under the same conditions ($\lambda_{ex} = 532$ nm and $I_{ex} = 80$ kW/cm^2) (39). The average fluorescence intensity of FND particles is around 50 counts/ms, which is almost eight times greater than that of DsRed-Monomer, as shown by the particle distribution histogram (Figure 6.5b). The estimated molar extinction coefficient, quantum yield, and observed number of NV^- centers in the 11-nm FND particle, coupled with the estimated eightfold change in fluorescence intensity, all accord nicely. A single DsRed molecule may release around 105 photons under 514-nm stimulation before photobleaching, according to Heikal *et al.* (40). FND, on the other hand, emits an infinite number of photons.

Hui *et al.* showed that by properly encapsulating particles in a lipid layer, the internalization of 40-nm FND particles in the cytoplasm of HeLa cells can be boosted by more than 10 times (41). To examine transmembrane signaling dynamics in live cells, Liu *et al.* combined FCS and SPT, grafted the transforming growth factor beta (TGF-β) onto FND's surface, and monitored the effects of small molecule inhibitors on the 3-D motions of TGF-β receptors in human lung cancer cells

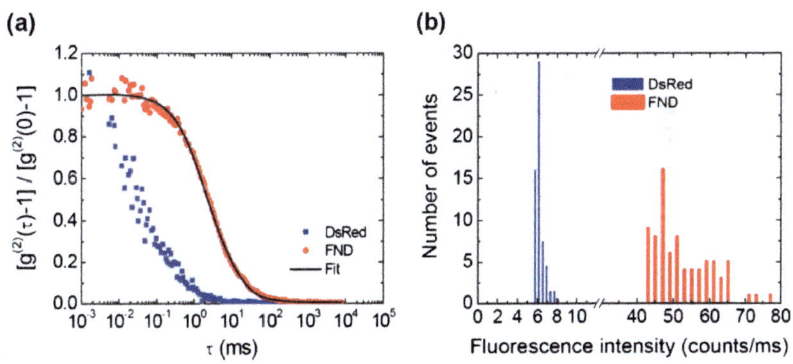

Figure 6.5 FCS spectroscopic analysis of 11-nm FNDs and DsRed-Monomers. Normalized autocorrelation functions of intensity variations and fluorescence intensities histograms of the particles are shown in (a) and (b), respectively. *Source:* (40) Mohan et al., 2010 / John Wiley & Sons.

(42). Their findings demonstrated that TGF-β conjugated FND had no discernible impact on the outcome of transmembrane signaling dynamics in live cells and that immobilized TGF-β was required for active signaling. In conclusion, these investigations indicate that researchers can examine the dynamics of FND bioconjugate in aqueous solutions and intricate biological environments in both the short- and long-term by combining FCS and SPT.

6.4 Two-Photon Fluorescence Imaging

Two-photon excitation (TPE) microscopy, a more sophisticated imaging technology in biomedicine, contains numerous advantages over one-photon excitation (OPE) microscopy. For instance, with TPE, there is (1) no out-of-focus absorption, and the picture contrast of target specimens is boosted dramatically. The technique also considerably decreases autofluorescence and scattering signals when applying red and NIR light (2). Furthermore, optical sectioning is reduced to a depth of a few hundred micrometers (43, 44). However, excitation requires an ultrafast laser. Wee *et al.* utilized a picosecond mode-locked laser (at 1064-nm wavelength) as the stimulating source to study the TPE fluorescence spectroscopy of NV^- centers in bulk diamond (45). They found the two-photon absorption cross section of the NV^- center at 1064 nm to be 0.45×10^{-50} $cm^4{\cdot}s$ per photon based on the comparable one-photon absorption cross section of 3.1×10^{-17} cm^2 at 532 nm. This measured two-photon absorption cross section is about 1/13 that of a dye molecule like rhodamine B. Notably, this difficulty can be readily overcome by increasing the quantity of NV^- centers in the particles (46).

TPE imaging using FNDs as the contrast agents in living cells was first demonstrated by Chang and coworkers (33). Better image contrast of FNDs in the cell was seen in TPE as it outperformed OPE in diminishing cell autofluorescence (Figure 6.6). Furthermore, in this experiment, the excited volume was minimal due to the fluorescence intensity's quadratic dependency on the incident light intensity. Consequently, photoinduced cell damage was significantly decreased. In another investigation, Hui *et al.* used a femtosecond Nd:glass laser conducting at 1064 nm to TPE lipid-encapsulated FNDs in living HeLa cells (47). Using this approach, they were successful in detecting single FND particles as small as 40 nm in the cytoplasm due to the high picture contrast. In summary, SPT enables the detection of individual FNDs in living cells and tissues and provides high-resolution imaging with minimal photodamage. The combination of FNDs and TPE microscopy provides a powerful tool for high-resolution imaging and tracking of individual particles in living cells and tissues, making it a valuable technique in biomedicine and other applications.

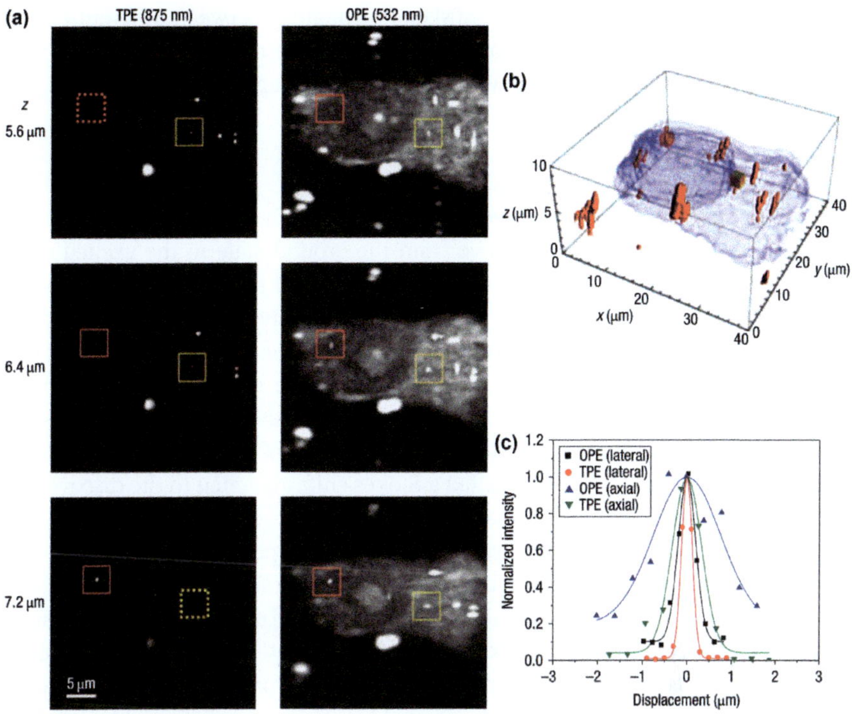

Figure 6.6 (a) Typical OPE and TPE images from the same cell. *Source:* Reproduced from ref. (34) / Springer Nature (b) 3-D TPE image of cellular internalized FND particles. (c) Lateral and axial cross sections of FND are labeled with a red box in (a). *Source:* (34) Chang et al., 2008 / Springer Nature.

6.5 Fluorescence Lifetime Imaging

6.5.1 Introduction to Fluorescence Lifetime Imaging Microscopy

Fluorescence lifetime imaging microscopy (FLIM) is one of the most used optical microscopies, especially in the living cell (48). The average amount of time a molecule spends in an excited state before going back to its ground state is known as the fluorescence lifetime. The time component of the fluorescence signal is the main advantage of FLIM compared with other intensity-based fluorescence microscopies. The basic principle of FLIM is generating and detecting fluorescence lifetime decays in biomolecules (49). It has proved itself not only in 2-D or 3-D fluorescence imaging but also for mining structural and chemical information on the nanometer scale in a single cell (50). FLIM is a powerful tool for studying biological processes, such as protein-protein interactions, enzyme kinetics, and intracellular signaling, as it provides additional information about the microenvironment of the fluorophore. Additionally, FLIM can distinguish between different fluorophores with overlapping emission spectra but different fluorescence lifetimes, allowing for multicolor imaging without spectral overlap.

6.5.2 Principles of FLIM

Fluorescent processes occur between the ground and first electronic states. When photons become excited, they move from the ground state to the excited state, quickly relaxing to the lowest vibrational level of their first electronic state – i.e., internal conversion. Usually, the time of internal conversion occurs in under a picosecond, but fluorescence is on the nanosecond scale (51–53). This means that the process of internal conversion is finished before the emission that occurs when photons move from the excited to the ground state. However, the emission band is red shifted with a longer wavelength than that of the excitation band. Such a loss of photon energy is often called a Stokes shift, referring to the discovery of the Irish physicist George G. Stokes.

Quantum yield and fluorescence lifetime are two important intrinsic characteristics of fluorescent materials. Quantum yield is the ratio of photons emitted to those absorbed, and this ratio is always less than one. Interestingly, the quantum yield of an NV center of FND is close to one (54). The other characteristic, fluorescence lifetime, is defined as the emission time of a photon as it moves from excited state to ground state. This factor is affected by environmental variables, such as pH and binding. FNDs are also inert, which can be necessary for future biological studies and for cell tracking markers. Furthermore, the lifetime of FNDs is 12 ns, which is longer than that of a biological sample (55).

A fluorescence lifetime is a form of exponential decay, given as

$$\tau = \frac{1}{(K_r + K_{nr})}$$

where τ is the fluorescence lifetime, K_r is the radiative decay rate, and K_{nr} is the nonradiative decay rate. The nonradiative decay rate is dependent on the local environment, which includes dynamic fluorophore interactions, such as fluorescence quenching and energy transfer.

The pulse-mode light exciting the fluorescent sample marks the initial time of the measurement. To obtain a fluorescence lifetime image, we must attribute the photons to different pixels. This is done by storing the absolute arrival times of the photons relative to the arrival times in respect to the laser pulse. Signals of line and frame markers from the scanner of the confocal microscope are also recorded, allowing us to sort the time stream of the photons into the different pixels. Next, we can acquire the probability distribution of a fluorescent photon's appearance. The curve is equal to the intensity decay curve (39) of fluorescence with time. Depending on the sensitivity of avalanche photodiode (APD) and the sophistication of the electronic equipment, the time accuracy can reach 20 ps. If the fluorescence lifetime distributions are combined with an instrument response function (IRF), a photon should achieve a shorter production time in the excited state, then decaying quickly.

In summary, the excited light has the strongest impact on fluorescence measurement. For instance, the measured time window depends on the repetition of the laser, and the intensity of fluorescence is based on the excitation laser's power. FLIM is sensitive to fluorescent photons, which can detect low intensities of excited light in a time-resolved spectrum. Furthermore, FILM can be used to understand the behaviors of different intensities of light. This form of single-photon detection possesses a high dynamic range and sensitivity; thus, it is a promising technique for researching the structure of quantum states.

6.5.3 Time-Correlated Single-Photon Counting

A typical FLIM system consists of a pulsed laser source for excitation to generate fluorescence and imaging optics to distinguish background photons, including tissue autofluorescence, from real fluorescence lifetime signals. A laser scanning or wide-field microscope configuration is typically used in conjunction with FLIM. Time-correlated single-photon counting (TCSPC) is the common setup for time-domain FLIM. Measurement of the temporal interval between a fluorophore's excitation by a light source and the subsequent emission of a single photon indicative of the fluorophore's de-excitation is the fundamental concept of TCSPC (Figure 6.7). Figure 6.8 shows a schematic set up for FLIM with TCSPC. The technique begins with light exciting the sample, which simultaneously triggers the detector. Next, the emission is coupled to the other detector. TCSPC detects the different timing and time-correlated signals, transferring the photon signals into a visual distribution of photon appearance. The key point of TCSPC is that it correctly measures the period between the excited light and a single photon's appearance. The results provide the distribution of fluorescence photons and time-correlated probability, which is also called the fluorescence lifetime. The light power is quite weak, so many pulses lack the strength to excite fluorophores. In this condition, the photon counting will continue to record until no photon is detected. This measurement is a statistical analysis, which requires a large quantity of data to plot the distribution of the fluorescence lifetime.

6.5.4 FND for FLIM

In a bulk diamond, the average fluorescence decay lifetime of an NV^- defect is 12 ns (56), whereas the lifetime extends to ~20 ns for the nanocrystals spin-coated on a glass substrate (57), suspended in water (46), or injected into living organisms (58). This difference stems from the alteration in the refractive index of the surrounding

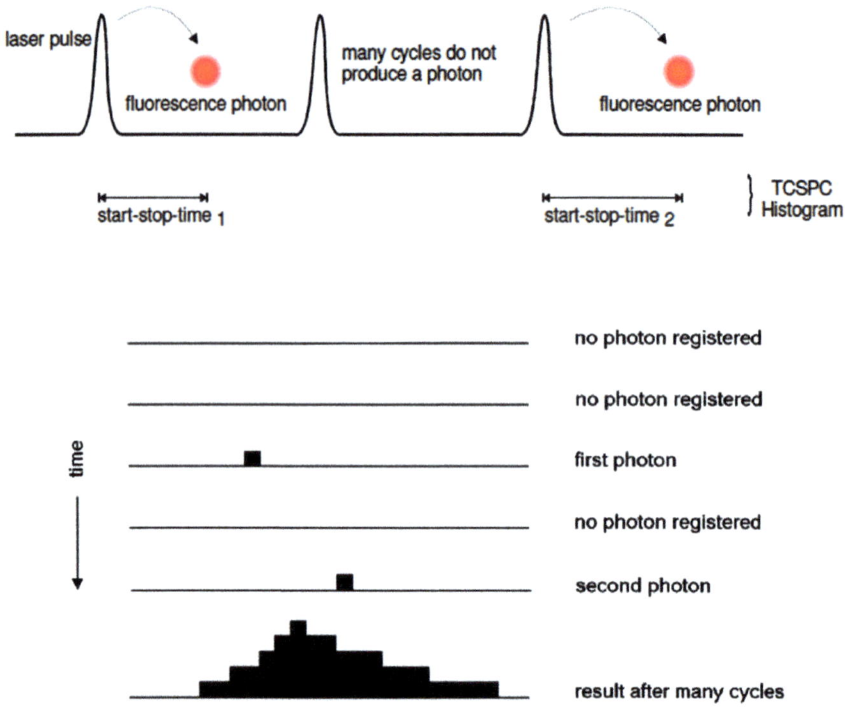

Figure 6.7 Time-correlated single-photon counting. *Source:* PicoHarp 300/http://ridl.cfd.rit.edu/products/manuals/PicoQuant/PicoHarp%20300%20v2.3/manual/Manual.pdf last accessed March 23, 2023.

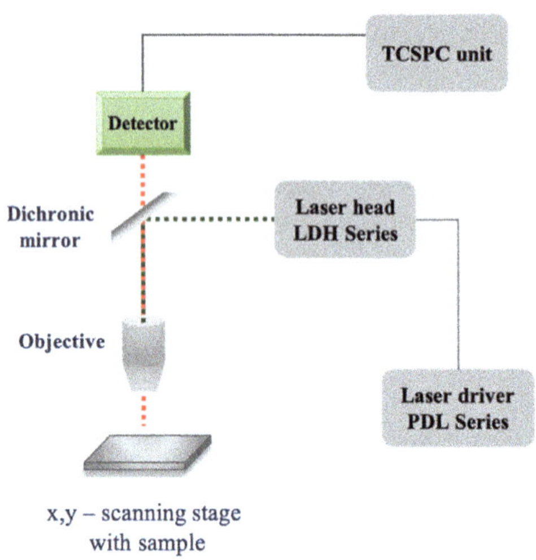

Figure 6.8 Generalized set up for a fluorescence lifetime imaging microscope.

medium when switching out a bulk diamond for nanocrystals (59). Furthermore, the average fluorescence lifetime of the NV^- centers in FND is substantially longer than the fluorescence lifetimes of commonly used endogenous and exogenous fluorophores (0.3–6.8 ns). Whether using a bulk diamond or an ND, the lifetime of the color center is significantly longer than that of dye molecules and cell autofluorescence. Thus, FND emissions can be isolated from the huge autofluorescence background of cells and tissue to enhance the image contrast by utilizing a variety of time-gating techniques.

Faklaris *et al.* were the first to leverage this technique to capture confocal fluorescence images of sub-50-nm FNDs in HeLa cells that contained only single NV^- centers, as shown in Figure 6.9 (17). The picture contrast was significantly improved by employing a picosecond pulsed laser as the excitation source and detecting photons between 15 and 53 ns following the excitation pulses. In another report, Hui *et al.* used a subnanosecond laser and an intensified charge-coupled device to excite and identify FND-labeled cancer cells in whole blood (47). Consequently, they acquired single-cell fluorescence pictures of FND-labeled HeLa cells enveloped within a 0.1-mm-thick chicken breast. Individual FND-labeled HeLa cells were also detected in blood flowing via a microfluidic device at a frame rate of up to 23 Hz. Finally, they extended their research by using the same technique to locate and trace FND-labeled lung cancer cells in a mouse ear's blood vessels.

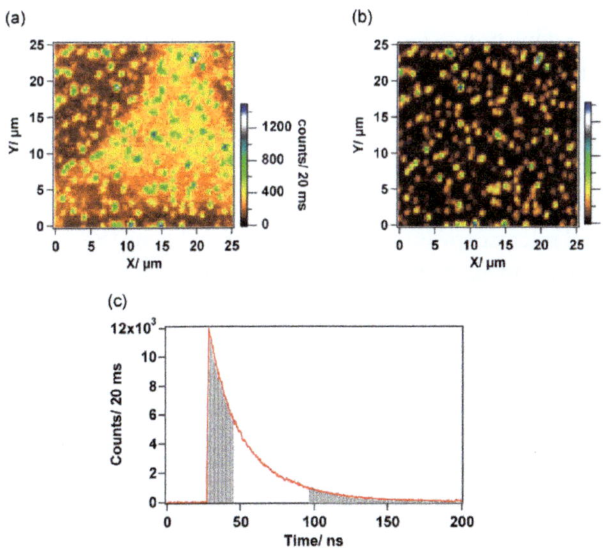

Figure 6.9 Time-gating scanning of internalized FND particles in a HeLa cell. (a) Raster scan showing FNDs and the cell's autofluorescence. (b) Time-resolved raster scan constructed from photons recorded between 15 and 53 ns following the laser excitation pulse. *Source:* Reproduced from ref. (17) / John Wiley & Sons (c) Fluorescence time decay from one of the FNDs. *Source:* (17) Faklaris et al., 2008 / John Wiley & Sons.

The result highlighted the advantages of fluorescent nanodiamonds (FNDs) over other types of fluorescent markers due to their exceptional characteristics such as superior photostability, biocompatibility, and ability to emit multicolor fluorescence from crystal defects. More important, the prolonged fluorescence lifetime of the NV^- center in FNDs enables the use of time-gating techniques to distinguish FND emission from the inherent fluorescence of cells/tissues and achieve background-free detection of FND-labeled cells *in vivo* using wide-field imaging. These unique photophysical properties of FNDs make them ideal for time-gated FLIM techniques, which can improve the detection sensitivity and specificity of FND-labeled cells *in vivo*. While further research is needed to better understand the behavior of FNDs in biological environments, the results thus far demonstrate their potential as a useful tool for FLIM studies and other biomedical applications.

6.6 Super-Resolution Fluorescence Imaging

Fluorescence microscopy has profoundly increased our understanding of cell biology. However, light diffraction poses the most significant challenge to uncovering the detailed structure of cell organelles. Super-resolution fluorescence microscopy techniques have been developed over the last decade to overcome this issue, which have advanced our understanding of cell biology dramatically. NV^- centers are well-suited for use as contrast agents in super-resolution imaging using stimulated emission depletion (STED) microscopy since they are perfectly photostable without photobleaching and blinking (60).

6.6.1 Introduction to STED Microscopy

The most important quality of microscopy is its resolution at the smallest distance between two spots. Optical microscopy is limited due to diffraction on the numerical aperture of the objective and excited light wavelength. Therefore, if the resolution is not sufficient to separate different spots under the diffraction limit, detailed information is lost. With near-field development, a single atom has already been observed clearly through microscope technology such as atomic force microscopy (AFM) (61), the transmission electron microscope (TEM) (62), or the scanning tunneling microscope (STM) (63). Near-field microscopes are required if we are to observe samples closely under the visible wavelength. Nevertheless, these tools have trouble displaying the structures of proteins inside a biosample. Thus, far-field microscopes are more promising in biological applications. Far-field microscopes are noninvasive and show specific fluorescence labels in living cells. However, there are some disadvantages to far-field microscope images, including their resolution (64). Sample preparation is another issue with far-field techniques. Samples must be fixed and embedded by epoxy or frozen.

Therefore, a fixed sample must be put under a vacuum condition, such as TEM imaging (62). Furthermore, STM, AFM, and TEM directly scan over a surface, possibly damaging soft biosamples.

Researchers have developed novel optical and mathematic techniques to resolve some of the issues associated with biosample microscopy. Examples include STED (51) and structure illustration microscopy (SIM) (65), which are fundamental optics. SIM and STED provide an approach to overcome the limits of stochastic super-resolution microscopy, such as overlapping emitters in a single emission spot or overcounting during the reconstruction process (66). Photoactivated localization microscopy (PALM) (52) and stochastic optical reconstruction microscopy (STORM) (53) combine optics and mathematic techniques. STED microscopy has demonstrated potential in achieving high-resolution imaging of fluorescent molecules, with the ability to approach atomic scale resolution. The use of STED microscopy has been extensively explored in the investigation of diamond nanoparticles with NV defects (55, 60, 67).

6.6.2 Principle of STED

STED microscopy employs two laser beams of varying wavelengths (61). During the measurement, STED's main laser beam brings the fluorophore of interest to its excited state. The second laser beam, the "depletion" beam, is shaped like a doughnut at the objective lens' focus and simultaneously quenches the fluorescence of all molecules except those in the middle of the nanometer-sized volume. The resultant "fluorescent volume" appears to be smaller than the diffraction limit. The smaller the "fluorescent volume," the greater the image resolution. STED is uncomplicated compared with other super-resolution approaches and does not require postprocessing mathematics. In principle, it is an ideal method for creating real-time and 3-D images of living cells.

In STED, photons are excited to a higher vibrational state (S_1^*) of an excited state (S_1) and relaxed to the ground state (S_0). Then, photons on S_1 instantaneously drop to S_0 through stimulated light emissions. Normally, this occurs before the photons release their fluorescence. In fact, the diffraction beam of excitation activates the diffraction range of fluorophores, but at the same time, the doughnut beam of STED power is strong enough to return the molecules to the ground state. The fluorescent molecule is effectively switched off.

We can say that the fluorescent spot becomes smaller than the diffraction limit, called h_{exc}.

$$h_{effect}(r) = h_{exc}(r) \cdot \left[h_{det}(r) \otimes p(r)\right]$$

where h_{effect} is the response for a point source;
$p(r)$ = the detection point spread function (PSF) of the pinhole function;

The process of depletion is called h_{det}. The point spread function is still in existence, but $h_{effect} = h_{exc} \cdot h_{det}$ is of insignificant magnitude. When h_{det} is close to zero in the center region, the intensity of the STED beam approaches zero. Still, using the STED beam leads to a better resolution in the nondepletion fluorescent region. However, the excitation beam cannot separate two or more fluorophores below the diffraction-limit spots, and it can only detect the signals of a single PSF in the position scanner. In addition, the depletion beam can be more precise due to the position of the piezo scanner with the ultrahigh laser power of STED light, so the pinhole can be ignored. Moreover, h_{exc} is nonzero and is assumed to be constant over the small region. The confocal resolution of PSF is still controlled by the excitation light. If it is applied to the STED beam (creates the center of dark spot), the resolution depends on the depletion beam to break the diffraction limitation.

Another important factor about fluorophores is the intensity of STED, I_{STED}. When using STED to inhibit fluorescence, the probability of an excited state becomes $\exp(-I_{STED}/I_{sat})$; I_{sat} is usually at $10^6\ \mathrm{W/cm^2}$. Such an experiment always uses an ultrahigh STED power, I_{STED} greater than I_{sat}, so most of the molecules are stimulated to the ground state and not emitted. The higher the intensity of the doughnut-shaped beam, the higher the depletion efficiency for this super-resolution technique.

In STED microscopy, a doughnut-shaped depletion beam is also used as a second laser in addition to the excitation laser. It provides a super imaging resolution. A typical fluorescence transition takes place in the fluorophore from the ground state to the excited state when the fluorophore in STED is solely stimulated by the excitation laser. The excited fluorophore can be quenched by STED when the STED beam is on, and a doughnut-shaped beam is subsequently applied, resulting in the reduction of the size of the emission region (66).

STED beams become more powerful in their ability to switch off the fluorescent probability. The signal in the center of the doughnut is much smaller than that outside the doughnut, implying that the center of detection can be smaller than the diffraction limit. Theoretically, the resolution (d) is based on the exciting wavelength and a numerical aperture NA of the objective. It is in the form of the famous Abbe diffraction limit equation:

$$d \approx \frac{\lambda}{2NA}.$$

In addition to the factor of a wavelength λ and numerical aperture NA in the equation, the intensity of the STED beam should also be considered

$$d \approx \frac{\lambda}{2NA\sqrt{1 + {max(I_{STED})}/{I_{sat}}}}$$

where $\max(I_{STED})$ is the maximum intensity of the STED beam.

The max ratio of $(I_{STED})/I_{sat}$ is always larger than 1. In biological applications, the confocal microscope is a powerful tool with which to see the microworld in time. However, STED takes it one step further by displaying the nanoworld. The resolution is around 20 nm for the dye molecules or fluorescent proteins and even close to 1 nm for diamond materials (67). With STED, photons are brought to an excited state and returned to the ground state with fluorescence.

Because the periphery of a photon is emitted earlier than the fluorescence lifetime by the stimulated emission process, only the doughnut center can detect the fluorescent signals. The specific wavelength of the STED doughnut-shaped beam is λ_{STED}, which depends on its specific emission spectra to reach a real inhibition effect to dark. No inhibited molecules are spontaneously emitted in the center of the doughnut, a fact that contributes to isolating the detected signals from the stimulated molecules.

The critical point of the stimulated technology is that the rate constant of STED must be much larger than that of fluorescence, i.e., rate constant $k_S >> k_F$. Therefore, most of the fluorescent molecules are stimulated to the ground state, and only a few fluorescent photons detect them. The excited molecules are far away from the doughnut center, allowing for a super-resolution image in the center.

After optical excitation from the ground state S_0 to the first excited state S_1, a molecule can take one of two possible paths to return to the ground state: First, in the case of STED, the electron gets stimulated down into the ground state, and no fluorescence occurs. Second, without STED, the molecule returns to the ground state due to the emission of a fluorescence photon.

6.6.3 FND-Enabled STED

STED was utilized to identify individual NV^- centers in bulk diamond with a noteworthy resolution of 6 nm (60). The same method was also used to effectively get high-resolution images of 35-nm FNDs spin coated on a glass plate (55). Accordingly, a resolution of ~40 nm could be obtained, which is theoretically restricted by particle size. No evidence of photobleaching is observed, even under intensive irradiation with a STED laser (up to 160 mW). This remarkable photostability positions FND as an ideal candidate for STED imaging. As a practical application, this feature has been leveraged to perform STED for FND imaging of HeLa cells (68), shown in Figure 6.10. FNDs were noncovalently coated with BSA and then endocytosis into the cell cytoplasm to avoid aggregation of particles in the cell medium. Using green light for excitation and a doughnut-shaped 740-nm laser beam for depletion, they successfully reached a spatial resolution close to 40 nm. The individual FND particles in cells were also isolated and distinguished from the FND aggregates confined in the endosomes. FND has single-particle sensitivity, and it would be advantageous to have a nanometric resolution. Meanwhile, a similar STED technique was used to reveal the merging of endosomes into late endosomes and measured the amount of drug uptake by

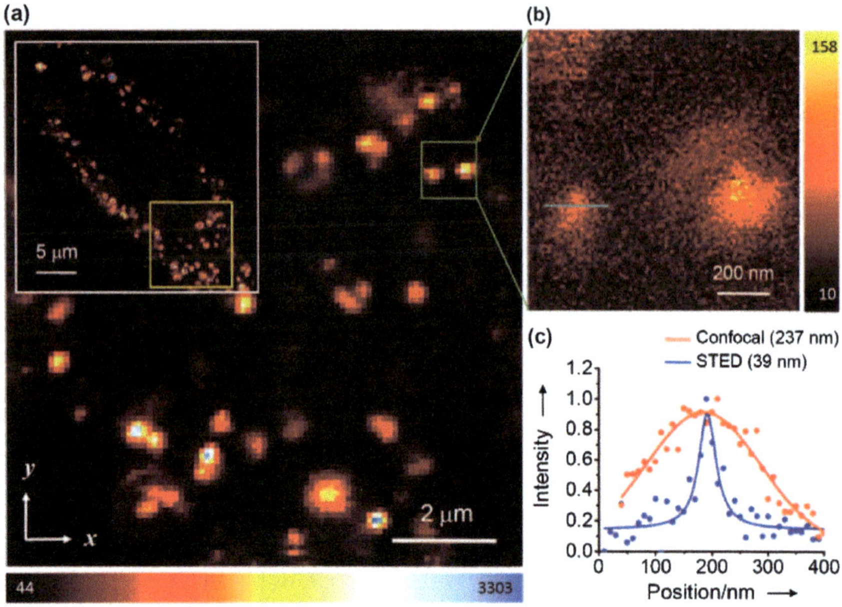

Figure 6.10 (a) Confocal image and fluorescence image (inset) of a HeLa cell labeled with BSA-conjugated FNDs. (b) STED image of FND bioconjugate particles indicated within the green box in (a). *Source:* Reproduced from ref. (73) / John Wiley & Sons (c) The FND particle's confocal and STED fluorescence intensity profiles are shown in (b) with a blue line. *Source:* (73) Tzeng et al., 2011 / John Wiley & Sons.

a single cell (69). Spatial resolution can be readily improved by a factor of 5 by using a high-power STED laser to deplete NV^- fluorescence. This advancement is crucial for optically revealing the properties of biological macromolecular complexes, cellular organelles, and subcellular compartments. FND has considerable promise and potential as the "brightest" contender among the nanocarbon series for bioimaging, displaying unparalleled resolution and accuracy.

Super-resolution live imaging with FND has not been achieved by STED because of the photodamage from the high-power doughnut STED beam. This occurs with the transfer of the negative NV center to the neutral NV center by a low-power laser at 637 nm (70). When using low laser power, a doughnut beam can possibly deplete the population of the NV^- center, except at the center of the beam. Thus, super-resolution imagery for bioimaging with the low-power doughnut beam can be further achieved. With a resolution of 4.1 nm, charge-state depletion microscopes can resolve the NV centers in bulk diamond and are anticipated to be more suitable for super-resolution live imaging with FNDs.

Finally, STORM, the application of sequential activation of photoswitchable fluorophores for time-resolved localization, can quickly image live cells at a super-resolution (with a frame rate approaching 1 Hz). Chen *et al.* have developed "deterministic emitter switch microscopy" (DESM) to resolve two negatively charged NV centers 55 nm apart. They use a bulk diamond and have achieved an acquisition time of 1.44 s (71). In addition, the fluorescence of the negatively charged NV centers was manipulated by microwave radiation, which allowed for the reconstruction of the super-resolution image.

A low-power 594-nm laser can also induce frequent switching between the NV^- center and NV^0 center for STORM imaging (72). Pfender *et al.* have devised a single-spin STORM with a low-power 594-nm laser to resolve the NV centers in bulk diamond. Single-spin STORM can completely switch off the fluorescence of the NV^- center, while DESM merely reduces the fluorescence intensity by a maximum of 30%. Hence, single-spin STORM outperforms DESM in resolving spin-coated nanodiamond particles hosting single NV^- centers on a cover glass. The FND with a single NV center is the best labeling agent for STORM in the fast, super-resolution imaging of live cells.

6.6.4 Combining STED with other Techniques for Bioimaging

3-D confocal microscopy is one of the most suitable microscopy techniques to confirm the presence of FND inside cells. This way of electroporation differs from the one of endocytosis for the bare-FND and BSA-FND attaching to the cell membrane. Figure 6.11 shows that HeLa cells with bare FNDs via endocytosis and that

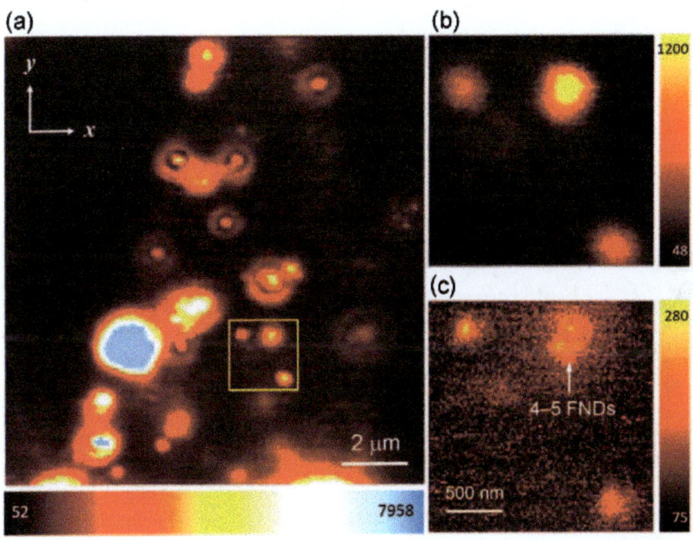

Figure 6.11 Confocal (a,b) and STED image (c) of a bare-FND-labeled HeLa cell. (b) Enlarged views of the bare-FND particles, shown by the yellow box in (a). (c) STED image of the same FND particles in (b). The cluster in (c), shown by the white arrow, is made up of up to 5 FND particles. *Source:* Reproduced from ref. (73) / John Wiley & Sons.

both confocal and STED systems produce colocalization images (68). Here, five FND nanoparticles were found in the same endosome, which could be differentiated through STED microscopy.

Prabhakar *et al.* employed a novel approach for correlative microscopy that uses FNDs as dual contrast probes (Figure 6.12) (73). In their work, they demonstrated that FNDs could be used as dual contrast labels with the automatic image registration tool SuperTomo and for precise image correlation in high-resolution STED/confocal and TEM correlative microscopy experiments. This research showed that FNDs could be utilized as probes in experiments on both live and fixed cells (73, 74).

Moffitt *et al.* have developed subdiffraction level imaging techniques such as gated stimulated emission depletion (g-STED) (Figure 6.13) (75). The time a fluorophore spends in the excited state is reduced using g-STED microscopy. This approach enables the user to control the portion of light decay that will be used in generating and optimizing the final image. In conclusion, combining STED

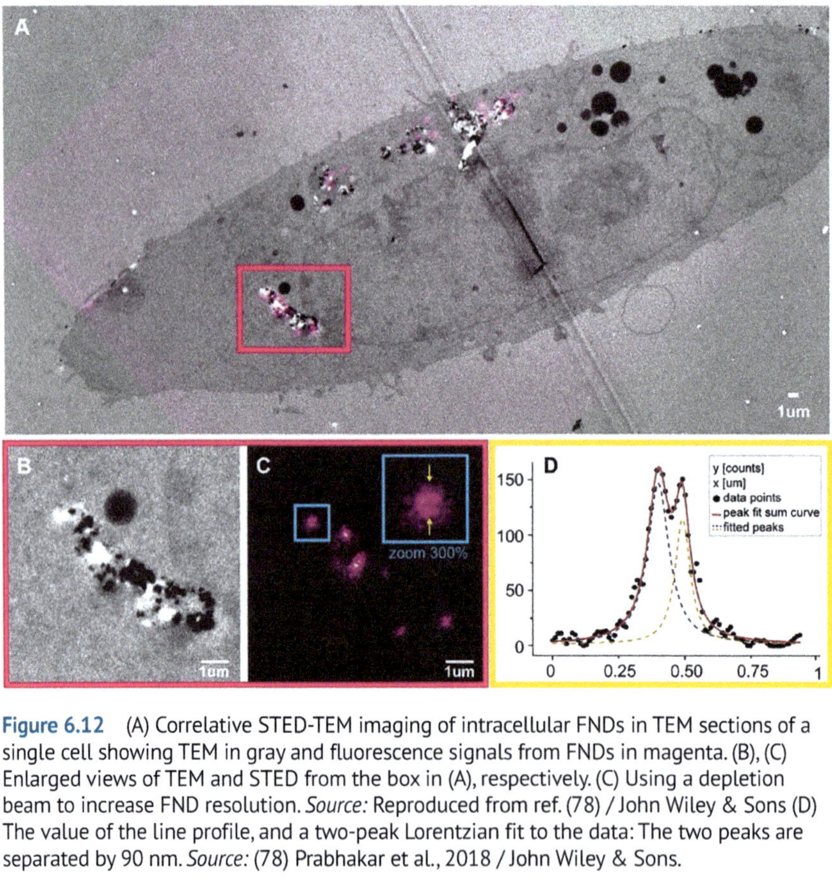

Figure 6.12 (A) Correlative STED-TEM imaging of intracellular FNDs in TEM sections of a single cell showing TEM in gray and fluorescence signals from FNDs in magenta. (B), (C) Enlarged views of TEM and STED from the box in (A), respectively. (C) Using a depletion beam to increase FND resolution. *Source:* Reproduced from ref. (78) / John Wiley & Sons (D) The value of the line profile, and a two-peak Lorentzian fit to the data: The two peaks are separated by 90 nm. *Source:* (78) Prabhakar et al., 2018 / John Wiley & Sons.

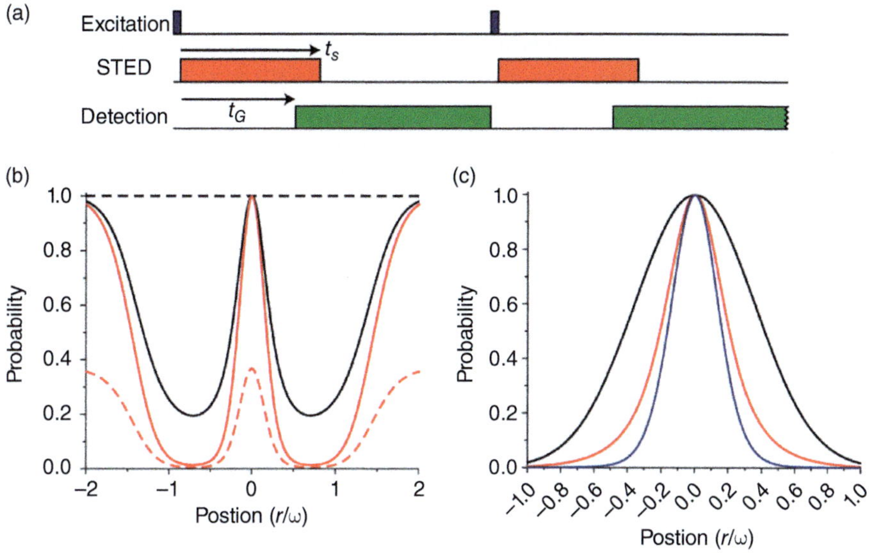

Figure 6.13 Improving spatial resolution with time gating: (a) Timing diagram for pulsed STED in which an excited beam of negligible duration (in blue) is followed by a depletion pulse of duration t_S (red), and the detection of spontaneous emission occurs during the time window beginning t_G after the completion of the first excitation pulse and continuing up until the beginning of the next excitation pulses to create the final image (green). (b) The probability of emitting via normal fluorescence (function of position) (in black), a uniform field of excited fluorophore (dashed black) is exposed to the pulsed STED beam. Probability of emitting via spontaneous emission after the STED pulse (in dashed red) and probability corrected for the decrease in brightness (red, solid). (c) The PSFs for confocal, STED, and time-gated STED detection. *Source:* This figure is adapted from ref. (76).

microscopy with other techniques has been shown to be an effective approach for imaging FNDs in cells. Confocal microscopy is suitable for confirming the presence of FNDs within cells, while STED microscopy allows for high-resolution imaging. FNDs can also be utilized as dual contrast probes for correlative microscopy experiments. These approaches offer great potential for using FNDs as probes in both live and fixed cell experiments.

6.7 Optically Detected Magnetic Resonance Imaging

Due to its spin state, the NV^- center is amenable to ultrasensitive detection on the nanometric scale under ambient conditions with the ODMR technique. As discussed in **Section 2.2**, the ODMR effect of an NV center has enabled many imaging applications of FND. Balasubramanian *et al.* scanned a nanoscale magnetic tip over an immobilized ND with a single NV^- center and were able to locate the position of the particle with nanometric accuracy (77). They also verified the imaging and resolving power of the technique by detecting two FND particles, separated by 100 nm, and obtained a spatial resolution of 20 nm – significantly less than the diffraction limit of optical microscopy. The method is potentially useful for monitoring the function of a single ion channel on a cell membrane (78).

The diamond crystal field detaches the $m_s = 0$ ground state from the degenerate $m_s = \pm1$ ground state of the NV^- center by 2.87 GHz; thus, this energy splitting defines a quantization axis, i.e., the $\langle 111 \rangle$ crystallographic axis, and provides an intrinsic compass. By applying microwave fields and pairs of coils to produce well-aligned magnetic fields with which to examine the spin splitting, researchers have decoded the spatial coordinates and the orientation of the NV^- axis based on the ODMR spectra. McGuinness *et al.* conducted orientation tracking of single FND particles within a living cell using ODMR peak positions (Figure 6.14). They binned the data acquired over a few hours correctly and resolved the rotating motion of FNDs within cells in a millisecond timeframe (34). By identifying single NV^- in each particle, super-resolution imaging of FNDs can be achieved (71). Besides, the ODMR effect on NV centers also plays an important role in other applications such as tissue and organism imaging, and quantum sensing. These applications will be discussed in the following chapters.

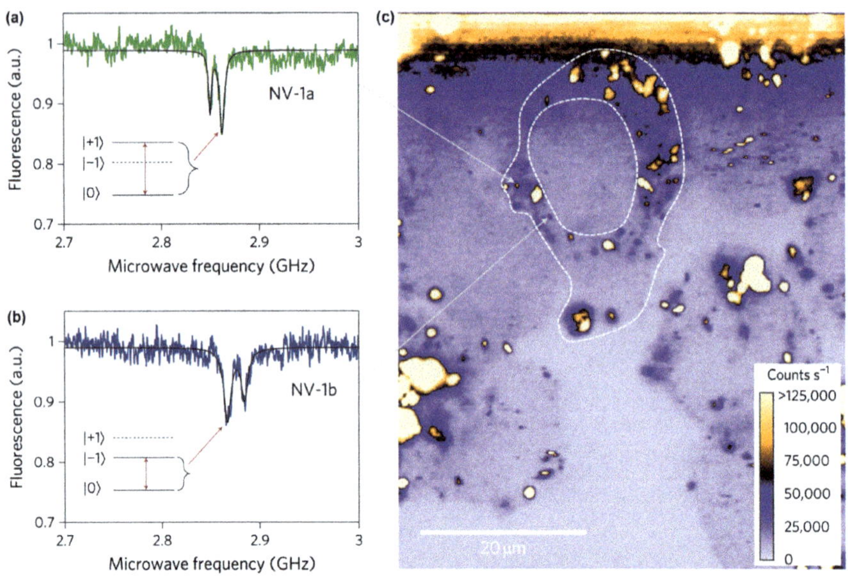

Figure 6.14 The ODMR spectra of NV-1a (a) and NV-1b (b) show the different strain splitting between the two centers in HeLa cell. *Source:* (35) McGuinness et al., 2011 / Springer Nature (c). The dashed lines indicate the nucleus and cell membrane. *Source:* Reproduced from ref. (35) / Springer Nature.

6.8 Cathodoluminescence Imaging

CL is the photon emission caused by an electron impacting a luminous substance. It is commonly used as a method for characterizing materials with nanometric precision. However, applying the approach to living tissue is difficult since the unavoidable destruction of the specimen generally results in low fluorescence intensity and quick signal deterioration. Nonetheless, some countermeasures can be taken, such as strong nanoparticles carrying photostable and spectrally distinct fluorophores. Glenn *et al.* explored the use of both blue and red FNDs for CL imaging (79). They acquired CL or secondary electron images that were well-correlated for individual particles from 40 to 80 nm. Unlike photoluminescence imaging, CL imaging has a spatial resolution of more than 5 nm and is not restricted by light diffraction. In a similar investigation, Tizei *et al.* examined the spectrally and spatially resolved CL of FNDs in a scanning transmission electron microscope (STEM) employing a focused electron beam (80, 81). They carefully analyzed the CL spectra of individual FNDs and identified different color centers, including H3 and NV^0. Their work has verified the feasibility of STEM for ultra-high-resolution CL imaging of FNDs. Meuret *et al.* recorded lifetime maps using time-resolved CL in an ultrafast TEM with a spatial resolution of 12 nm and a subnanosecond precision of 100 nm FND (Figure 6.15) (82). However, no CL was

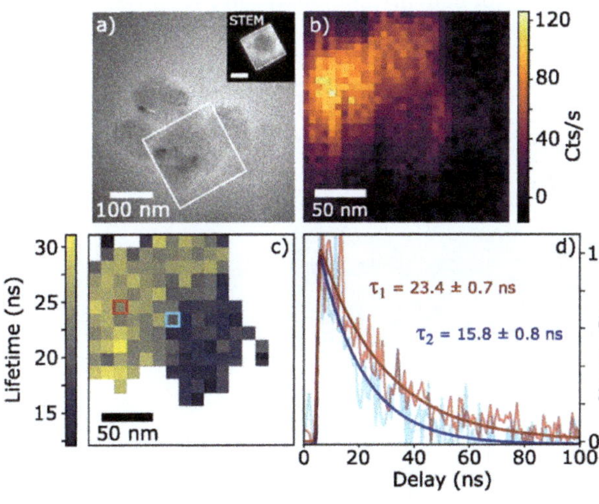

Figure 6.15 (a) TEM images of the nanodiamond cluster; the white square indicates the scanned region. (b) Intensity map derived from the decay trace map by adding all the histogram bins of a 500-ns frame with a pixel size of 6.3 nm. (c) Lifetime map with four neighboring pixels binned (effective pixel size 12.4 nm). A decay trace fit was performed for each pixel, and the retrieved lifespan was reported on the map. Pixels with insufficient signal to derive a lifespan are masked. *Source:* Reproduced from ref. (87) / AIP Publishing LLC (d) Decay traces and fitting at the places indicated by the blue and red squares in (c). *Source:* (87) Meuret et al., 2021 / AIP Publishing.

seen from NV^- in any of these studies, which is most likely due to the charge-state shift of the centers during electron irradiation from NV^- to NV^0.

In another study, direct electron beam excitation-assisted (D-EXA) fluorescence microscopy was used to provide multicolor, high spatial resolution imaging of FNDs in live HeLa cells (83). Green and red light-emitting FNDs were both directly stimulated with a concentrated electron beam, and the resulting CL was recorded with nanometer-scale resolution. Two-color imaging indicated that these two forms of FNDs can be observed in cells at the same time (Figure 6.16). Such an approach might be used for multicolor immunostaining to reveal distinct cell activities after properly conjugating FNDs with antibodies. Researchers will be able to achieve molecular localization during drug administration as well as show structural details within live cells or even species.

Plasmonic waveguides made up of chains of metal nanoparticles may localize and direct light far below the diffraction limit. Still, they experience substantial propagation losses due to the lithographically constrained space between big particles. At a 62-nm ($\lambda/10$) deep subwavelength confinement, Gür *et al.* demonstrated that self-assembly of monocrystalline gold nanoparticles (AuNPs) using DNA origami allowed the interparticle distance to be reduced to ~2 nm, lowering propagation loss to 0.8 dB per 50 nm (84). Electron energy loss spectroscopy was used to characterize individual waveguides with nanometer-scale resolution. CL imaging spectroscopy was used to directly visualize light propagation toward FND at the single device level, allowing for light manipulation and energy conversion at

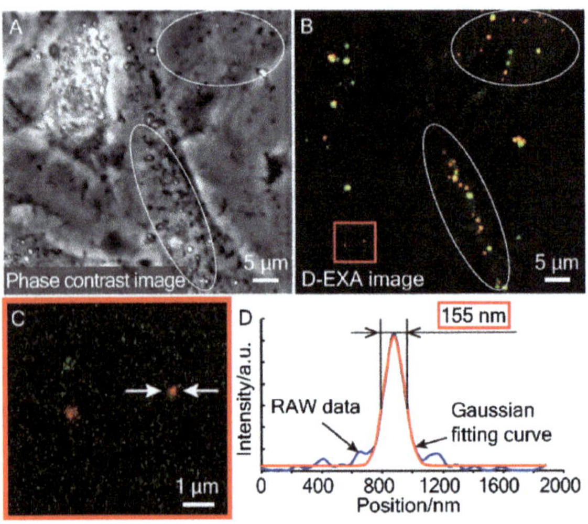

Figure 6.16 Internalized green and red FNDs were seen in live HeLa cells using (a) phase-contrast microscopy and (b) D-EXA fluorescence microscopy. The enlarged D-EXA picture of the red square in (c) exhibits a spatial resolution of 155 nm *Source:* Reproduced from ref. (88) / John Wiley & Sons (d). *Source:* (88) Nawa et al., 2014 / John Wiley & Sons.

the nanoscale. CL measurements on complete and incomplete waveguides that have been isolated close to a single FND are shown by Gür *et al.* In a diamond lattice, FND exhibits an NV point defect, where nitrogen replaces the carbon atom next to the vacancy, resulting in a characteristic zero-phonon line (ZPL) at 576 nm for the neutral charge state (NV^0) and 637 nm for the negative charge state (NV^-). A distinct CL spectrum with high-intensity NV^0 emission is observed when the electron beam is positioned on the FND. When any of the AuNPs, including the particle farthest away from the FND, is removed from the waveguide, this typical FND spectrum is detected (Figure 6.17). After CL measurements, high-resolution

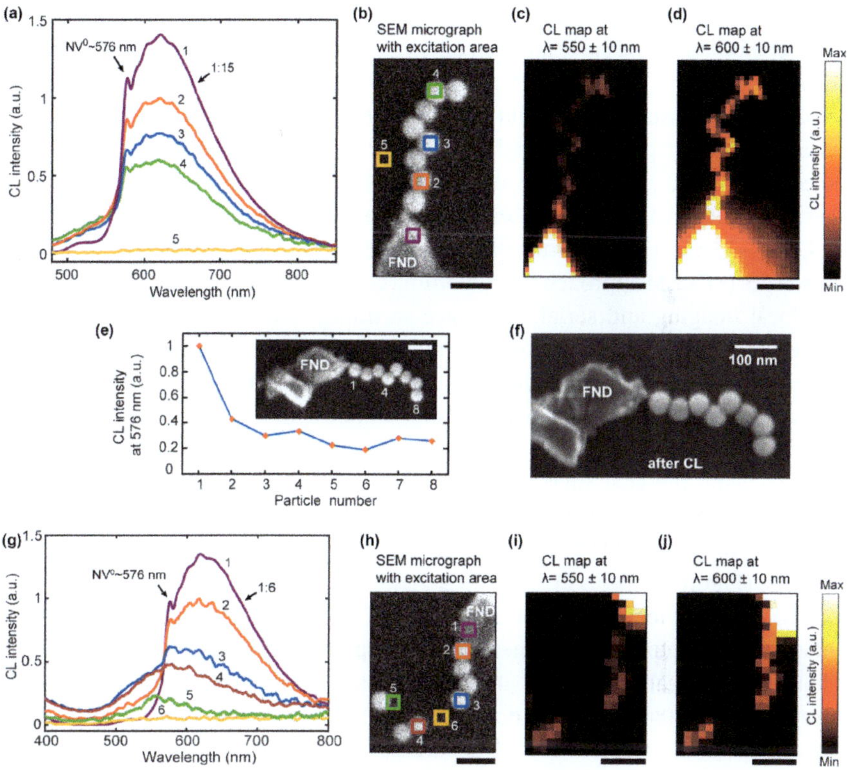

Figure 6.17 Plasmonic waveguides are used to transport energy. When waveguides are illuminated near an FND, the spectra of CL are seen. Complete (a) and incomplete (g) waveguide CL spectra measured on SEM micrographs, at electron beam excitation points represented by squares in correlating hues (b,h). For clarity, a measure of the FND emission's intensity (purple) using direct e-beam illumination is lowered 15-fold (a) or 6-fold (g). Intensity of CL as a function of electron beam excitation positions, spatially resolved CL maps at detection wavelengths of 550 nm (10 nm) (c,i) and 600 nm (10 nm) (d,j). (e) The amount of CL produced as a function of excitement position in the whole waveguide at 576 nm (equivalent to NVO). (f) After the CL measurements, a SEM micrograph of the same waveguide with an FND was acquired at high resolution with scale bars of 100 nanometers. *Source:* Reproduced from ref. (89) / American Chemical Society.

SEM photos of the waveguide demonstrated that the waveguide remained intact and that no AuNP fusion had occurred (84).

6.9 Correlative Light and Electron Microscopy

The properties of two different instruments are combined in correlative light and electron microscopy (CLEM). As a result, this imaging technique can analyze the kinetics and localization of macromolecules and proteins using live-cell light microscopy (LM) and then use electron microscopy (EM) to examine the ultrastructural morphology of relevant cells. QDs, dye-doped polystyrene beads, and dye-labeled nanogold have all been employed as fluorescent fiducial markers for CLEM. However, these nanomaterials deteriorate during sample preparation, making them insufficiently stable for standard CLEM analysis. For instance, the time-consuming process of high-pressure freezing is required to preserve the specimens. FND appears to be a potential fluorescent fiducial marker for CLEM because of its great stability, intense fluorescence, and high electron density.

Using ~35-nm FND as intracellular correlation indicators, Prabhakar *et al.* proposed a 3-D CLEM approach for combining cell volume information from live-cell confocal imaging and serial block-face scanning electron microscopy (SB-EM) (85). The 3-D CLEM procedure was initiated by incubating human breast adenocarcinoma (MDA-MB-23) cells with 35-nm FND. Following the independent acquisition of specific cell images using confocal microscopy and 3-D SB-EM (with a 15-kV electron beam), image correlation was performed on the collected images using the eC-CLEM plugin on the Icy bioimage analysis platform. The MitoTracker signal displayed in the EM image stack had good colocalization with mitochondria since FND had mapped mitochondrial locations across the cell's volume (Figure 6.18).

FND labeling techniques can be used for TEM in addition to SEM. After TEM photos were captured, researchers collected specimen fluorescence signals with a delay period greater than 3 ns after pulsed laser illumination to cancel out background signals (86). This was done to take LM images. Researchers were able to distinguish individual 100-nm FND particles on the surface of the cells using this time-gating approach. The distinctive optical characteristics of FNDs allow for the identification of CD44 antigens with a 50-nm resolution. It is possible to locate antigens on the surface of cells with amazing precision and accuracy by combining LM and EM analyses and using smaller FND.

In summary, by combining the advantages of both light and electron microscopy techniques, FNDs can provide nanoscale resolution for imaging biological structures in living cells and tissues. In CLEM, FNDs can be used as fiducial markers for aligning the light and electron microscopy images, enabling the precise localization

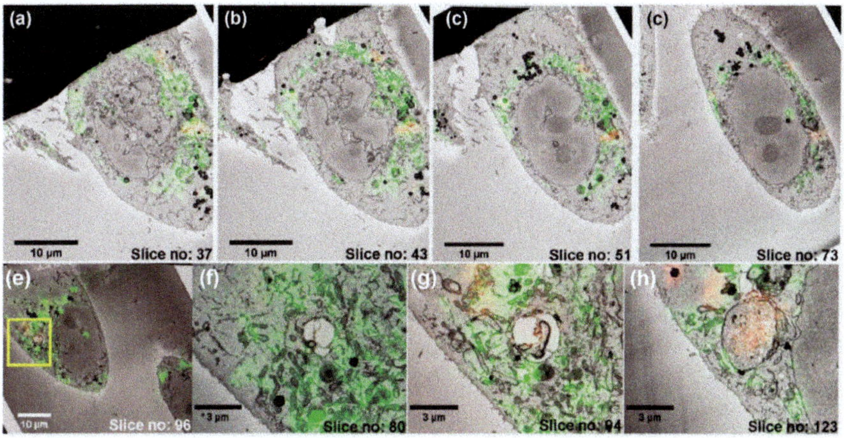

Figure 6.18 (a–d) Multiplane CLEM pictures with rising z-axis. (e) CLEM image's region of interest. (f) Slice no. 80 has a single high-resolution picture that displays several mitochondria (green) and one empty vesicle. (g) Slice no. 94 shows a vesicle, FND, and mitochondria in different colors. (h) Slice no. 123 shows an FND-filled vesicle in red and a decreased number of mitochondria in green. Inside the vesicle, the black spots are FNDs. *Source:* Reproduced from ref. (91) / MDPI/CC BY 4.0.

of fluorescent signals within the ultrastructural context. Furthermore, FNDs can be used as contrast agents for electron microscopy, as they can generate CL upon electron irradiation. CL imaging can provide information on the spatial distribution and spectral characteristics of FNDs with sub-nanometer resolution. Combining FNDs with CLEM and CL imaging can enable the visualization of cellular structures and molecular interactions at unprecedented detail and accuracy.

References

1 Yan JL, Estevez MC, Smith JE, Wang KM, He XX, Wang L, *et al.* Dye-doped nanoparticles for bioanalysis. *Nano Today*. 2007;2(3):44–50.

2 Kim Y, Kim HS, Yang Y, Baek K, Choi N, Park M-H. Quantitative analysis of hydrogen peroxide using ratiometric fluorescent probe-doped silica nanoparticles. *Toxicol Environ Health Sci*. 2017;9(2):108–15.

3 Medintz IL, Uyeda HT, Goldman ER, Mattoussi H. Quantum dot bioconjugates for imaging, labelling and sensing. *Nat Mater*. 2005;4(6):435–46.

4 Zheng J, Nicovich PR, Dickson RM. Highly fluorescent noble-metal quantum dots. *Annu Rev Phys Chem*. 2007;58:409–31.

5 Aldewachi H, Chalati T, Woodroofe M, Bricklebank N, Sharrack B, Gardiner P. Gold nanoparticle-based colorimetric biosensors. *Nanoscale*. 2018;10(1):18–33.

6 Lin B, Yan Y, Guo M, Cao Y, Yu Y, Zhang T, *et al.* Modification-free carbon dots as turn-on fluorescence probe for detection of organophosphorus pesticides. *Food Chem.* 2018;245:1176–82.

7 Melnychuk N, Klymchenko AS. DNA-functionalized dye-loaded polymeric nanoparticles: Ultrabright FRET platform for amplified detection of nucleic acids. *J Am Chem Soc.* 2018;140(34):10856–65.

8 Derfus AM, Chan WCW, Bhatia SN. Probing the cytotoxicity of semiconductor quantum dots. *Nano Lett.* 2004;4(1):11–18.

9 Ang LY, Lim ME, Ong LC, Zhang Y. Applications of upconversion nanoparticles in imaging, detection and therapy. *Nanomedicine-Uk.* 2011;6(7):1273–88.

10 Vaijayanthimala V, Chang HC. Functionalized fluorescent nanodiamonds for biomedical applications. *Nanomedicine-Uk.* 2009;4(1):47–55.

11 Hsiao WW-W, Le T-N, Chang H-C. Applications of fluorescent nanodiamond in biology. *Encyclopedia of Analytical Chemistry.* Nazare MH and Neves AJ, editor. Stevenage, United Kingdom: The Institution of Engineering and Technology; 2022. pp. 1–43.

12 Yu S-J, Kang M-W, Chang H-C, Chen K-M, Yu Y-C. Bright fluorescent nanodiamonds: No photobleaching and low cytotoxicity. *J Am Chem Soc.* 2005;127(50):17604–5.

13 Wu T-J, Tzeng Y-K, Chang W-W, Cheng C-A, Kuo Y, Chien C-H, *et al.* Tracking the engraftment and regenerative capabilities of transplanted lung stem cells using fluorescent nanodiamonds. *Nat Nanotechnol.* 2013;8(9):682–9.

14 Fu CC, Lee HY, Chen K, Lim TS, Wu HY, Lin PK, *et al.* Characterization and application of single fluorescent nanodiamonds as cellular biomarkers. *Proc Natl Acad Sci USA.* 2007;104(3):727–32.

15 Haziza S, Mohan N, Loe-Mie Y, Lepagnol-Bestel A-M, Massou S, Adam M-P, *et al.* Fluorescent nanodiamond tracking reveals intraneuronal transport abnormalities induced by brain-disease-related genetic risk factors. *Nat Nanotechnol.* 2017;12(4):322–8.

16 Neugart F, Zappe A, Jelezko F, Tietz C, Boudou JP, Krueger A, *et al.* Dynamics of diamond nanoparticles in solution and cells. *Nano Lett.* 2007;7(12):3588–91.

17 Faklaris O, Garrot D, Joshi V, Druon F, Boudou JP, Sauvage T, *et al.* Detection of single photoluminescent diamond nanoparticles in cells and study of the internalization pathway. *Small.* 2008;4(12):2236–9.

18 Faklaris O, Joshi V, Irinopoulou T, Tauc P, Sennour M, Girard H, *et al.* Photoluminescent diamond nanoparticles for cell labeling: Study of the uptake mechanism in mammalian cells. *ACS Nano.* 2009;3(12):3955–62.

19 Weng MF, Chiang SY, Wang NS, Niu H. Fluorescent nanodiamonds for specifically targeted bioimaging: Application to the interaction of transferrin with transferrin receptor. *Diam Relat Mater.* 2009;18(2–3):587–91.

20 Zhang BL, Li YQ, Fang CY, Chang CC, Chen CS, Chen YY, *et al.* Receptor-mediated cellular uptake of folate-conjugated fluorescent nanodiamonds: A combined ensemble and single-particle study. *Small.* 2009;5(23):2716–21.

21 Chang B-M, Lin H-H, Su L-J, Lin W-D, Lin R-J, Tzeng Y-K, *et al.* Highly fluorescent nanodiamonds protein-functionalized for cell labeling and targeting. *Adv Funct Mater.* 2013;23(46):5737–45.

22 Rehor I, Slegerova J, Kucka J, Proks V, Petrakova V, Adam MP, *et al.* Fluorescent nanodiamonds embedded in biocompatible translucent shells. *Small.* 2014;10(6):1106–15.

23 Slegerova J, Hajek M, Rehor I, Sedlak F, Stursa J, Hruby M, *et al.* Designing the nanobiointerface of fluorescent nanodiamonds: Highly selective targeting of glioma cancer cells. *Nanoscale.* 2015;7(2):415–20.

24 Fang CY, Chang CC, Mou CY, Chang HC. Preparation and characterization of ion-irradiated nanodiamonds as photoacoustic contrast agents. *J Nanosci Nanotechno.* 2015;15(2):1037–44.

25 Le XL, Lepagnol-Bestel A-M, Adam M-P, Thomas A, Dantelle G, Chang C-C, *et al.*, editors. *Fluorescent diamond nanoparticle as a probe of intracellular traffic in primary neurons in culture*; 2012.

26 Lee JW, Lee S, Jang S, Han KY, Kim Y, Hyun J, *et al.* Preparation of non-aggregated fluorescent nanodiamonds (FNDs) by non-covalent coating with a block copolymer and proteins for enhancement of intracellular uptake. *Mol Biosyst.* 2013;9(5):1004–11.

27 Chu Z, Zhang S, Zhang B, Zhang C, Fang C-Y, Rehor I, *et al.* Unambiguous observation of shape effects on cellular fate of nanoparticles. *Sci Rep.* 2014;4:4495.

28 Hsu T-C, Liu K-K, Chang H-C, Hwang E, Chao J-I. Labeling of neuronal differentiation and neuron cells with biocompatible fluorescent nanodiamonds. *Sci Rep.* 2014;4:5004.

29 Huang YA, Kao CW, Liu KK, Huang HS, Chiang MH, Soo CR, *et al.* The effect of fluorescent nanodiamonds on neuronal survival and morphogenesis. *Sci Rep.* 2014;4:6919.

30 Hui YY, Hsiao WW-W, Haziza S, Simonneau M, Treussart F, Chang H-C. Single particle tracking of fluorescent nanodiamonds in cells and organisms. *Curr Opin Solid State Mater Sci.* 2017;21(1):35–42.

31 Saxton MJ, Jacobson K. Single-particle tracking: Applications to membrane dynamics. *Ann Rev Biophys Biomol Struct.* 1997;26(1):373–99.

32 Levi V, Gratton E. Exploring dynamics in living cells by tracking single particles. *Cell Biochem Biophys.* 2007;48(1):1–15.

33 Chang Y-R, Lee H-Y, Chen K, Chang C-C, Tsai D-S, Fu C-C, *et al.* Mass production and dynamic imaging of fluorescent nanodiamonds. *Nat Nanotechnol.* 2008;3(5):284–8.

34 McGuinness LP, Yan Y, Stacey A, Simpson DA, Hall LT, Maclaurin D, *et al.* Quantum measurement and orientation tracking of fluorescent nanodiamonds inside living cells. *Nat Nanotechnol.* 2011;6(6):358–63.

35 Epperla CP, Mohan N, Chang C-W, Chen C-C, Chang H-C. Nanodiamond-mediated intercellular transport of proteins through membrane tunneling nanotubes. *Small*. 2015;11(45):6097–105.

36 Knox SL, Steinauer A, Alpha-Cobb G, Trexler A, Rhoades E, Schepartz A. Quantification of protein delivery in live cells using fluorescence correlation spectroscopy. *Meth Enzymol*. 2020;641:477–505.

37 Qian H, Sheetz MP, Elson EL. Single-particle tracking – Analysis of diffusion and flow in 2-dimensional systems. *Biophys J*. 1991;60(4):910–21.

38 Krichevsky O, Bonnet G. Fluorescence correlation spectroscopy: The technique and its applications. *Rep Prog Phys*. 2002;65(2):251–97.

39 Mohan N, Tzeng YK, Yang L, Chen YY, Hui YY, Fang CY, *et al*. Sub-20-nm fluorescent nanodiamonds as photostable biolabels and fluorescence resonance energy transfer donors. *Adv Mater*. 2010;22(7):843–7.

40 Heikal AA, Hess ST, Baird GS, Tsien RY, Webb WW. Molecular spectroscopy and dynamics of intrinsically fluorescent proteins: Coral red (dsRed) and yellow (Citrine). *Proc Natl Acad Sci USA*. 2000;97(22):11996–2001.

41 Hui YY, Zhang B, Chang Y-C, Chang C-C, Chang H-C, Hsu J-H, *et al*. Two-photon fluorescence correlation spectroscopy of lipid-encapsulated fluorescent nanodiamonds in living cells. *Opt Express*. 2010;18(6):5896–905.

42 Liu W, Yu F, Yang J, Xiang B, Xiao P, Wang L. 3D single-molecule imaging of transmembrane signaling by targeting nanodiamonds. *Adv Funct Mater*. 2016;26(3):365–75.

43 Xu C, Zipfel W, Shear JB, Williams RM, Webb WW. Multiphoton fluorescence excitation: New spectral windows for biological nonlinear microscopy. *Proc Natl Acad Sci USA*. 1996;93(20):10763–8.

44 Helmchen F, Denk W. Deep tissue two-photon microscopy. *Nat Methods*. 2005;2(12):932–40.

45 Wee TL, Tzeng YK, Han CC, Chang HC, Fann W, Hsu JH, *et al*. Two-photon excited fluorescence of nitrogen-vacancy centers in proton-irradiated type ib diamond. *J Phys Chem A*. 2007;111(38):9379–86.

46 Su L-J, Fang C-Y, Chang Y-T, Chen K-M, Yu Y-C, Hsu J-H, *et al*. Creation of high density ensembles of nitrogen-vacancy centers in nitrogen-rich type Ib nanodiamonds. *Nanotechnology*. 2013;24(31):315702.

47 Hui YY, Su LJ, Chen OY, Chen YT, Liu TM, Chang HC. Wide-field imaging and flow cytometric analysis of cancer cells in blood by fluorescent nanodiamond labeling and time gating. *Sci Rep*. 2014;4:5574.

48 Fujita H, Zhong C, Arai S, Suzuki M. Bright dots and smart optical microscopy to probe intracellular events in single cells. *Front Bioeng Biotechnol*. 2018;6:204.

49 Becker L, Janssen N, Layland SL, Mürdter TE, Nies AT, Schenke-Layland K, *et al*. Raman imaging and fluorescence lifetime imaging microscopy for diagnosis of cancer state and metabolic monitoring. *Cancers*. 2021;13(22):5682.

50 Sharpe J. Optical projection tomography as a new tool for studying embryo anatomy. *J Anat.* 2003;202(2):175–81.

51 Willig KI, Rizzoli SO, Westphal V, Jahn R, Hell SW. STED microscopy reveals that synaptotagmin remains clustered after synaptic vesicle exocytosis. *Nature.* 2006;440(7086):935–9.

52 Betzig E, Patterson GH, Sougrat R, Lindwasser OW, Olenych S, Bonifacino JS, *et al.* Imaging intracellular fluorescent proteins at nanometer resolution. *Science.* 2006;313(5793):1642–5.

53 Rust MJ, Bates M, Zhuang XW. Sub-diffraction-limit imaging by stochastic optical reconstruction microscopy (STORM). *Nat Methods.* 2006;3(10):793–5.

54 Kurtsiefer C, Mayer S, Zarda P, Weinfurter H. Stable solid-state source of single photons. *Phys Rev Lett.* 2000;85(2):290–3.

55 Han KY, Willig KI, Rittweger E, Jelezko F, Eggeling C, Hell SW. Three-dimensional stimulated emission depletion microscopy of nitrogen-vacancy centers in diamond using continuous-wave light. *Nano Lett.* 2009;9(9):3323–9.

56 Batalov A, Zierl C, Gaebel T, Neumann P, Chan IY, Balasubramanian G, *et al.* Temporal coherence of photons emitted by single nitrogen-vacancy defect centers in diamond using optical Rabi-oscillations. *Phys Rev Lett.* 2008;100(7):077401.

57 Beveratos A, Brouri R, Gacoin T, Poizat JP, Grangier P. Nonclassical radiation from diamond nanocrystals. *Phys Rev A.* 2001;64(6):061802(R).

58 Kuo Y, Hsu T-Y, Wu Y-C, Hsu J-H, Chang H-C, editors. *Fluorescence lifetime imaging microscopy of nanodiamonds in vivo*; 2013.

59 Tisler J, Balasubramanian G, Naydenov B, Kolesov R, Grotz B, Reuter R, *et al.* Fluorescence and spin properties of defects in single digit nanodiamonds. *Acs Nano.* 2009;3(7):1959–65.

60 Rittweger E, Han KY, Irvine SE, Eggeling C, Hell SW. STED microscopy reveals crystal colour centres with nanometric resolution. *Nat Photon.* 2009;3(3):144–7.

61 Carpick RW, Salmeron M. Scratching the surface: Fundamental investigations of tribology with atomic force microscopy. *Chem Rev.* 1997;97(4):1163–94.

62 Mohler WA, Simske JS, Williams-Masson EM, Hardin JD, White JG. Dynamics and ultrastructure of developmental cell fusions in the Caenorhabditis elegans hypodermis. *Curr Biol.* 1998;8(19):1087–90.

63 Tersoff J, Hamann DR. Theory of the scanning tunneling microscope. *Phys Rev B.* 1985;31(2):805–13.

64 Abbe E. Beiträge zur Theorie des Mikroskops und der mikroskopischen Wahrnehmung. *Archiv für mikroskopische Anatomie.* 1873;9(1):413–68.

65 Anderson TF. Techniques for the preservation of three-dimensional structure in preparing specimens for the electron microscope. *Trans N Y Acad Sci.* 1951;13:130–4.

66 Xu Y, Xu R, Wang Z, Zhou Y, Shen Q, Ji W, *et al.* Recent advances in luminescent materials for super-resolution imaging via stimulated emission depletion nanoscopy. *Chem Soc Rev.* 2021;50(1):667–90.

67 Wildanger D, Patton BR, Schill H, Marseglia L, Hadden JP, Knauer S, *et al.* Solid immersion facilitates fluorescence microscopy with nanometer resolution and sub-angstrom emitter localization. *Adv Mater.* 2012;24(44):Op309–13.

68 Tzeng YK, Faklaris O, Chang BM, Kuo Y, Hsu JH, Chang HC. Superresolution imaging of albumin-conjugated fluorescent nanodiamonds in cells by stimulated emission depletion. *Angew Chem Int Ed.* 2011;50(10):2262–5.

69 Prabhakar N, Näreoja T, von Haartman E, Karaman DŞ, Jiang H, Koho S, *et al.* Core–shell designs of photoluminescent nanodiamonds with porous silica coatings for bioimaging and drug delivery II: Application. *Nanoscale.* 2013;5(9):3713–22.

70 Chen XD, Zou CL, Gong ZJ, Dong CH, Guo GC, Sun FW. Subdiffraction optical manipulation of the charge state of nitrogen vacancy center in diamond. *Light Sci Appl.* 2015;4.

71 Chen EH, Gaathon O, Trusheim ME, Englund D. Wide-field multispectral super-resolution imaging using spin-dependent fluorescence in nanodiamonds. *Nano Lett.* 2013;13(5):2073–7.

72 Pfender M, Aslam N, Waldherr G, Neumann P, Wrachtrup J. Single-spin stochastic optical reconstruction microscopy. *Proc Natl Acad Sci USA.* 2014;111(41):14669–74.

73 Prabhakar N, Peurla M, Koho S, Deguchi T, Näreoja T, Chang HC, *et al.* STED-TEM correlative microscopy leveraging nanodiamonds as intracellular dual-contrast markers. *Small.* 2018;14(5):1701807.

74 Castelletto S, Boretti A. Color centers in wide-bandgap semiconductors for subdiffraction imaging: A review. *Adv Photonics.* 2021;3(05):054001.

75 Moffitt JR, Osseforth C, Michaelis J. Time-gating improves the spatial resolution of STED microscopy. *Opt Express.* 2011;19(5):4242–54.

76 Reineck P, Gibson BC. Near-infrared fluorescent nanomaterials for bioimaging and sensing. *Adv Opt Mater.* 2017;5(2):1600446.

77 Balasubramanian G, Chan I, Kolesov R, Al-Hmoud M, Tisler J, Shin C, *et al.* Nanoscale imaging magnetometry with diamond spins under ambient conditions. *Nature.* 2008;455(7213):648–51.

78 Hall LT, Hill CD, Cole JH, Stadler B, Caruso F, Mulvaney P, *et al.* Monitoring ion-channel function in real time through quantum decoherence. *Proc Natl Acad Sci USA.* 2010;107(44):18777–82.

79 Glenn DR, Zhang H, Kasthuri N, Schalek R, Lo PK, Trifonov AS, *et al.* Correlative light and electron microscopy using cathodoluminescence from nanoparticles with distinguishable colours. *Sci Rep.* 2012;2:865.

80 Tizei LH, Kociak M. Spectrally and spatially resolved cathodoluminescence of nanodiamonds: Local variations of the NV(0) emission properties. *Nanotechnology*. 2012;23(17):175702.

81 Tizei LHG, Meuret S, Nagarajan S, Treussart F, Fang CY, Chang HC, *et al.* Spatially and spectrally resolved cathodoluminescence with fast electrons: A tool for background subtraction in luminescence intensity second-order correlation measurements applied to subwavelength inhomogeneous diamond nanocrystals. *Phys Status Solidi A*. 2013;210(10):2060–5.

82 Meuret S, Tizei LHG, Houdellier F, Weber S, Auad Y, Tencé M, *et al.* Time-resolved cathodoluminescence in an ultrafast transmission electron microscope. *Appl Phys Lett*. 2021;119(6):062106.

83 Nawa Y, Inami W, Lin S, Kawata Y, Terakawa S, Fang CY, *et al.* Multi-color imaging of fluorescent nanodiamonds in living HeLa cells using direct electron-beam excitation. *Chemphyschem*. 2014;15(4):721–6.

84 Gür FN, McPolin CPT, Raza S, Mayer M, Roth DJ, Steiner AM, *et al.* DNA-assembled plasmonic waveguides for nanoscale light propagation to a fluorescent nanodiamond. *Nano Lett*. 2018;18(11):7323–9.

85 Prabhakar N, Belevich I, Peurla M, Heiligenstein X, Chang H-C, Sahlgren C, *et al.* Cell volume (3D) correlative microscopy facilitated by intracellular fluorescent nanodiamonds as multi-modal probes. *Nanomaterials*. 2021;11(1):14.

86 Chang H-C, Hsiao WW-W, Su M-C *Fluorescent Nanodiamonds*. Hoboken, NJ, USA: John Wiley & Sons; 2018.

7

In Vivo Bioimaging of Fluorescent Nanodiamonds

Wesley Wei-Wen Hsiao, Yun-Yu Chen, Thi-Hong-Hanh Le, Yuen Yung Hui, Stefanny Angela, and Trong-Nghia Le**

* *Corresponding authors*

In vivo imaging involves non-invasive visualization of living organisms, which can aid in monitoring drug and therapeutic distribution, improving drug and gene treatment, monitoring tumor growth, and tracking disease progression in both animals and plants. This chapter addresses *in vivo* imaging applications of fluorescent nanodiamonds (FNDs) and the biological interactions between the model organism and this novel nanomaterial. The exceptional brightness, photostability, and non-toxicity of FNDs have made them ideal for imaging and tracking in various biological models using techniques like wide-field fluorescence imaging, time-gated fluorescence imaging, photoacoustic imaging, magnetic modulation fluorescent imaging, and microwave modulation fluorescent imaging.

7.1 Wide-Field Fluorescence Imaging

A variety of biological models have been used to assess the application of fluorescent nanodiamonds (FNDs) for wide-field fluorescence imaging, including *Paramecium caudatum* (1), *Tetrahymena thermophila* (1), *Caenorhabditis elegans* (2–4), *Drosophila melanogaster* (5), as well as zebrafish, mice, and rats (4, 6). *C. elegans* is the first organism that FNDs have been applied to for *in vivo* imaging. This is because of its tiny size, transparency, clear morphology, simplicity of culture, and quick generation time. Particles are introduced to the worms by either feeding or microinjection (2, 3). Fluorescence imaging has indicated that FNDs

Nanodiamonds in Analytical and Biological Sciences: Principles and Applications, First Edition.
Edited by Wesley Wei-Wen Hsiao and Trong-Nghia Le.
© 2023 John Wiley & Sons Ltd. Published 2023 by John Wiley & Sons Ltd.

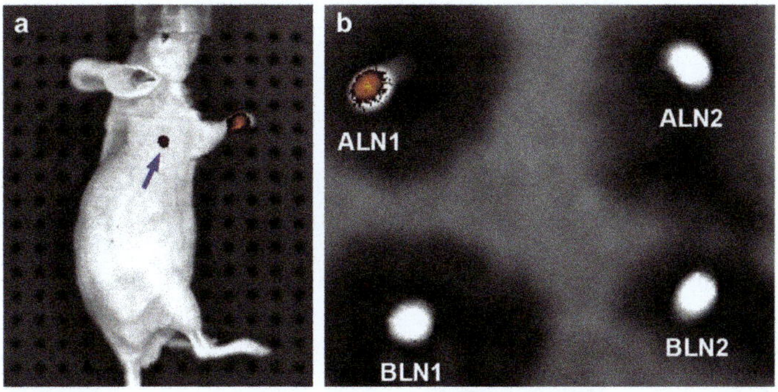

Figure 7.1 (a) *In vivo* and (b) *ex vivo* lymph node photos of a nude mouse post FND injection intradermally. The lymph nodes ALN1 and ALN2 are placed in the right and left axilla, respectively, while the lymph nodes BLN1 and BLN2 are positioned in the right and left brachial areas, respectively. *Source:* Reproduced from ref. (6) / Elsevier.

were taken up by feeding predominantly stayed in the intestinal lumen. In contrast, FNDs injected into the gonad were transported to the organisms' embryos and could be found within the hatched larvae of the following generation. This finding enabled researchers to image the whole digestive system and follow the biological and developmental processes of the living organism for a couple of days. FNDs have also been used in sentinel lymph node imaging (6), which is a critical and typical feature of cancer therapy. In this procedure, mice were first intradermally injected with FNDs and then supervised by a standard *in vivo* imaging system. To reduce interference from background tissue autofluorescence, specimens were stimulated at 605 nm, and their fluorescence was recorded at 780 nm. For the contrast improvement, the primary pictures were modified by removing the scaled background images obtained with 430-nm stimulation. On day 8, following injection, accumulation of the FND particles in the right axillary lymph node was plainly apparent. Figure 7.1 presents the findings of *in vivo* and *ex vivo* lymph node imaging on a naked mouse following the FNDs' intradermal injection. These findings suggest that FNDs have the potential to revolutionize the field of bioimaging, allowing for non-invasive and high-resolution imaging of biological processes in living organisms.

7.2 Time-Gated Fluorescence Imaging

Although background subtraction is a viable means of improving image contrast, effects can be significantly enhanced with the aid of fluorescence time gating. *C. elegans* serves as an excellent model organism to study this phenomenon.

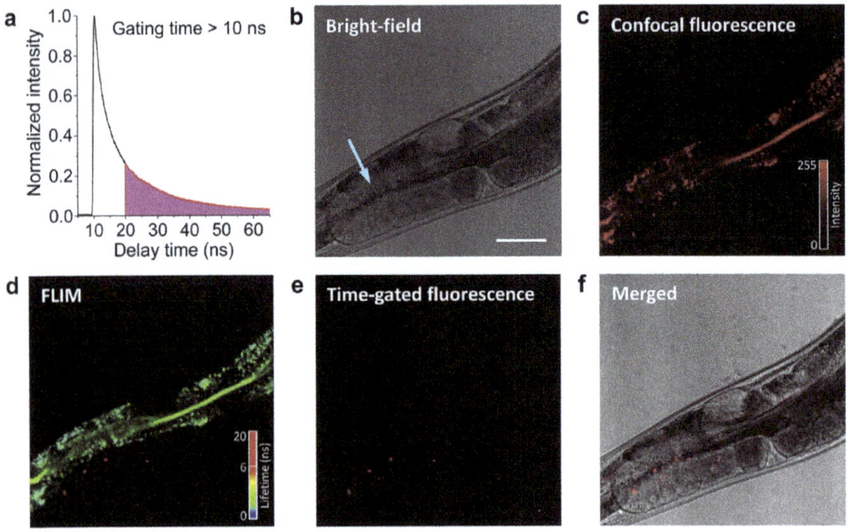

Figure 7.2 Fluorescence imaging of FNDs in *C. elegans*. (a) Fluorescence decay time trace of 100-nm FNDs suspended in water. *Source:* (7) Kuo Y et al., 2013 / with Permission of Elsevier. (b) Bright-field, (c) confocal fluorescence, (d) FLIM, (e) time-gated fluorescence, and (f) merged images of FND microinjected in an adult worm. *Source:* Reproduced from ref. (7) / Elsevier.

Figure 7.2 compares confocal fluorescence images of worms after the microinjection of FNDs into the distal gonad with or without time gating (7). Without time gating, background fluorescence signals of intestinal cells are observed and dominate the confocal fluorescence image. Selective detection of the late photons in the fluorescence emission markedly lowered background signals, simultaneously heightening the signal-to-noise ratio. Thanks to the excellent photostability of FND, a higher image contrast could be achieved, allowing for precise localization of nanoparticles within living organisms. Kuo *et al.* exploited this strategy to track the intercellular transport of yolk lipoproteins coated on FNDs in the nematode (7). FND is an ideal nanocarrier for intercellular transport and long-term imaging uses *in vivo*, as demonstrated by real-time imaging, which tracked worms for 50 minutes.

More recently, Chang's group has used the fluorescence lifetime imaging method for tracking the engraftment and regenerative capabilities of transplanted FND-labeled stem cells in mice (Figure 7.3). They reported that FNDs are capable of identifying transplanted lung stem/progenitor cells *in vivo* by combining immunostaining with fluorescence-activated cell sorting and fluorescence lifetime imaging for *ex vivo* tissue sections (8). FNDs with the biological inertness property have no effect on the abilities of cells to regenerate and differentiate into type I and type II pneumocytes. Moreover, FND-labeled lung stem/progenitor cells are more likely

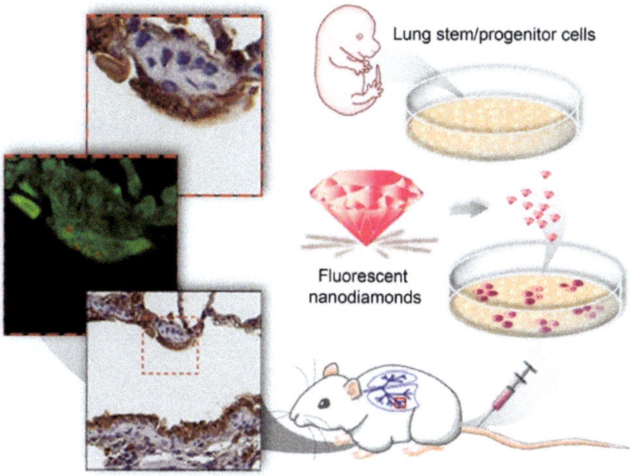

Figure 7.3 Tracking the engraftment and regenerative capabilities of transplanted lung stem/progenitor cells using FNDs in a lung-injured mouse model. *Source:* Reproduced from ref. (8) / Springer Nature.

to remain in the terminal bronchioles of the lungs a week after intravenous treatment. This study revealed that mice's injured lung cells recovered quickly after being transplanted with FND-labeled cells. It also elucidated the factors that inhibit stem cell adoption and the mechanisms by which they regenerate within a host. Thus, a high-potential technique for monitoring the homing of stem cells such as bone marrow stem cells and mesenchymal stem cells in larger animals is expected. Overall, the photostability of FNDs, combined with fluorescence time gating and fluorescence lifetime imaging methods, has enabled precise localization of nanoparticles within living organisms and improved image contrast. These findings highlight the versatility and promise of FNDs for *in vivo* imaging applications.

7.3 Photoacoustic Imaging

Photoacoustic (PA) imaging is a hybrid imaging approach that has currently gained traction in the field of biomedicine (9). PA imaging was first proposed by Alexander Graham Bell in 1880 and is based on a notion known as the PA effect (Figure 7.4).

Because ultrasonic waves scatter less than optical waves in biological tissue, the PA technology gives deep tissue imaging with a spatial resolution that is superior to fluorescence approaches (11). Normally, in PA imaging, a sample is irradiated with brief laser pulses. The presence of ultrasonic emissions from specimens is detected and combined to create a 3-D image. Near-infrared

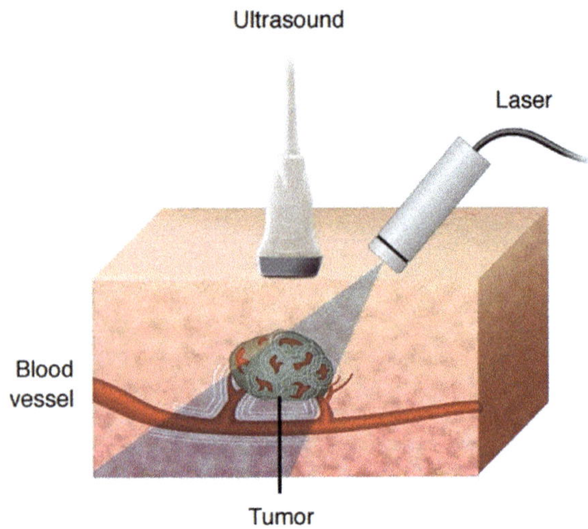

Figure 7.4 The photoacoustic effect occurs when electromagnetic radiation, such as a high-energy laser pulse is transformed into ultrasonic waves, and is the basis of photoacoustic imaging. *Source:* (10) Valluru et al., 2016 / Korean Society of Ultrasound in Medicine.

nanodiamonds (NDs) with a high optical absorbance (700–1000 nm) have been utilized as contrast agents for PA imaging (12). Ion-irradiated nanodiamond (IND) particles could have a high-density ensemble of neutral vacancy centers (V^0) (up to 3000 ppm) or even partly graphitized inside the crystal matrix due to extensive ion irradiation during manufacturing (13). After being subcutaneously injected into the lower back of a mouse, INDs were photoacoustically scanned. Figure 7.5 shows the PA pictures, which clearly reveal the injection site and path along which the needle was withdrawn against the tissue backdrop. NDs showed no signs of decay, even after using a high-power nanosecond laser for continuous illumination.

The same research group has also leveraged INDs in their attempts to use molecularly tailored PA contrast agents for the high-resolution and phenotype-specific detection of breast cancer (14). First, acid-treated INDs were coated with polyethylene glycol (PEG) to enhance stability and blood circulation duration. The PEGylated materials were subsequently coupled with antihuman epidermal growth factor receptor-2 peptides for selective targeting. According to PA imaging, the bioconjugated INDs mainly aggregated in tumors, wholly demarcating the whole tumor after 10 hours. This study demonstrates that when paired with IND contrast ants, hybrid imaging techniques could be a helpful strategy for guiding drug administration in deep tissues.

Cui *et al.* proposed a single nanoplatform (HPNDIC) based on hydrophobic and electrostatic noncovalent interactions by combining multicomponent hyaluronic acid (HA), protamine (PS), NDs, curcumin (Cur), and IR780. The HPNDIC may be utilized for ternary collaboration dual-modal fluorescence/photoacoustic imaging directed Cur-therapy/photothermal/photodynamic treatment of

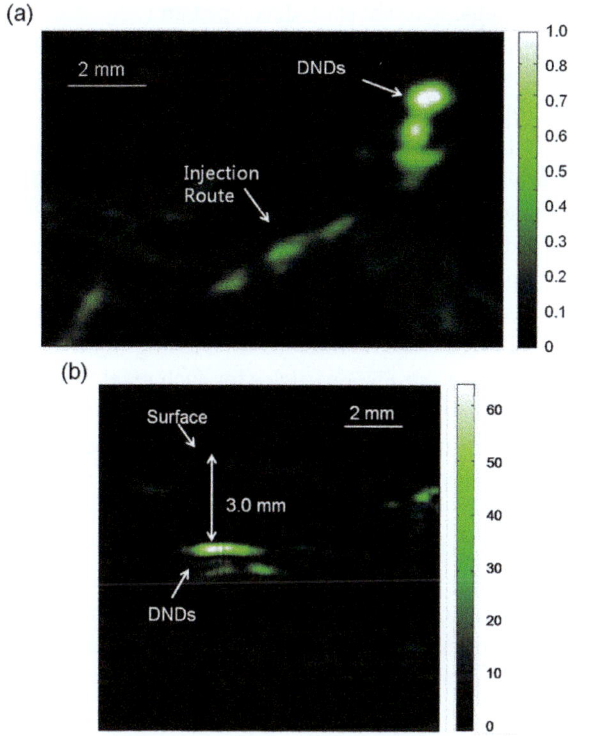

Figure 7.5 Photoacoustic images of a mouse after IND injection in the (a) rear and (b) ventral sides of the thigh. *Source:* Reproduced from ref. (12) / SPIE.

triple-negative breast cancer (TNBC) (15). To achieve this goal, a two-step coordinated assembly technique was used. The resulting HPNDIC particles were homogeneous in size, had a high drug-loading capacity, and demonstrated outstanding colloidal stability. IR780 might be activated to demonstrate photothermal therapy (PTT)/photodynamic therapy (PDT) dual-pattern treatment effects under near-infrared irradiation, resulting in improved Cur chemotherapeutic efficiency *in vitro* and *in vivo* with excellent biocompatibility. The effective accumulation of HPNDIC at tumor locations was further demonstrated using *in vivo* fluorescence and PA imaging. Researchers created a multicomponent theranostic platform to guide triple-collaborative Cur-therapy/photothermal/photodynamic against TNBC using dual-modal fluorescence/photoacoustic imaging (15).

7.4 Tissue Imaging by Microwave Modulation

Microwave modulation is a relatively new imaging technique that offers several advantages over traditional imaging techniques. Firstly, microwave modulation can penetrate deeper into tissues than visible light, making it a useful tool for imaging internal structures. Additionally, microwave modulation is non-invasive, so it does not damage tissues or cells. Igarashi *et al.* developed an early selective imaging approach to increase FND image contrast *in vitro* and *in vivo* by using the NV^- center's spin resonance signal (16). The zero-field energy splitting between the bright state ($m_s = 0$ sublevel) and dark state ($m_s = \pm1$ sublevel) without an external magnetic field is 2.87 GHz, which matches the microwave frequency. These researchers then microwaved the organisms at 2.87 GHz and captured wide-field fluorescence photos when the microwave was on or off (Figure 7.6). When the microwave was turned on and is resonate with the transition, electrons in the triplet ground state migrated from the bright state to the dark state. Compared to the bright state, electrons in the dark state can easily go to the intermediate state and then shift to the bright state via intersystem crossing. Subtraction was used to create selected pictures of the FND particles for each pixel. Because alternate microwave irradiation only modified the NV centers' fluorescence intensity, subtracting these two pictures successfully eliminated the specimen's autofluorescence background and increased image contrast. This approach works on single cells in entire animals like *C. elegans* (17) and mice. These studies were able to identify FNDs aggregates (170 nm in diameter and 260 centers) intradermally injected into the flank of a nude mouse for hours.

In conclusion, FNDs have been successfully applied in tissue imaging using microwave modulation. This technique offers advantages such as deeper tissue penetration and non-invasiveness, making it a useful tool for imaging internal structures. By utilizing the NV center's spin resonance signal and selective imaging approaches, FND image contrast can be significantly increased, eliminating autofluorescence background and enabling the identification of FND aggregates *in vivo*. These findings suggest that FNDs have great potential in tissue imaging and could be a valuable addition to the toolkit of researchers and clinicians studying various diseases and disorders.

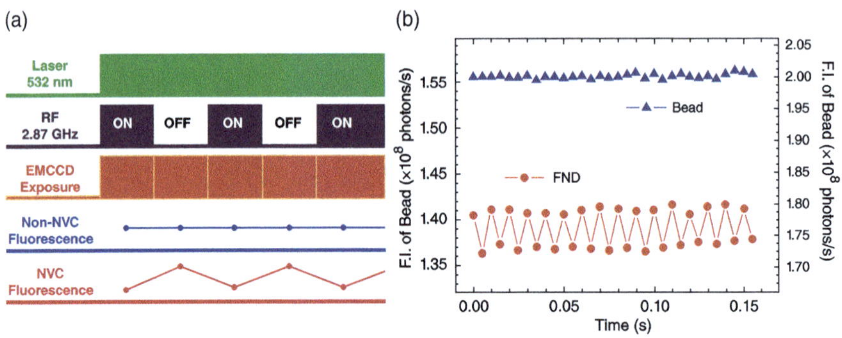

Figure 7.6 (a) Time chart of laser excitation, microwave irradiation, and image acquisition using non-NV^- center (non-NVC) and NVC for selective imaging of FNDs. (b) Time trace of observed fluorescence intensities of FND and fluorescent beads. *Source:* (16) Igarashi et al., 2012 / American Chemical Society.

7.5 Magnetically Modulated Fluorescence Imaging

In contrast to microwave radiation, a magnetic field can reduce FND's fluorescence intensity since the spin sublevels of the NV^{-} center are mixed, leading to remove the selectivity of intersystem crossover (18). In other words, the magnetic field mechanism for FND darkening differs from the microwave process described in the previous subsection. With no magnetic field applied during the continuous excitation, the intersystem crossover from the singlet state to the $m_s = 0$ sublevel of the ground state is preferred compared to the $m_s = \pm 1$ sublevels. Thus, the NV center is polarized, and the bright state is mostly occupied. The NV center has a higher probability of nonradiative transition, which stems from the loss of its polarization. FNDs, therefore, become 10–20% darker in the presence of a magnetic field than without one. Using the image subtraction technique, the confounding background signals, such as autofluorescence and scattered light, can be eliminated, resulting in background-free FND signals. Applying this subtraction technique, Chapman *et al.* improved FND image contrast *in vitro* from the strong autofluorescence background (18). Sarkar *et al.* conducted their research by injecting intact mouse foot pads with silica-coated FNDs and imaging the animals' sentinel lymph nodes (19). As shown in Figure 7.7, FND particles were

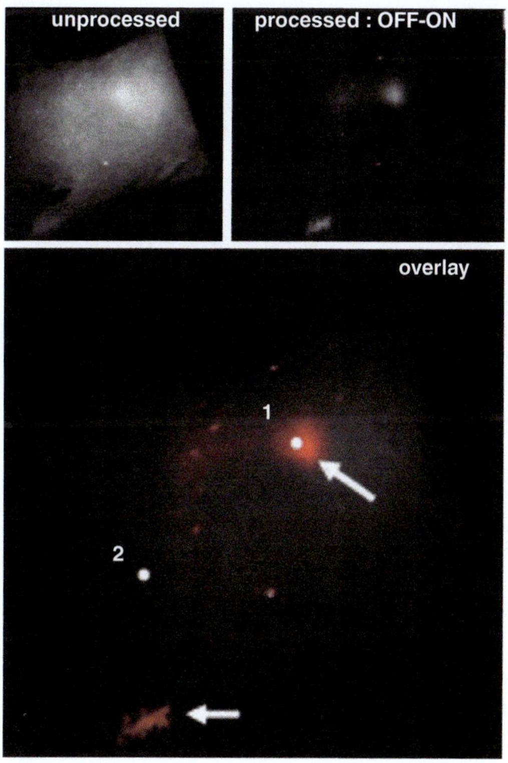

Figure 7.7 FNDs were used to image the sentinel lymph nodes of a mouse *in vivo*. The image was generated using pairwise subtraction with and without the magnetic field, averaged over 475 photos. The corrected image (top right inset, red in overlay) was superimposed over an unprocessed image acquired with the magnetic field turned off (top left inset, white in overlay). The white arrows indicate the injection site in the footpad and the position of the auxiliary lymph node. Signals from FNDs in the lymph node were clearly recognized. *Source:* Reproduced from ref. (19) / Optica Publishing Group.

observed in the proximal axillary sentinel lymph nodes, and FND signals were strongly improved using computational lock-in detection (19).

Jones *et al.* captured the background-subtracted nanoparticles swallowed by *C. elegans*, using magnetic fields to regulate the fluorescence of NV NDs (20). In a wide-field imaging configuration, the fluorescence of 40-nm NV NDs could be modified by 10% along with the time-modulated magnetic fields generated by an affordable "hobby" electromagnet. Furthermore, differential magnetic imaging was applied to observe and separate fluorescence from NDs within the stomach of the organism *C. elegans* (Figure 7.8). As we can see, this method allows researchers to

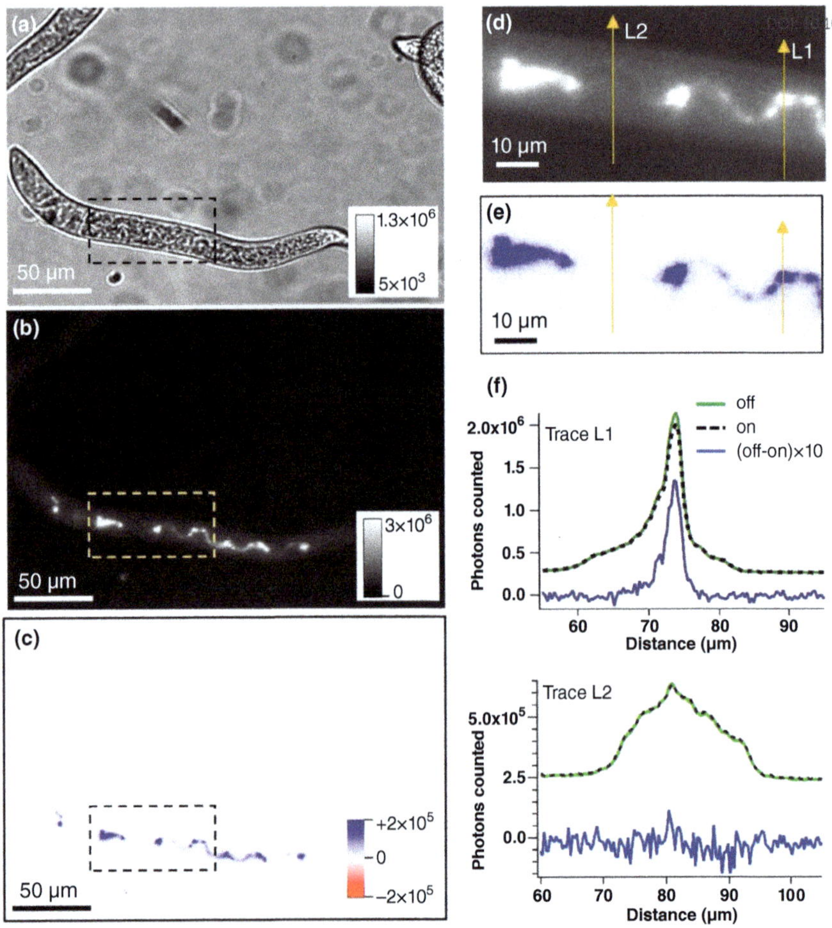

Figure 7.8 (a) Bright-field image and (b) Fluorescence image of NV ND-exposed *C. elegans*. (c) Image resulting from image subtraction. (d), (e) Selected-area images of fluorescence and difference images. (f) Line traces across the regions indicated in (d) and (e). *Source:* Reproduced from ref. (20) / Royal Society of Chemistry.

monitor nanoparticle uptake in organisms as well as evaluate their mobility and interactions in biological models.

7.6 Tissue Imaging Combining Microwaves and Quadruple Coils

Hegyi *et al.* used the optically detected magnetic resonance (ODMR) approach with a quadruple magnetic coil to acquire high-resolution images of FNDs in tissue (Figure 7.9) (21). In the center of this 2-D imaging system, magnetic fields cancel each other out. This is called the "field-free line." FND only resonates with the 2.87-GHz microwave at the field-free line. However, FNDs will not resonate at 2.87 GHz in regions with nonvanishing magnetic fields because the resonance frequency of the transition is split by the magnetic field. The microwave is switched on and off at random throughout the experiment (Figure 7.9b). When the field-free line strikes FND, the intensity of FND fluorescence is regulated by the presence of the periodic microwave and detected with the lock-in amplifier, which is proportional to FND concentration at the field-free line. Researchers sought to modulate FND signal by sweeping the field-free line throughout the

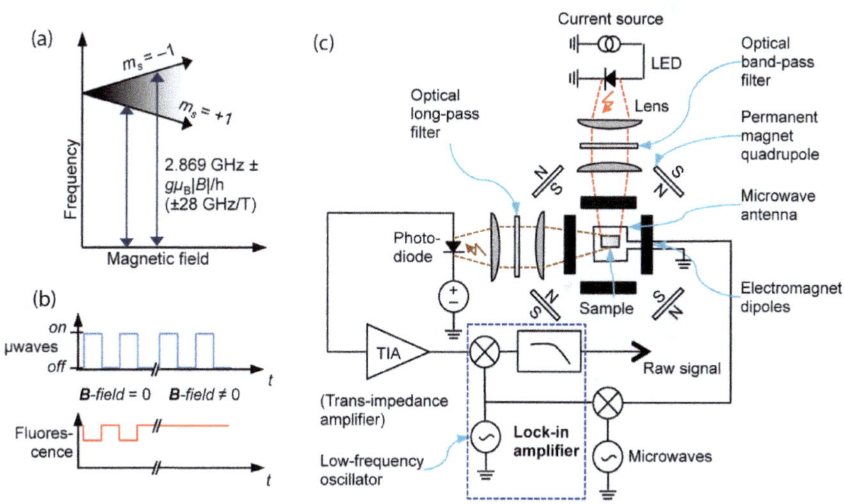

Figure 7.9 ODMR-based imaging of FND's NV^- centers. (a) The spin transition frequency is 2.869 GHz at a zero magnetic field, but as the magnetic field increases, the upper spin sublevels split. (b) A microwave at 2.869 GHz reduces the observed fluorescence from FND when $|B| = 0$ but does not regulate fluorescence in a strong B-field ($|B| > 0$). (c) The "magnetic well" is constructed by the quadruple coils, which have a nonvanishing magnetic field everywhere but in the imaging system's center. *Source:* (21) Hegyi et al., 2013 / American Chemical Society.

whole region. As a result, a map of FND concentration was produced. This method improved the resolution of deep tissue optical imaging substantially. To exemplify this, researchers photographed big letters written with FND covered by 5-mm-thick chicken breast (22).

FND's with an NV^- center have a long coherence time in ambient environments, and its ODMR spectrum is sensitive to changes in magnetic field, electric field, and temperature, as well as mechanical stress. These features have led to many breakthroughs in quantum sensing and imaging. Its application in biology is still in progress; however, we anticipate that the study of the outstanding coherence properties of the NV center will lead to a Nobel Prize in Physics in the very near future.

In summary, FNDs have shown great potential for *in vivo* bioimaging of small animals such as zebrafish and mice, where they can be delivered via intravenous injection or inhalation due to their excellent photostability, biocompatibility, and low toxicity. In addition, FNDs have also been used for imaging of plant cells and tissues, demonstrating their potential for a broad range of biological applications. However, further research is needed to optimize the delivery and imaging protocols and to fully understand the long-term effects of FNDs on living organisms.

Acknowledgments

This work was supported by the National Science and Technology Council (NSTC), Taiwan, under Grant No. 112-2221-E-011-031.

References

1 Lin YC, Perevedentseva E, Tsai LW, Wu KT, Cheng CL. Nanodiamond for intracellular imaging in the microorganisms *in vivo*. *J Biophotonics*. 2012;5(11–12):838–47.

2 Mohan N, Chen C-S, Hsieh H-H, Wu Y-C, Chang H-C. In vivo imaging and toxicity assessments of fluorescent nanodiamonds in Caenorhabditis elegans. *Nano Lett*. 2010;10(9):3692–9.

3 Mohan N, Zhang B, Chang C-C, Yang L, Chen C-S, Fang C-Y, *et al*. Fluorescent nanodiamond – A novel nanomaterial for in vivo applications. *Mater. Res. Soc. Symp. Proc*. 2011;1362:601.

4 Yanagi T, Kaminaga K, Suzuki M, Abe H, Yamamoto H, Ohshima T, *et al*. All-optical wide-field selective imaging of fluorescent nanodiamonds in cells, in vivo and ex vivo. *ACS Nano*. 2021;15(8):12869–79.

5 Simpson DA, Thompson AJ, Kowarsky M, Zeeshan NF, Barson MSJ, Hall LT, *et al.* In vivo imaging and tracking of individual nanodiamonds in drosophila melanogaster embryos. *Biomed Opt Express.* 2014;5(4):1250–61.

6 Vaijayanthimala V, Cheng P-Y, Yeh S-H, Liu K-K, Hsiao C-H, Chao J-I, *et al.* The long-term stability and biocompatibility of fluorescent nanodiamond as an in vivo contrast agent. *Biomaterials.* 2012;33(31):7794–802.

7 Kuo Y, Hsu T-Y, Wu Y-C, Chang H-C. Fluorescent nanodiamond as a probe for the intercellular transport of proteins in vivo. *Biomaterials.* 2013;34(33):8352–60.

8 Wu T-J, Tzeng Y-K, Chang W-W, Cheng C-A, Kuo Y, Chien C-H, *et al.* Tracking the engraftment and regenerative capabilities of transplanted lung stem cells using fluorescent nanodiamonds. *Nat Nanotechnol.* 2013;8(9):682–9.

9 Wang LHV, Hu S. Photoacoustic tomography: In vivo imaging from organelles to organs. *Science.* 2012;335(6075):1458–62.

10 Valluru KS, Willmann JK. Clinical photoacoustic imaging of cancer. *Ultrasonography.* 2016;35(4):267–80.

11 Ntziachristos V. Going deeper than microscopy: The optical imaging frontier in biology. *Nat Methods.* 2010;7(8):603–14.

12 Zhang T, Cui H, Fang C-Y, Su L-J, Ren S, Chang H-C, *et al.* Photoacoustic contrast imaging of biological tissues with nanodiamonds fabricated for high near-infrared absorbance. *J Biomed Opt.* 2013;18(2).

13 Fang CY, Chang CC, Mou CY, Chang HC. Preparation and characterization of ion-irradiated nanodiamonds as photoacoustic contrast agents. *J Nanosci Nanotechno.* 2015;15(2):1037–44.

14 Zhang T, Cui H, Fang CY, Cheng K, Yang X, Chang HC, *et al.* Targeted nanodiamonds as phenotype-specific photoacoustic contrast agents for breast cancer. *Nanomedicine (Lond).* 2015;10(4):573–87.

15 Cui X, Deng X, Liang Z, Lu J, Shao L, Wang X, *et al.* Multicomponent-assembled nanodiamond hybrids for targeted and imaging guided triple-negative breast cancer therapy via a ternary collaborative strategy. *Biomater Sci.* 2021;9(10):3838–50.

16 Igarashi R, Yoshinari Y, Yokota H, Sugi T, Sugihara F, Ikeda K, *et al.* Real-time background-free selective imaging of fluorescent nanodiamonds in vivo. *Nano Lett.* 2012;12(11):5726–32.

17 Yoshinari Y, Mori S, Igarashi R, Sugi T, Yokota H, Ikeda K, *et al.* Optically detected magnetic resonance of nanodiamonds in vivo implementation of selective imaging and fast sampling. *J Nanosci Nanotechno.* 2015;15(2):1014–21.

18 Chapman R, Plakhoitnik T. Background-free imaging of luminescent nanodiamonds using external magnetic field for contrast enhancement. *Opt Lett.* 2013;38(11):1847–9.

19 Sarkar SK, Bumb A, Wu X, Sochacki KA, Kellman P, Brechbiel MW, *et al.* Wide-field in vivo background free imaging by selective magnetic modulation of nanodiamond fluorescence. *Biomed Opt Express*. 2014;5(4):1190–202.

20 Jones ZR, Niemuth NJ, Robinson ME, Shenderova OA, Klaper RD, Hamers RJ. Selective imaging of diamond nanoparticles within complex matrices using magnetically induced fluorescence contrast. *Environ Sci Nano*. 2020;7(2):525–34.

21 Hegyi A, Yablonovitch E. Molecular imaging by optically detected electron spin resonance of nitrogen-vacancies in nanodiamonds. *Nano Lett*. 2013;13(3):1173–8.

22 Hegyi A, Yablonovitch E. Nanodiamond molecular imaging with enhanced contrast and expanded field of view. *J Biomed Opt*. 2014;19(1):011015.

8

Quantum Sensing of Fluorescent Nanodiamonds

*Yuen Yung Hui, Miranda Liu, Stefanny Angela, Thi-Hong-Hanh Le, and Wesley Wei-Wen Hsiao**

* *Corresponding author*

Being the "brightest" candidate among the nanocarbon family, fluorescent nanodiamond (FND) has increasingly been applicable in ultrahigh-resolution and precision bioimaging. The negative nitrogen-vacancy (NV^-) center in FND has a triplet ground state exhibiting a crystal field splitting of 2.88 GHz, separating the $m_s = 0$ and ± 1 sublevels. Interestingly, microwave radiation may be applied to detect and alter the spin sublevels optically, which is commonly called optically detected magnetic resonance (ODMR). Additionally, the NV^- spin has a very long coherence time, making FND suitable for ultrasensitive measurements for temperature, magnetic field, and mechanical stress at the nanoscale. This chapter describes the applications of fluorescent nanodiamond for quantum sensing.

8.1 Quantum Coherence

Quantum mechanics predicts many counterintuitive phenomena. One such phenomenon is superposition. A quantum particle can simultaneously exist in several different states. We do not normally think of a die as being both "here" and "there" simultaneously. However, a "quantum die" can be described as being in two places at once. The position of a quantum die is a superposition of being "here" and "there" simultaneously.

This concept can be more accurately illustrated by a two-level system (1). Suppose that an atom has two eigenstates with discrete energy levels: the ground state $|0\rangle$ and the excited state $|1\rangle$. The wave function of the atom is a

Nanodiamonds in Analytical and Biological Sciences: Principles and Applications, First Edition.
Edited by Wesley Wei-Wen Hsiao and Trong-Nghia Le.
© 2023 John Wiley & Sons Ltd. Published 2023 by John Wiley & Sons Ltd.

superimposition of $|0>$ and $|1>$ and is described by $\psi = a|0> + b|1>$. a and b are the probability amplitudes for $|0>$ and $|1>$, respectively. Hence, $|a|^2$ and $|b|^2$ are the probabilities of finding the atom at the $|0>$ and the $|1>$ states, respectively, constrained by $|a|^2 + |b|^2 = 1$. Furthermore, the atom is now excited by continuous light. Quantum mechanically speaking, the atom is perturbed by a small, local electromagnetic field. Which state is the atom in now? ($|0>$ or $|1>$?) Generally speaking, the atom should be in a state of either $|0>$ or $|1>$. To find the atom in a state of both $|0>$ and $|1>$ at the same time would be impossible. However, quantum mechanics allows and predicts this two-level system to undergo a so-called Rabi oscillation (2). The probability of finding the atom in the state of $|0>$ changes with time as a perfect sinusoidal wave and oscillates forever, as shown in Figure 8.1. In this ideal case, the $|0>$ state is described as having an infinitely long "coherence time." The phase difference between $|0>$ and $|1>$ is always 180° and remains the same forever. If the probability amplitudes of a and b are modified in a predictable manner with the tiny perturbation, the atom can serve as a sensitive probe for external perturbation.

In everyday life, we do not observe quantum phenomena because atoms (or molecules) are rarely isolated and usually interact strongly with the environment. An uncontrolled environment can easily destroy the "coherence" of a system. Preparing an isolated atom with a long "coherence time" is problematic. For instance, Serge Haroche prepared electrically charged atoms and ions in an ion trap for quantum phenomena. These ions were first kept in a vacuum at an extremely low temperature to isolate them from environmental heat and radiation. In addition, a laser was applied to suppress the thermal motion of the ions in the trap, enabling the study of quantum phenomena with the trapped ion. Later, Serge Haroche shared the Nobel Prize in physics with David J. Wineland in 2012 for "ground-breaking methods that enable measuring and manipulation of individual quantum systems."

The nitrogen-vacancy (NV) center in a diamond is like an artificial atom with a long spin coherence time at room temperature (T_2 ~ ms). They have a paramagnetic

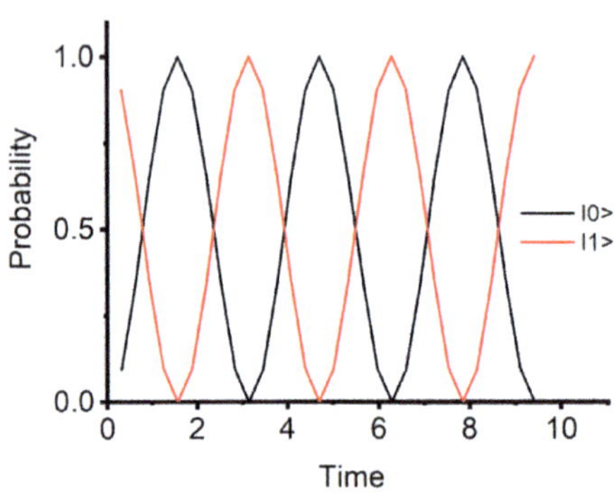

Figure 8.1 Probability of finding an atom at $|0>$ and $|1>$ during Rabi oscillation.

triplet ground state, which can be manipulated by light, magnetic fields, and microwaves. These centers offer outstanding magnetic field sensitivity. Furthermore, as NV centers are close to the diamond's surface, they can be applied to a magnetic sample at a distance of a few nanometers for nanoscale magnetic imaging (3, 4). These sensing capabilities are maintained under ambient conditions, which is crucial for biological applications. Apart from being sensitive to magnetic fields, NV centers are also sensitive to variations in electric fields, temperatures, and pressures. Hence, FNDs have also been applied as nanoscale sensors. In the following sections, we focus on a new field of research on the basis of a biophysical groundwork for quantum sensing (5). Here, we highlight several diverse and exciting applications enabled by these novel quantum sensors, including magnetic field sensing, orientation tracking, nanothermometry, and nanoscopic spin probes.

8.2 NV^- Centers for Quantum Sensing

The presence of a vacancy site close to one substitutional nitrogen atom in a diamond crystal (with C_{3v} symmetry) can form the NV center. These vacancies in diamonds are generally formed by radiation damage via bombardment with high-energy particles such as electrons, neutrons, and gamma rays (6). (In defectless diamond, the valence electrons are entirely occupied by sp^3 hybridized orbitals, and each orbital is filled by a couple of electrons.) When a carbon atom is eliminated from the diamond matrix, four valence electrons along with a +4-charged ion core, consisting of six protons and two inner shell electrons in the 1s orbital, are properly eliminated. These neutral vacancies are called GR1 centers, and they contain four holes in the previously filled orbitals. When a nitrogen atom replace the carbon atom close to the vacancy, the additional valence electron fills one hole to form a neutral NV defect center with zero net charge (NV^0) (7). However, NV defect centers near an electron-rich source, such as adjacent defects or crystal surface, can capture one extra electron, leading to another hole being altered (8). In these cases, an NV^- center is formed. NV^- centers possess two holes with the net charge of −1 and, hence, induce an integral electron spin ($S = 1$) of the ground state ^3A (Figure 8.2). Although both NV^0 and NV^- center in nanoparticles have similar optical properties for bioimaging, only NV^- centers contain the triplet ground and excited states for quantum sensing.

The NV^0 center has a distinct sharp zero-phonon line (ZPL) at 575 nm, enabling researchers quickly differentiate from the NV^- centers which have a ZPL at 637 nm (9, 10). Both NV defect centers have a broad phonon sidebands that are red-shifted by roughly 50 nm in their emission spectra. Additionally, diamond lattices containing both NV centers are photostable with other organic fluorophores. Aside from their exceptional photostability, as discussed in Chapter 5,

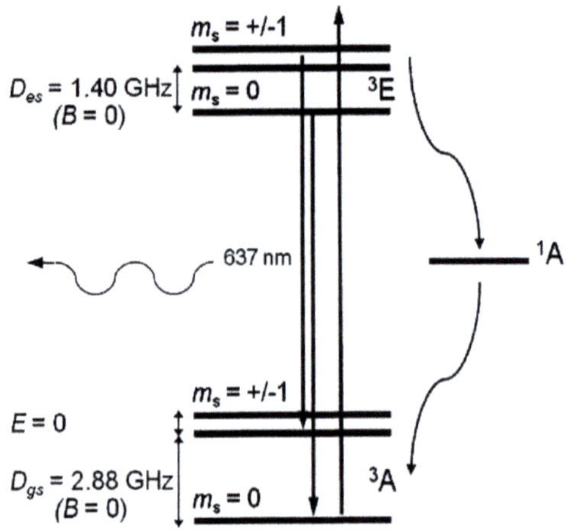

Figure 8.2 Energy level diagram of the NV^- center. *Source:* (11) Bradac et al., 2013 / John Wiley & Sons.

nanodiamonds (NDs) also offer superior biocompatibility both *in vitro* and *in vivo*. With these unique characteristics, NV centers in NDs are usually applied for long-term, risk-free fluorescence tracking in living organisms.

The NV^- center in diamond exhibits unique spin properties with the triplet ground state (^3A) as well as excited state (^3E) and an intermediate singlet state (^1A). Much research in both experimental and theoretical models have proposed an energy level diagram of the NV^- center (12). Figure 8.2 depicts a simplified energy level diagram of the NV^- center. The energy for the transition from ^3A to ^3E is 1.95 eV, which corresponds to the ZPL of the NV^- center at 637 nm. Both ground and excited states of the NV^- center are separated into the spin sublevels with $m_s = 0$ and $m_s = \pm 1$. As a result, 2.87 GHz is revealed as an energy splitting of the ground state, which correlates to the microwave excitation.

Additionally, the spin state of the NV^- center processes properties similar to the polarization of light. It can be unpolarized or polarized (13). At a thermal equilibrium, the population of the ground state in both $m_s = 0$ and $m_s = \pm 1$ sublevels follow the Boltzmann distribution. Since the $m_s = 0$ and $m_s = \pm 1$ sublevels of the ground state are close to each other, they both contain the same electron population, as shown in Figure 8.3a. The NV^- center is "unpolarized." However, optical excitation can polarize the NV^- center, producing a non-Boltzmann steady-state spin alignment in its ground state. The $m_s = 0$ sublevel of the ground state can be preferentially populated.

Figure 8.3 shows how the NV^- center is initially unpolarized. When a green laser is used to stimulate the NV^- center, the population of ^3A shifts to the corresponding ^3E sublevels. Since the optical transition is spin-conserving, the quantum spin number remains the same, as shown in Figure 8.3b. However, the $m_s = 1$ sublevel of the excited state reverts to the ground state through a radiative or a nonradiative transition. The radiative transition emits a fluorescence with a ZPL

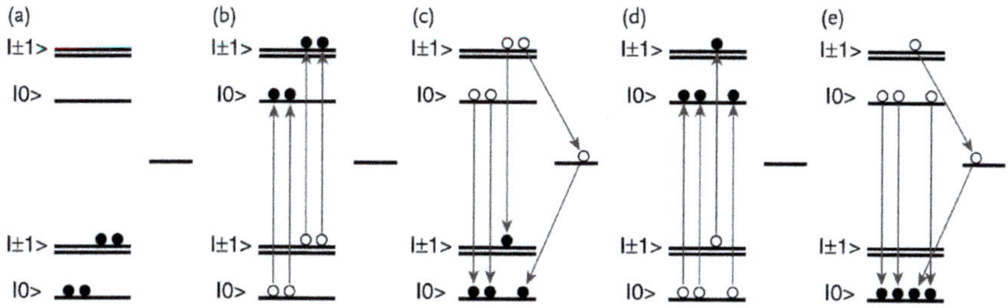

Figure 8.3 Polarization mechanism of NV^- center in diamond. *Source:* (9) Bradac et al., 2013 / John Wiley & Sons.

of 637 nm, whereas the nonradiative transition involves an intersystem crossover over the singlet state and hence is not spin-conserving. According to the spin–orbital interaction (14), the population is shifted from the $m_s = \pm 1$ sublevel of ^3E, then to ^1A, and eventually to the $m_s = 0$ sublevel of ^3A by the nonradiative decay path (Figure 8.3c). After a few excitation cycles, the $m_s = 0$ sublevel of the ground state tends to be populated (Figure 8.3e). The spin state of NV^- centers is commonly described as polarized. (Note: The electron distribution of the polarized NV^- center is different from the unpolarized NV^- center.)

Because the NV^- center spin state couples to the environment, it cannot stay polarized forever. The coherence of the NV^- center is restricted by the spins around it, like paramagnetic defects and ^{13}C nuclear spin (15). The center stays in a certain spin state for a certain relaxation time and then becomes unpolarized. The NV^- center can undergo different relaxation processes and have different relaxation times, generally indicated by T_1 and T_2. T_1 is the spin–lattice or longitudinal relaxation time, while T_2 is generally known as spin–spin or transverse dephasing relaxation time – also commonly known as "coherence time." We will further discuss its application in Section 8.3.3.

With its distinctive spin properties, the NV^- center is gaining more and more attention from researchers. Recent works have revealed that the NV^- center ground state sublevel population can be optically monitored as well as optically controlled with a 2.87-GHz resonant microwave (16). For instance, a polarized NV^- center with the $m_s = 0$ sublevel of the ground state can produce the maximum fluorescence intensity. The application of resonant microwave on the center, however, increases the electron population in the $m_s = \pm 1$ sublevel of the ground state. If a green laser is applied to probe the NV^- center, its fluorescence intensity will drop owing to nonradiative decay via the intermediate singlet state. Thus, the $m_s = 0$ sublevel of the ground state is known as the "bright state," while the $m_s = \pm 1$ sublevel of the ground state is the "dark state." As a result, analyzing the NV^- center's fluorescence intensity can offer information about the ground state sublevel population and hence the spin state. As illustrated in Figure 8.4, to

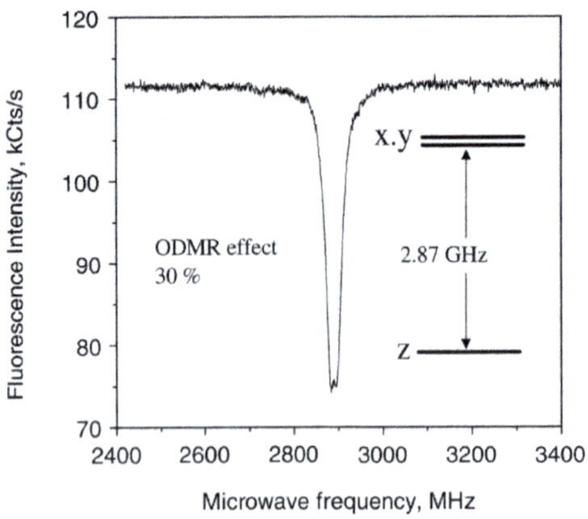

Figure 8.4 ODMR spectrum of a single defect recorded with 514-nm irradiation at an ambient temperature and without an external magnetic field. *Source:* (16) Jelezko et al., 2006 / John Wiley & Sons.

optically probe the spin state of the NV^- center, a diamond sample is further slowly scanned by an auxiliary microwave to produce an ODMR spectrum. (It is noted that this ODMR effect has only been found in a handful of molecules.)

8.3 Emerging Sensing Technologies Using FND

8.3.1 Magnetic Field Sensor

Due to their high responsivity to external magnetic fields, NV^- centers have been widely used for magnetic sensing. NV sensors are also ideal for bio-oriented nanoscale magnetic sensing and imaging owing to their biocompatibility (either bulk diamond or NDs). Based on the sophisticated sensing procedures, NV centers can detect a broad bandwidth magnetic fields from static (~Hz) up to dynamic (~GHz). In the case of unperturbed electronic energy, NV^- centers contain a magnetically sensitive triplet ground state with the $m_s = 0$ sublevel and $m_s = \pm 1$ sublevel. The latter two sublevels degenerate at a zero external magnetic field and are energetically higher than the $m_s = 0$ sublevel by an electron spin resonance (ESR) frequency of 2.8694 GHz. When NV^- centers are put in a magnetic field, they split the $m_s = \pm 1$ sublevel, thereby changing the center's ODMR peaks (17). If a magnetic field of $B = D / \gamma_e$ is introduced, the ODMR spectrum of the single NV^- center is divided into two components with the frequencies $f_\pm$:

$$f_\pm = D + \frac{3\gamma_e^2 B^2}{2D} \sin^2\theta_B \pm \gamma_e B \cos\theta_B \sqrt{1 + \frac{\gamma_e^2 B^2}{4D^2} \tan^2\theta_B \sin^2\theta_B}, \qquad (1)$$

where D is the crystal field splitting; $D = 2869.4$ MHz; θ_B is the angle between the magnetic field and the center's major symmetry axis; and $\gamma_e = g_e \mu_B / h = 28.03$ MHz/mT with $g_e \sim 2.003$, μ_B being the Bohr magneton, and h being the Planck

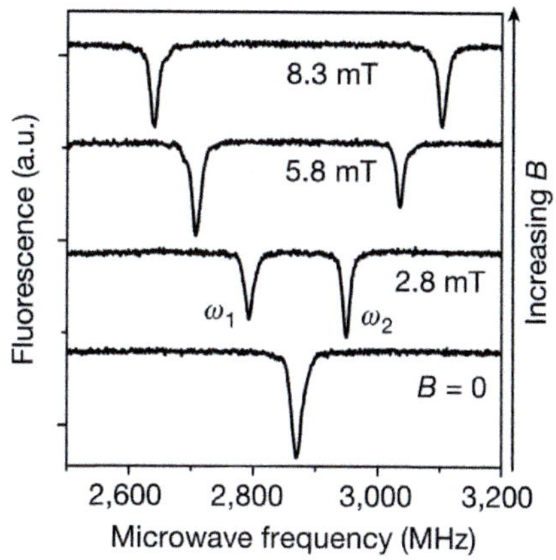

Figure 8.5 A single NV^- center ODMR spectra were recorded in an increasing magnetic field condition. *Source:* (18) Balasubramanian et al., 2008 / Springer Nature.

constant. The above energy splitting (Zeeman splitting) reflects the magnetic field strength in Figure 8.5. Furthermore, the magnetic field sensitivity (a minimum detectable field within 1s acquisition time) depends on the coherence time of the NV^- center (3), and the NV^- center is applicable as a sensitive magnetic probe.

Gruber *et al.* optically read the ODMR spectrum of an NV^- center in a bulk diamond at the single-molecule level (13). The variation in the fluorescence signal between the ^3A and ^3E transitions was efficiently used to detect the sublevel population of the triplet ground state. Techniques based on the ODMR of NV^- centers have been further developed in nanoscale magnetic resonance imaging to identify single electronic spin states with extraordinary sensitivity and speed under ambient circumstances. For example, Balasubramanian *et al.* integrated FND (~40 nm) with a single NV^- center into an atomic force microscope (AFM) cantilever and then used it as a scanning probe magnetometer with the ODMR setup to acquire subwavelength imaging resolution (18). NDs hosting a single NV spin have also been widely applied to probe a very weak magnetic field (19, 20). Remarkably, nanoscale magnetic imaging was described by Rondin *et al.*, with comparable all-optical and large off-axis magnetic fields sensitivity (20). Without microwave control, this technique significantly extends the operating range of diamond-based magnetometry (21). However, the coupling of NV centers to paramagnetic entities like other NV centers, ^{13}C, and ^{14}N nuclei in FND limits the maximum sensitivity of these magnetic field measurements to about 20 μT (22).

Lim *et al.* mapped the magnetic field of suspended objects in a microfluidic system with high nanometer spatial resolution and high sensitivity with a fixed FND. They manipulated a single magnetic particle in a microfluidic system with a planar electroosmotic flow control and a vertical magnetic actuation (23). A fixed ND with a single NV color center was further applied to probe the local magnetic field (Figure 8.6). With the ODMR setup, the authors were able to simultaneously measure the magnetic field and map out the field distribution with nanoscale-level accuracy. The conjunction of NV magnetic sensors with

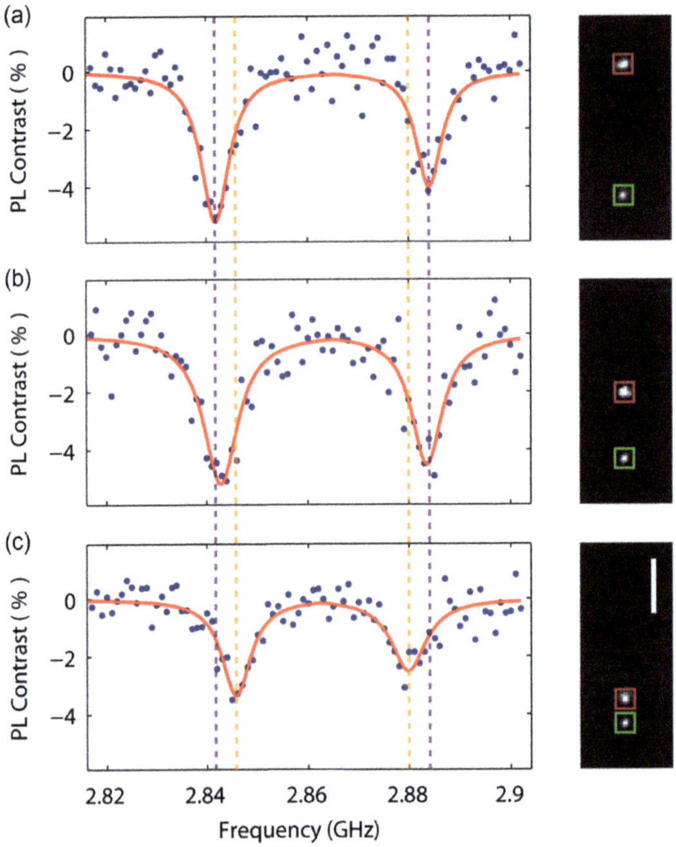

Figure 8.6 ODMR spectra (left side) and camera images (right side) of the magnetic particle (red square) and NV emission (green square). ODMR spectra were obtained with different distances between the magnetic particle and the FND particle (a) 7.25 μm, (b) 3.62 μm, and (c) 1.50 μm. The measured data are fitted using the Lorentzian function, as represented by the red line. The purple and orange dashed lines are plotted according to the center frequencies of the Lorentzian fits in (a) and (c). *Source:* Reproduced from ref. (23) / American Chemical Society.

microfluidic devices opens up new possibilities for real-time magnetic sensing in biological and chemical systems.

Horowitz *et al.* controlled the three dimension motion of the FND particle in solution with an infrared laser as an optical tweezer (24). By simultaneously applying a green laser and microwave, the ground state ESR transition of the NV^- centers were revealed. The fluorescence intensity decreased significantly (10%) when the operating frequency varied over 2.87 GHz. Hence, optically trapped FNDs can map the three-dimensional magnetic field in solution and also probe the complex environments like microfluidic channels. Furthermore, individual

NDs hosting NV centers were deterministically trapped and spatially manipulated by Geiselmann *et al.* for local magnetic field measurement (25). In addition to the laser tweezer, anti-Brownian electrokinetic (ABEL) traps also provide a novel nonperturbative alternative for trapping and manipulating single fluorescent molecules in solution. The operating principle of this technique is tracking the Brownian motion with a time-dependent electric field to produce an electrokinetic drift cancelling the Brownian motion. Because Brownian motions of freely diffusing particles are uncorrelated, the applied potential can select and trap a single molecule with a certain Brownian motion. Kayci *et al.* demonstrated that ABEL trapping could control an FND particle in both ambient and aqueous conditions for magneto-optical spin detection (26).

By modulating the infrared (1064-nm) trapping laser during spectroscopic measurements, Russell *et al.* (27) substantially improved the ODMR contrast compared to continuous wave trapping (Figure 8.7). They acquired dual-sensing (ODMR and T_1) measurements and exact Brownian motion observation on sub-50-nm diamonds confined under intense trapping conditions. This research

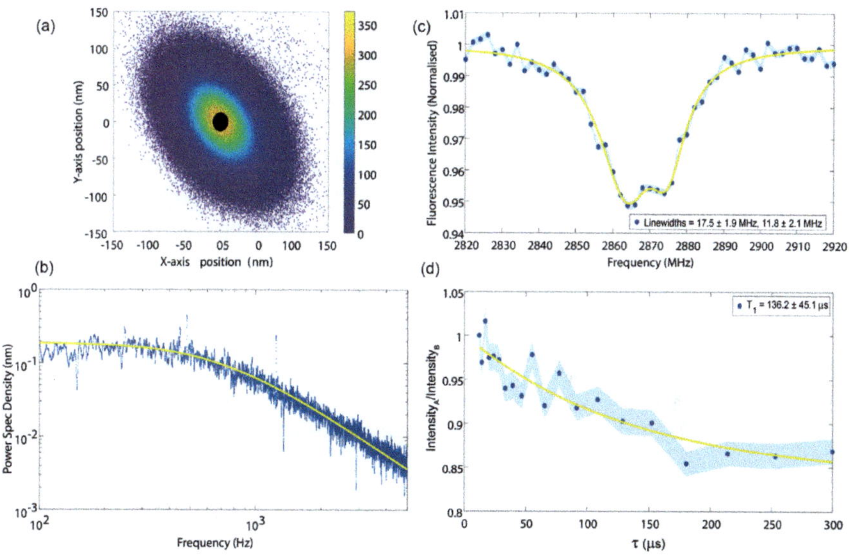

Figure 8.7 Single trapped sub-50-nm ND with multimodal sensing capability. (a) A 2-D lateral position histogram of the confined ND's stochastic motion. (b) Power spectrum's log–log plot and a Lorentz fit of the power spectrum of a particle trajectory in the 0.1–5 kHz range. Particle size is estimated to be on the order of 21 nm in X and 26 nm in Y, assuming a Stokes' drag relation (as seen by the size of the black ellipse in (a)). (c) Pulsed ODMR and (d) Spin-lattice (T1) relaxation measurement from the same ND showing dual sensing capability of the system. *Source:* (27) Russell et al., 2021 / American Chemical Society.

further investigated the DC magnetic field sensing capabilities of trapped 100-nm-sized diamonds, resulting in a one order of magnitude increase in sensitivity. Similarly, several studies have shown that NV centers in FND or near diamond surface (28), with the ODMR spectroscopy, are good probes for magnetic field mapping in challenging environment. For instance, it has been difficult to reach the inside of live cells and microfluidic channels with traditional scanning probe techniques (29–31). But NV centers provide an alternative probe for nanoscale magnetic field sensing.

8.3.2 Orientation Tracker for NV Axis

The ground state's $m_s = 0$ sublevel and the degenerate $m_s = \pm 1$ sublevel are separated by the diamond crystal field's ESR frequency of 2.87 GHz. The angle between the NV axis and the external magnetic field changes the energy splitting (32). When exposed to the magnetic field, a single NV^- center behaves as an intrinsic compass. The orientation tracking of the NV axis was reported by McGuinness *et al.* by using the microwave field to measure the energy splitting of the NV^- center (33). Then, a uniform magnetic field was used to determine the ODMR spectra's peak positions in live HeLa cells. As a result, the millisecond-scale spin movement of FND particles in cells was successfully resolved. This research was also extended to continuously track the orientation of the NV^- center in a HeLa cell for more than 16 hours.

The orientation of the NV axis was also identified by Rondin *et al.* (34). The azimuthal angle of the NV axis was determined by varying the azimuthal angle of the magnetic field in the xy-plane, producing a dataset of ODMR spectra. The polar angle of the NV axis was then computed using another data set of ODMR spectra generated by adjusting the magnetic field's polar angle. This technique has not been widely applied due to its time-consuming nature and the requirement of two sets of ODMR spectra. However, homogeneous and accurate magnetic field calibrations are sorely required. To overcome the above shortcomings, Fukushige *et al.* suggested an alternate strategy to determine the orientation of the NV axis (35). This approach consists of a single ODMR dataset obtained applying the magnetic field in a single direction and rotating the magnetic field in a plane. This shortened the operation time by half.

Feng *et al.* reported for the first time six-dimensional (6-D) motion tracking composing of synchronized 3-D translation and 3-D rotation tracking (36). Based on NV center sensing, the study was carried out on single diamond particles attached to a giant plasma membrane vesicle. These researchers figured out that the geometric effect on diamond particles' motion could be eliminated during quantitative analysis, while determining the net rotation on the vesicle at the same time. This approach was also tested on live HeLa cells by monitoring the translation and rotation of single 200-nm NDs. Within the 500-second recording period, the most likely instantaneous speed of the ND was found to be ~0.03 $\mu m\ s^{-1}$, and the angular speed is proximately $1°\ s^{-1}$. However, under various controlled physiological conditions, NDs' rotation is related to cell metabolic activity. The 6-D tracking approach opens up new avenues for studying local environment dynamics, such as plasma membrane fluctuation, cargo delivery in cells, and rheology at the nanoscale.

8.3.3 Nanoscopic Spin Probe

In ambient environments, Kaufmann *et al.* employed an ND with a single NV^- center to find Gd spin labels in a synthetic cell membrane (37). They first labeled the lipid molecules in a lipid bilayer supported with Gd^{3+}, a prominent magnetic resonance imaging contrast agent hosting a spin of 7/2. Characteristic magnetic fluctuations were produced by motional diffusion of individual Gd-labeled lipids, cross-lipid Gd–Gd spin dipole interactions, and intrinsic Gd spin relaxation effects. By optically polarizing the $m_s = 0$ state, the relaxation time T_1 of the NV^- center in the lipid bilayer was determined, and the possibility P(t) of discovering the NV^- center in the $m_s = 0$ state was subsequently measured. Relaxation time was found to be shortened if the synthetic membrane was marked with 10% Gd spin-labeled lipids, as shown in Figure 8.8. T_1 relaxation time decreased by $74 \pm 6\%$ compared to the reference value, and close individual (4 ± 2) proximal Gd labels were detected by the ND probe. Hence, this strategy is useful for detecting a finite amount of spins in biological models. The gap between traditional ESR ensemble-based approaches and few-spin nanoscale detection under biological environment has been efficiently bridged. Moreover, a door for the *in situ* detection of dynamical processes in biology at the nanoscale level has been opened through this effective monitoring of limited numbers of spins.

Ferritin has a minor part of uncompensated Fe^{3+} spin with a magnetic moment of 300 μ_B, and its content in blood serum correlates with body iron storage. Thus, ferritin assessments in physiological settings are critical for disease pathophysiology and diagnosis. Ermakova *et al.* detected individual ferritin molecules by applying diamond–ferritin complexes (38). Ferritin molecules were first attached to 20-nm NDs via noncovalent binding. Then, the relaxation time T_1 of free NDs was compared to the diamond–ferritin combination and found to be much shorter. In comparison to other ferritin detection studies, this technique demonstrates an extraordinary degree of sensitivity, simplicity and, most importantly, the absence of the need for expensive antibodies or sophisticated immunochromatography equipment. Moreover, this method also overcomes the inconsistent reproducibility of electrochemical immunoassays and, therefore, creates a channel for single protein detection that can easily be applied to cellular models.

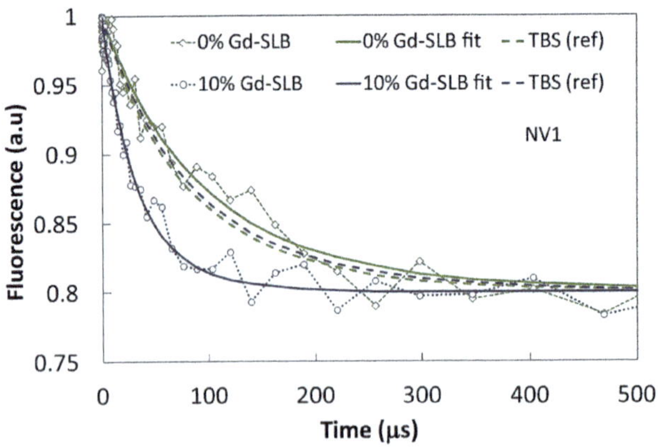

Figure 8.8 The relaxation time T_1 of the NV^- center in an ND in a supported lipid bilayer was obtained in conditions of no spin labels (green) and 10% Gd spin labels (blue). *Source:* (36) Kaufmann et al., 2013 / National Academy of Science.

Acknowledgments

This work was supported by the National Science and Technology Council (NSTC), Taiwan, under Grant No. 112-2221-E-011-031.

References

1 Acosta V, Hemmer P. Nitrogen-vacancy centers: Physics and applications. *Mrs Bull.* 2013;38(2):127–33.

2 Feynman RP, Leighton RB, Sands M. *The Feynman Lectures on Physics.* Boston, US: Addison-Wesley publishing Company Inc.; 1965.

3 Hong S, Grinolds MS, Pham LM, Le Sage D, Luan L, Walsworth RL, *et al.* Nanoscale magnetometry with NV centers in diamond. *Mrs Bull.* 2013;38(2):155–61.

4 Hsiao WW-W, Le T-N, Chang H-C. Applications of fluorescent nanodiamond in biology. In: Meyers, RA (Ed.), *Encyclopedia of Analytical Chemistry.* Hoboken, NJ, USA: John Wiley & Sons; 2022. pp. 1–43. https://doi.org/10.1002/9780470027318.a9776.

5 Balasubramanian G, Lazariev A, Arumugam SR, Duan DW. Nitrogen-vacancy color center in diamond – emerging nanoscale applications in bioimaging and biosensing. *Curr Opin Chem Biol.* 2014;20:69–77.

6 Campbell B, Choudhury W, Mainwood A, Newton M, Davies G. Lattice damage caused by the irradiation of diamond. *Nucl Instrum Methods Phys Res A.* 2002;476(3):680–5.

7 Maze JR, Gali A, Togan E, Chu Y, Trifonov A, Kaxiras E, *et al.* Properties of nitrogen-vacancy centers in diamond: The group theoretic approach. *New J Phys.* 2011;13(2):025025.

8 Schirhagl R, Chang K, Loretz M, Degen CL. Nitrogen-vacancy centers in diamond: Nanoscale sensors for physics and biology. *Annu Rev Phys Chem.* 2014;65:83–105.

9 Chang Y-R, Lee H-Y, Chen K, Chang C-C, Tsai D-S, Fu C-C, *et al.* Mass production and dynamic imaging of fluorescent nanodiamonds. *Nat Nanotechnol.* 2008;3(5):284–8.

10 Faklaris O, Joshi V, Irinopoulou T, Tauc P, Sennour M, Girard H, *et al.* Photoluminescent diamond nanoparticles for cell labeling: Study of the uptake mechanism in mammalian cells. *Acs Nano.* 2009;3(12):3955–62.

11 Bradac C, Gaebel T, Rabeau JR. Nitrogen-vacancy color centers in diamond: Properties, synthesis, and applications. In: Mildren, RP and Rabeau, JR (Eds.), *Optical Engineering of Diamond.* Weinheim, Germany: Wiley-VCH Verlag GmbH & Co. KGaA; 2013. pp. 143–75.

12 Doherty MW, Manson NB, Delaney P, Jelezko F, Wrachtrup J, Hollenberg LCL. The nitrogen-vacancy colour centre in diamond. *Phys Rep.* 2013;528(1):1–45.

13 Gruber A, Dräbenstedt A, Tietz C, Fleury L, Wrachtrup J, Borczyskowski CV. Scanning confocal optical microscopy and magnetic resonance on single defect centers. *Science.* 1997;276(5321):2012.

14 Manson NB, Harrison JP, Sellars MJ. Nitrogen-vacancy center in diamond: Model of the electronic structure and associated dynamics. *Phys Rev B.* 2006;74(10):104303.

15 Childress L, Hanson R. Diamond NV centers for quantum computing and quantum networks. *Mrs Bull.* 2013;38(2):134–8.

16 Jelezko F, Wrachtrup J. Single defect centres in diamond: A review. *Phys Status Solidi A.* 2006;203(13):3207–25.

17 Doherty MW, Michl J, Dolde F, Jakobi I, Neumann P, Manson NB, *et al.* Measuring the defect structure orientation of a single NV- centre in diamond. *New J Phys.* 2014;16:063067.

18 Balasubramanian G, Chan I, Kolesov R, Al-Hmoud M, Tisler J, Shin C, *et al.* Nanoscale imaging magnetometry with diamond spins under ambient conditions. *Nature.* 2008;455(7213):648–51.

19 Taylor JM, Cappellaro P, Childress L, Jiang L, Budker D, Hemmer PR, *et al.* High-sensitivity diamond magnetometer with nanoscale resolution. *Nat Phys.* 2008;4(10):810–6.

20 Rondin L, Tetienne JP, Hingant T, Roch JF, Maletinsky P, Jacques V. Magnetometry with nitrogen-vacancy defects in diamond. *Rep Prog Phys.* 2014;77(5):056503.

21 Rondin L, Tetienne JP, Spinicelli P, Dal Savio C, Karrai K, Dantelle G, *et al.* Nanoscale magnetic field mapping with a single spin scanning probe magnetometer. *Appl Phys Lett.* 2012;100(15).

22 Aman H, Plakhotnik T. Accuracy in the measurement of magnetic fields using nitrogen-vacancy centers in nanodiamonds. *J Opt Soc Am B.* 2016;33(3):B19–B27.

23 Lim K, Ropp C, Shapiro B, Taylor JM, Waks E. Scanning localized magnetic fields in a microfluidic device with a single nitrogen vacancy center. *Nano Lett.* 2015;15(3):1481–6.

24 Horowitz VR, Aleman BJ, Christle DJ, Cleland AN, Awschalom DD. Electron spin resonance of nitrogen-vacancy centers in optically trapped nanodiamonds. *Proc Natl Acad Sci USA.* 2012;109(34):13493–7.

25 Geiselmann M, Juan ML, Renger J, Say JM, Brown LJ, de Abajo FJG, *et al.* Three-dimensional optical manipulation of a single electron spin. *Nat Nanotechnol.* 2013;8(3):175–9.

26 Kayci M, Chang HC, Radenovic A. Electron spin resonance of nitrogen-vacancy defects embedded in single nanodiamonds in an ABEL trap (vol 14, pg 5335, 2014). *Nano Lett.* 2014;14(10):6036-.

27 Russell LW, Dossetor EC, Wood AA, Simpson DA, Reece PJ. Optimizing optical tweezers experiments for magnetic resonance sensing with nanodiamonds. *ACS Photonics.* 2021;8(4):1214–21.

28 Xu Y, Yu Y, Hui YY, Su Y, Cheng J, Chang H-C, *et al.* Mapping dynamical magnetic responses of ultrathin micron-size superconducting films using nitrogen-vacancy centers in diamond. *Nano Lett.* 2019;19(8):5697–702.

29 Rehor I, Lee KL, Chen K, Hajek M, Havlik J, Lokajova J, *et al.* Plasmonic nanodiamonds: Targeted core–shell type nanoparticles for cancer cell thermoablation. *Adv Healthc Mater.* 2015;4(3):460–8.

30 Schlücker S. Surface-Enhanced Raman spectroscopy: Concepts and chemical applications. *Angew Chem Int Ed.* 2014;53(19):4756–95.

31 Aslan K, Gryczynski I, Malicka J, Matveeva E, Lakowicz JR, Geddes CD. Metal-enhanced fluorescence: An emerging tool in biotechnology. *Curr Opin Biotechnol.* 2005;16(1):55–62.

32 Maclaurin D, Hall LT, Martin AM, Hollenberg LCL. Nanoscale magnetometry through quantum control of nitrogen-vacancy centres in rotationally diffusing nanodiamonds. *New J Phys.* 2013;15.

33 McGuinness LP, Yan Y, Stacey A, Simpson DA, Hall LT, Maclaurin D, *et al.* Quantum measurement and orientation tracking of fluorescent nanodiamonds inside living cells. *Nat Nanotechnol.* 2011;6(6):358–63.

34 Rondin L, Tetienne J-P, Rohart S, Thiaville A, Hingant T, Spinicelli P, *et al.* Stray-field imaging of magnetic vortices with a single diamond spin. *Nat Commun.* 2013;4(1):1–5.

35 Fukushige, K, Kawaguchi, H, Shimazaki, K, Tashima, T, Takashima, H, Takeuchi, S. Identification of the orientation of a single NV center in a nanodiamond using a three-dimensionally controlled magnetic field. *Appl Phys Lett.* 2020;116(26):264002.

36 Feng X, Leong W-H, Xia K, Liu C-F, Liu G-Q, Rendler T, *et al.* Association of nanodiamond rotation dynamics with cell activities by translation-rotation tracking. *Nano Lett.* 2021;21(8):3393–400.

37 Kaufmann S, Simpson DA, Hall LT, Perunicic V, Senn P, Steinert S, *et al.* Detection of atomic spin labels in a lipid bilayer using a single-spin nanodiamond probe. *Proc Natl Acad Sci USA.* 2013;110(27):10894–8.

38 Ermakova A, Pramanik G, Cai JM, Algara-Siller G, Kaiser U, Weil T, *et al.* Detection of a few metallo-protein molecules using color centers in nanodiamonds. *Nano Lett.* 2013;13(7):3305–9.

9

Nanoscale Thermometry with Fluorescent Nanodiamonds

Thi-Hong-Hanh Le, Richard Hsin, Duc-Thang Vo, Yan-Kai Tzeng, Trong-Nghia Le, and Wesley Wei-Wen Hsiao**

* *Corresponding authors*

Temperature is an important factor in biochemical reactions, and understanding how cells sense and react to temperature is important for disease treatment. Fluorescent nanodiamonds (FNDs) are biocompatible and not affected by external conditions such as pH, ions, viscosity, and organic solvents, making them reliable thermometers for use in complex biological environments. The nitrogen-vacancy (NV) centers in FND have been widely used as stable nanothermometers due to the optically detected magnetic resonance (ODMR) thermal shift and the zero-phonon line (ZPL) resulting from thermally induced lattice strains. Additionally, diamonds have high thermal conductivity, which ensures that all NV centers are in thermal equilibrium with their surroundings. The chapter provides an overview of temperature sensing techniques in complex biological environments based on the magnetic and optical features of NV centers.

9.1 NV Thermometry in Biosystems

Temperature is a physical quantity that governs many biochemical reactions, including enzyme reactions and metabolism. Decades ago, research revealed that temperature is involved in cell differentiation, proliferation, death, protein functions, and expression of gene. Temperature changes could alter cells' life cycles, and thoroughly understanding the effects of the temperature on the cells turns out to be crucial for disease treatment. As a result, real-time monitoring of temperatures at a nanoscale spatial resolution has become a growing field of research (1, 2). Nanothermometry, in particular, is expected to greatly impact nanotechnology and molecular biology (3, 4). An ideal nanothermometer should be accurate and precise, enabling its use for a large range of temperatures in various practical applications. Moreover, the measure time must be fast while maintaining the precision. Many different measurement techniques have been developed

Nanodiamonds in Analytical and Biological Sciences: Principles and Applications, First Edition.
Edited by Wesley Wei-Wen Hsiao and Trong-Nghia Le.
© 2023 John Wiley & Sons Ltd. Published 2023 by John Wiley & Sons Ltd.

based on various sensor materials, including single-molecule fluorophores (5, 6), green fluorescent protein (GFP) (3), quantum dot (QD) (4), dye molecules covalently linked to a host matrix (7), probes with two different dyes covalently linked to one another (8, 9), and fluorescent nanodiamond (FND) (10–12).

Owing to the optical properties of its color centers, FND is uniquely capable of measuring local temperatures at the subcellular level with sensitivity down to the millikelvin degree (10, 13–17). The color centers referred to here are point defects embedded in sp^3 carbon diamond crystal and are commonly known as negative nitrogen-vacancy (NV^-) centers. This center possesses a thermal shift of the zero-field splitting at 2.87 GHz ($m_s = 0$ to $m_s = \pm 1$) or the ZPL at 637 nm leading to the temperature response of FNDs differs strongly from that of most fluorescence probes (18, 19). Temperature of nanodiamonds (NDs) can be probed by measuring optically detected magnetic resonance (ODMR) frequency shift because ODMR frequency is sensitive to temperature changes. However, environmental parameters, such as pH, ions, viscosity, and organic solvents, have no effect on the thermosensing capabilities of NV^- centers, making them trustworthy thermometers, even in complicated biological environments. Therefore, FNDs can serve as precise intracellular temperature probes. Owing to their biocompatibility and environmental sensitivity, they can also be used for ion concentration and even magnetic resonance imaging (MRI) with nanometer-scale resolution (20–23). This chapter provides an overview of recently developed temperature sensing techniques based on the magnetic and optical features of NV^- centers.

9.2 Ultrahigh Precision Temperature Measurement

NV^- centers have a unique spin state with an amazingly long coherence period of greater than 1 ms in nonisotope diamond at ambient conditions. This enables them to serve as ultrahigh sensitive temperature meters. In 2010, Acosta *et al.* utilized ODMR to explore the relationship between temperature and the spin resonance of NV^- centers (24). In this study, they measured the dependence of the temperature shift of the zero-field splitting in the range of 280–330 K for an ensemble of NV^- centers in bulk diamond (0.010–15 ppm). A thermal shift of $\Delta D/\Delta T = -75$ kHz K^{-1} and the red shift stemmed from the thermal expansion of the diamond crystal lattice were recorded at 300 K (Figure 9.1). Later studies of single NV^- centers over a wider temperature range showed a change in frequency. However, the relationship between frequency and temperature was not linear. The temperature dependence of D at 300–700 K was adequately reflected by a third-order polynomial (25):

$$D(T) = 2.8697 + 9.7 \times 10^{-5} T - 3.7 \times 10^{-7} T^2 + 1.7 \times 10^{-10} T^3$$

where D(T) is in gigahertz and T is in kelvin.

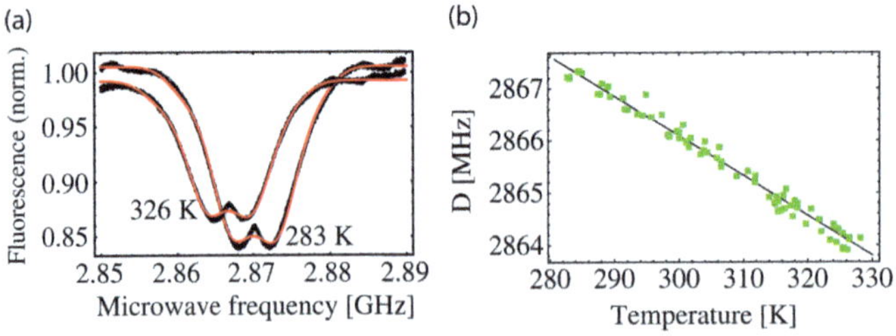

Figure 9.1 (a) ODMR spectra of NV^- centers recorded at different temperatures (283 and 326 K). (b) ODMR peak centers' frequency shifts in temperature range of 280–330 K. *Source:* (24) Acosta et al 2010 / American Physical Society.

Even though the amplitude of thermal change is small, it can still be measured precisely due to the abnormal stability of the ODMR spectra at room temperature.

To elucidate the thermal effect on NV fluorescence lifetime, Plakhotnik and Gruber heated 30-nm FNDs. They observed a fourfold decrease of the photoluminescence accompanied by a decrease in fluorescence lifetime about 2.7-fold of NV centers in the temperature range of 300–670 K (26). They also used the same FNDs to study the thermal shift of D and found similar findings as those of bulk diamond (27). FNDs, therefore, showed outperformance in terms of wider operating limits and better single-particle detection sensitivity.

In 2013, Kucsko and colleagues used FNDs to measure the temperature of a local thermal environment on a length scale of 200 nm in a living cell (10). Both nanoparticles of 100-nm FNDs and gold (GNPs) were injected into a single human embryonic fibroblast. GNP was excited by 532-nm laser to work as a heat source and was distinguished from NV fluorescence by 560±25 nm band pass filter. The electron spin resonance spectra were then detected by a continuous-wave laser. From that, the local temperature change was detected with the sensitive of 80 $\mu K\ Hz^{-1/2}$.

9.3 Time-Resolved Nanothermometry

In terms of temperature accuracy, FND surpasses other reported nanothermometers, such as CdSe QDs or GFPs. However, there are still many challenges associated with using FND in this context. This is because of the time to obtain the ODMR spectra and the additional data processing required following the test to determine peak positions. Many techniques have been developed to simplify and accelerate NV thermometry, including the three-point technique (28), frequency modulation schemes (16, 17), and coherent control methods (29) (e.g., Coop-D-Ramsey sequence). Tzeng *et al.* introduced a new method named "time-resolved temperature measurement." This technique is based on the independence of

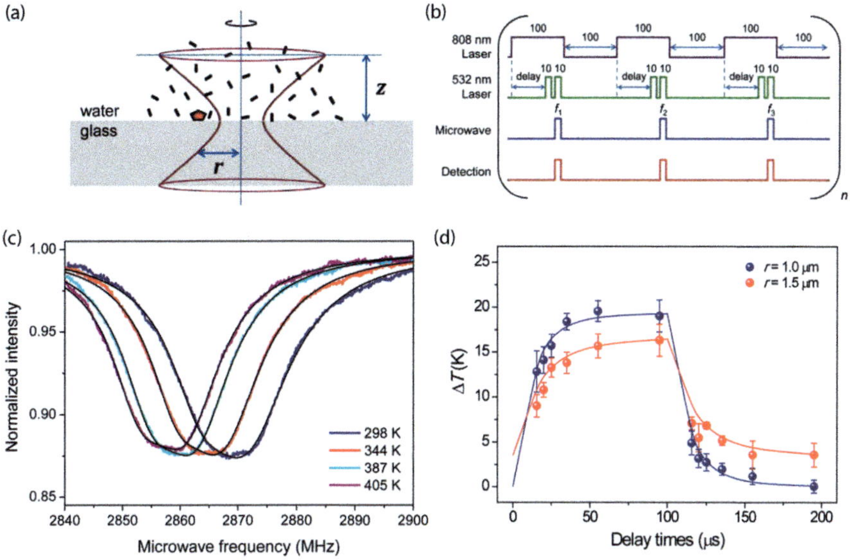

Figure 9.2 (a) Time-resolved nanothermometry of GNR denoted by the black rods, heat source is the 808-nm laser with FND. (b) Time sequences of the laser, microwave, and detection pulses in microseconds with the three-point method. (c) ODMR spectra of the heated GNR solution by the NIR laser (0–10 mW). (d) Time evolution of the GNR solution's heat dissipation at different radial locations (r = 1.0 and 1.5 μm). *Source:* (28) Tzeng et al 2015 / American Chemical Society.

ODMR width peaks of NV^- centers on T up to 600 K (28). 100-nm FND was employed as a single-particle temperature sensor, while gold nanorod (GNR) was used as a heater when introducing $\lambda = 808$ nm laser (Figure 9.2a). By recording FND fluorescence at three predetermined microwave frequencies, ODMR peak shifts were determined (Figure 9.2b). Remarkably, these researchers used a three-point sampling strategy to monitor differences in temperature throughout ±100 K in real time while simultaneously doing extensive research on nanoscale heat transfer with a temporal resolution of more than 10 μs (Figure 9.2c and 9.2d).

9.4 All-Optical Luminescence Nanothermometry

The thermal changes of NV^- spin resonances allow for sensitive nanoscale temperature sensing with excellent temporal and spatial accuracy. Despite the success of ODMR, difficulties remain with integrating this technique into real biological applications. For instance, in most ODMR research, microwave (MW) excitation is applied via a metal wire or nanostructure positioned close to the examination point (usually about 10 μm). The close proximity required for excitation has prevented ODMR from more practical uses. All-optical NV^- thermometry has been confirmed as a new method to address the abovementioned drawback of ODMR, involving the detection of the thermal shifts of NV centers' ZPL located at 637 nm (30, 31), the Debye–Waller factor (i.e., the ratio of the area under the ZPL versus the total emission band) (27), and the variation in ZPL height corresponding to the phonon sideband (30, 32). All-optical approaches are good candidates for biological applications since they do not require MW irradiation and are uncomplicated and easy to implement using any confocal fluorescence microscope combined with a charge-coupled device (CCD)-based spectrometer. Chen revealed that thermal shifts of the ZPL at 638 nm are equivalent to that of the spin transitions at 2.87 GHz (31). Chang and coworkers created a high density and uniform FND-embedded poly(2-hydroxyethyl methacrylate) (PHEMA) film to study the temperature dependence of ZPL positions and the heights of NV^- centers from 35 to 120 °C (Figure 9.3) (30). In this range, the sensitivity of the ZPL shifts and height changes is 0.46–1.1 and 0.15–0.62 K $Hz^{-1/2}$, respectively. These

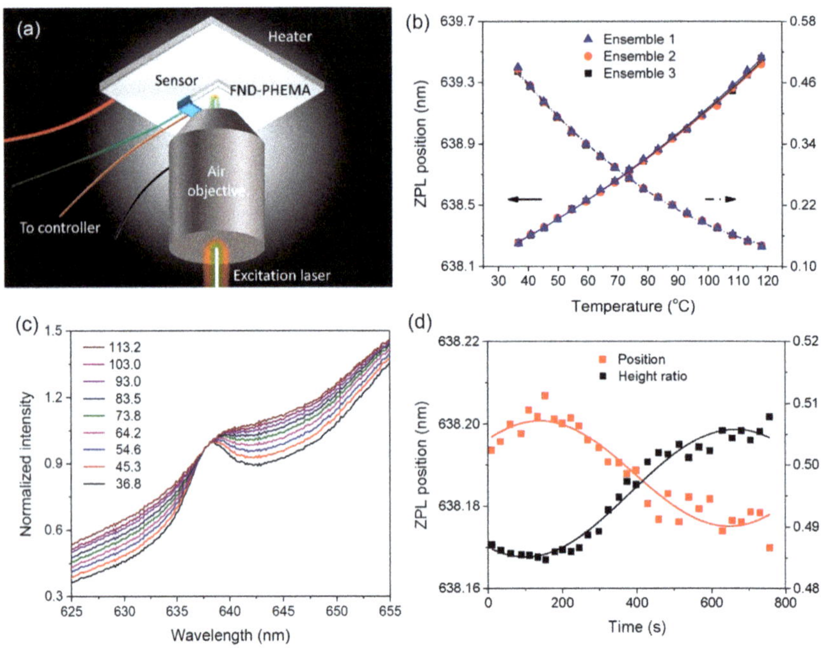

Figure 9.3 (a) Schematic diagram illustrating the setup for monitoring ZPL thermal changes in NV^- centers. (b) The changes of the ZPL location and height ratios at temperature of 35–120 °C for three typical PHEMA-embedded FND. (c) Fluorescence spectra of NV^- centers at the ZPL region of 36.8–113.2 °C. (d) Stability test of PHEMA-embedded FND. *Source:* (30) Hui et al 2019 / American Chemical Society.

polymer films could be mass-produced, making them ideal for commercial uses as all-optical thermometers.

While using ZPL for thermometry, Plakhotnik *et al.* measured the amplitude of the ZPL by equation fitting the fluorescence spectra instead of detecting temperature shifts in the ZPL of NV centers (27). 0.3 $K/Hz^{1/2}$ of the sensitivity could be obtained when a 590-nm laser was used for exciting 50-nm FND. Long-term stability greater than 0.6 K (peak-to-peak value) was further attained by a single FND spin-coated on a glass substrate. This approach is naturally more sensitive than detecting ZPL shifts. However, background (or baseline) signals make this method more vulnerable to interference. To reduce background interference, some researchers have successfully employed fluorescence lifetime gating or magnetic manipulation of the NV^- emission.

9.5 Scanning Thermal Imaging

Thermal mapping is a method for measuring an object's spatial temperature fluctuation. It is difficult to use single-particle FND nanothermometry for nanoscale thermal mapping. To get data on the spatial temperature distribution required for mapping, FND probes need to be properly swept across the sample. In a study by Tetienne *et al.*, a single FND nanoparticle containing ≈100 NV^- centers attached to the tip of an atomic force microscope (AFM) to serve as a thermal probe (33) could offer up to an order of magnitude gain in acquisition time. This technique was used together with ODMR to map local temperature changes caused by the laser irradiation of a single 40-nm GNP in water (Figure 9.4). Integrating these two methods maintained the sub-100-nm spatial resolution inherited from the scanning probe microscope for temperature sensing and topographic imaging. These findings open the door to novel uses for FND nanothermometry, such as thermal imaging and the operation of microelectronic devices.

9.6 Intracellular Temperature Sensing

Highly sensitive nanothermometers are needed to monitor and record changes in local temperatures that are unaffected by external conditions. FNDs have been reported for a few proof-of-concept applications that enable temperature measurements inside a living cell and other biological systems. Table 9.1 summarizes several studies that have used NV thermometry in biosystems. These studies range from temperature sensing the interiors of cells (10, 34) and laser-activating neurons (35), to intraneuronal cell temperature mapping (36), and detecting the absolute temperature of cells (12). Researchers have also used NV thermometry to

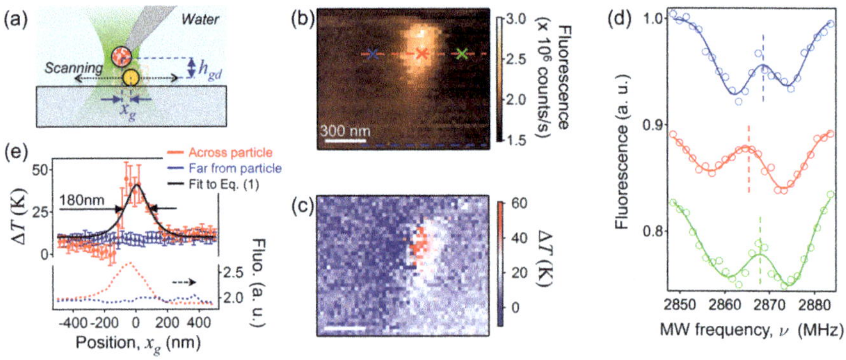

Figure 9.4 (a) Experimental setup for gold nanoparticle thermal imaging. AFM tip and the sample are submerged in an aqueous solution to achieve adequate thermal conductivity. Fluorescence image (b) and temperature map (c) recorded at the same time by scanning a 40-nm gold particle relative to the ND probe and its excitation laser. *Source:* Reproduced from ref. (33) / American Chemical Society (d) Optically detected spin resonance spectra for three separate scan pixels, as represented by symbol of crosses in (b) with matching colors. *Source:* (33) Tetienne et al 2016 / American Chemical Society.

measure HeLa cell's intracellular thermal conductivity (37), monitor and regulate *Caenorhabditis elegans* embryo division (38), measure living stem cells' intracellular temperature over an incubation period of several days (39), and monitor the thermogenic reaction of *C. elegans* worms to chemical stimuli of mitochondrial uncouplers in real time (29). As shown in Table 9.1, NV thermometry has poorer sensitivity in biosystems than under normal experiment settings (e.g., in just water or buffer solution) (10, 40), and the biosystem's temperature measurements are mainly based on ODMR line shift.

Kucsko *et al.* injected GNPs and ~100-nm FNDs into human embryonic fibroblast cells by cells via nanowire-assisted delivery (10). GNPs were used for heating, while FNDs (NV_1 and NV_2) were used for temperature sensing. The temperature fluctuations of two FND particles were monitored within the same cell, while an adjacent GNP was heated locally by a separate laser (Figure 9.5a). NV particles positioned close to the heat source showed a strong temperature dependence on the laser power (Figure 9.5b and 9.5c). They recorded a temperature shift of 0.5 K at the FND position and a change of ~10 K at the GNP region when using a laser power of 12 μW, and especially, the cell was still alive. The death ratio of cells increases alongside increases in laser power. This approach could thereby modulate cell viability at the nanoscale, allowing for single cell optimization of nanoparticle-based photothermal treatment.

Even though there is no doubt about NV center's capacity to be used for intracellular thermometric techniques, there have been only a few publications that have monitored the intracellular temperature of stem cells over a long period of

Table 9.1 NV-center-based nanothermometry in biosystems.

Methods	Biological model	Diamond size (technique)/ others	Temp. control	Measurable range	Sensitivity	Ref.
ODMR peak shift	Human embryonic fibroblast WS1 cells	NDs 100 nm; 500 NVs/ ND	Local heating: gold nanoparticle heated by green laser in cell	$\Delta T \sim 20$ K	9 mK $Hz^{-1/2}$ ultrapure bulk diamond 200 ml $Hz^{1/2}$	(10)
ODMR peak shift	HeLa cells	ND-GNRs size ~ 140 nm; (n.a.)	Local heating: GNRs heated by (near-IR) laser	$\Delta T \sim 20$ K	NA	(34)
ODMR peak shift	Mouse neurons	Diamond on fiber (diamond size 20–30 μm); (n.a.)	Global heating: IR laser-induced heating	22–37 °C	NA (accuracy of 0.1 °C)	(35)
ODMR peak shift	Mouse primary cortical neurons	NDs size ~ 170 nm (measured by Dynamic light scattering [DLS] in water); 500 NVs per diamond particles	Global heating incubator with temperature control	35.2–37.3 °C	1 K Hz–1/2	(36)
ZPL positions	Membrane nanotubes in human embryonic kidney cells	ND@GNRs size ~ 100 nm (DLS); 900 NVs per diamond particles	Local heating: GNRs heated by 594 nm laser	28–75 °C	2 °C Hz–1/2	(41)
ODMR peak shift	HeLa cells	NDs size – n.a.	Global heating: sample stage with temperature control	27–37 °C	NA (accuracy better than 0.1 °C)	(12)
ODMR peak shift	C. elegans embryos	NDs size ~ 50 nm – (n.a.)	Local heating: IR laser-induced heating	$\Delta T \sim 20$ K	In vivo sensitivity of ~2 K Hz–1/2	(38)

(Continued)

Table 9.1 (Continued)

Methods	Biological model	Diamond size (technique)/ others	Temp. control	Measurable range	Sensitivity	Ref.
ODMR peak shift	C. elegans worms	NDs size ~ 168 nm (DLS in water); 900 NVs per diamond particles	Local heating: chemical stimuli of mitochondrial uncoupler, i.e., FCCP	$\Delta T \sim 4–7\ °C$	1.4 °C Hz–1/2 (in vivo)	(39)
ODMR peak shift	Adipose tissue-derived stem cells	150 ± 0.8 nm (range: 80 to 300 nm) (DLS in water); 500 NVs per diamond particles	Local heating: intracellular temperature change	29.1–40.9 °C	2.2 °C Hz–1/2	(39)
ODMR peak shift	Hela cells	ND@PDA size ~ 100 nm; (DLS in water); 500 NVs per diamond particles	Local heating PDA heated by a 532 nm laser	$\Delta T \sim 5$ K	NA (accuracy of 1 K)	(37)
ODMR peak shift	HeLa cells	NDs size ~ 100 nm; 500 NVs per diamond particles ~	A continuous-wave 532-nm laser was used for optical excitation		1.2 K Hz–1/2	(15)
ODMR peak shift	HeLa cells	NDs size ~ 35 nm (DLS)	Local heating 532 nm excitation with a power of 110 μW in front of the objective	$\Delta T \sim 30\ °C$	1.2 mK Hz–1/2	(11)
ZPL position	HeLa cells	NDs size: 30; 70; 100 nm; (n.a.)	Global heating: F-3030 temperature bath	25–60 °C	From 5.73% ∘C−1 to 6.994% ∘C−1 (NV0) and from 4.14% ∘C−1 to 6.475% ∘C−1 (NV^-)	(42)

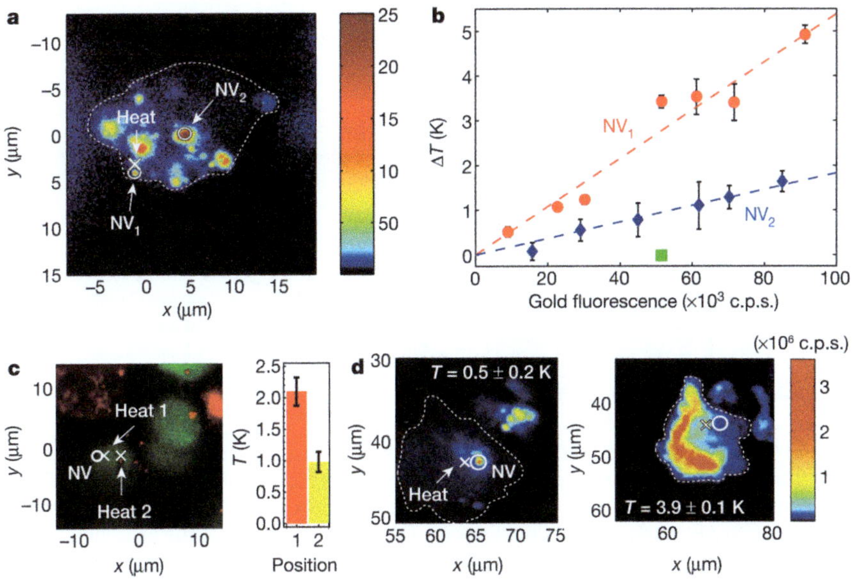

Figure 9.5 (a) Confocal scan of single cell is obtained by using 532-nm laser stimulation and collected at wavelengths longer than 638 nm. Crosses signify the position of GNPs and circle denotes the location of FNDs. The dotted line directs the viewer's attention and defines the cell membrane. Color bars show the fluorescence in the unit of counts per seconds. (b) Temperature shift measured at two centers as a function of laser power given to the GNPs. Dashed lines are linear fits to the data. Each point is made up of an average of five measurements, each lasting for 4 seconds. The error bars (1σ) are caused by fluctuations in GNPs laser heating. (c) Fluorescence scan of stained cells (live/dead assay) with excitation at 494 or 528 nm and emission at 515 nm (green; live cells) and 617 nm (red; dead cells). The bar graph depicts the temperature of a single ND (circle) with local heat applied at two different locations (crosses). (d) Individual cell confocal fluorescence images with variable light strength. The excitation wavelength is 532 nm and the collection wavelength is 630 nm. *Source:* Reproduced from ref. (10) / Springer Nature.

time. Yukawa *et al.* developed a unique quantum thermometric sensing and analysis system (QTAS) based on FNDs, as shown in Figure 9.6, to monitor the intracellular temperature of stem cells for several days (39). They also explored the effect of intracellular temperature on the abilities of adipose tissue-derived stem cells (ASCs) including growth factor synthesis and differentiation. ASCs were efficiently labeled by FNDs after 24 h of incubation. Over 95% of the cells were still alive even though FND concentration was in excess of 500 μg ml^{-1}. However, > 75% of cells were dead if incubated with an FND concentration of 750 μg ml^{-1}. The FNDs' peak was verified at roughly 2.87 GHz with typical NV centers' fluorescence spectra that could be observed by ODMR. A mean temperature dependence of $-(77.6 \pm 11.0)$ kHz $°C^{-1}$ was determined, and intracellular temperatures ranging from 24.0 to 40.9 °C could be accurately detected

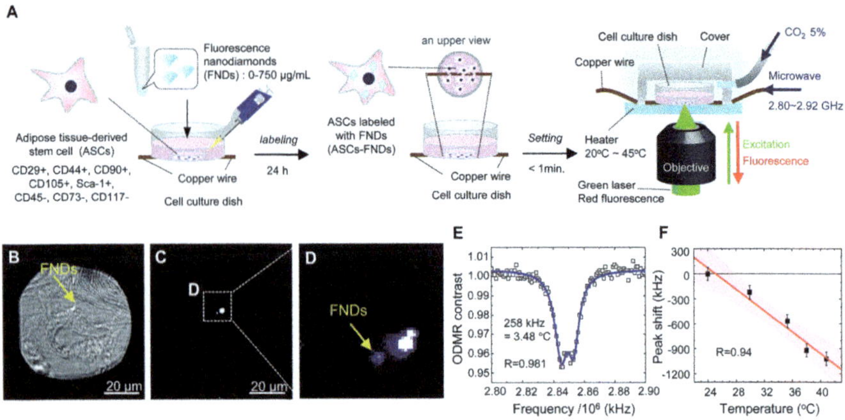

Figure 9.6 (A) FND-based intracellular temperature measurement of ASCs using the QTAS. Images of (B) living ASCs–FNDs and (C) FND fluorescence alone (white-light off). (D) Confocal fluorescence scanning image of the small box in (C). *Source:* Reproduced from ref. (39) / Royal Society of Chemistry (E) The ODMR spectrum of the FND indicated by a yellow arrow in (D) at 40.8 °C with fitted functions. (F) The observed peak shifts as a function of the dish temperatures. The solid line and the associated shaded area are the linear fit and 95% confidence interval. *Source:* (39) Yukawa et al 2020 / Royal Society of Chemistry.

using the QTAS for several days of culture. Furthermore, the effect of intracellular temperature on growth factor synthesis and differentiation into adipocytes and osteocytes was elucidated. This research highlighted the potential usefulness of QTAS for determining the link between intracellular temperature and cellular activities.

Improving NV thermometry sensitivity in biosystems is one of the most important issues. Wang *et al.* combined copper–nickel magnetic nanoparticle (MNP) alloy with NV centers to produce a material able to enhance the thermometry sensitivity under ambient conditions down to 11 mK $Hz^{-1/2}$ (43). This was accomplished by transforming temperature measurements into a more sensitive magnetic field using NV centers, particularly around the ferromagnetic–paramagnetic transition temperature, which is accompanied by massive magnetization shifts. In recent research, Liu *et al.* further improved sensitivity to 76 μK $Hz^{-1/2}$ by combining a single NV center in a diamond nanopillar with a single MNP (14). Adopting this method, Zhang *et al.* (44) created a hybrid nanothermometer by using temperature-responsive hydrogels as "spacers" between NDs (with NV centers) and MNP. When temperature changed, this hydrogel could experience the volume phase transition, resulting in a dramatic variation in distance between the MNPs and the NDs. However, owing to the inevitable breakdown of MNPs under complicated biological conditions, the entire sensor unit is quite intricate and may be challenging to deal with (45).

In summary, NV centers in NDs have emerged as important and promising nanoscale thermometers for myriad biological applications. With unprecedented precision and sensitivity, NV center-driven temperature sensing techniques could soon become a reality in our daily lives. Serving as a diverse platform for cellular/intracellular sensing and imaging, the aforementioned examples show NV center's promise in practical applications.

Despite the positive qualities of NV centers, diamond nanothermometry is not without its challenges. Temperature sensor size is an ongoing issue with the technology. Sub-20-nm FNDs are desirable for sustained fluorescence and detectable ZPL and ODMR signals within intracellular organelles. When FNDs are sized at equal or smaller than 20 nm, the average number of NV centers per particle approaches 1, and the fluorescence quantum yield drops by 80%. Surface modification, size, shape, crystal strain, purity concentration, and the number of NV centers in nanocrystals require additional study. Through additional technological advancements, a future "all-in-one" NV-driven treatment with nanoscale tracking, heating, and sensing capabilities is expected to be produced.

References

1 Albers AE, Chan EM, McBride PM, Ajo-Franklin CM, Cohen BE, Helms BA. Dual-emitting quantum dot/quantum rod-based nanothermometers with enhanced response and sensitivity in live cells. *J Am Chem Soc.* 2012;134(23):9565–8.

2 Brites CD, Lima PP, Silva NJ, Millán A, Amaral VS, Palacio F, *et al.* Thermometry at the nanoscale. *Nanoscale.* 2012;4(16):4799–829.

3 Donner JS, Thompson SA, Kreuzer MP, Baffou G, Quidant R. Mapping intracellular temperature using green fluorescent protein. *Nano Lett.* 2012;12(4):2107–11.

4 Maestro LM, Jacinto C, Silva UR, Vetrone F, Capobianco JA, Jaque D, *et al.* CdTe quantum dots as nanothermometers: Towards highly sensitive thermal imaging. *Small.* 2011;7(13):1774–8.

5 Shah JJ, Gaitan M, Geist J. Generalized temperature measurement equations for rhodamine B dye solution and its application to microfluidics. *Anal Chem.* 2009;81(19):8260–3.

6 Arai S, Lee S-C, Zhai D, Suzuki M, Chang YT. A molecular fluorescent probe for targeted visualization of temperature at the endoplasmic reticulum. *Sci Rep.* 2014;4(1):1–6.

7 Kumar R, Roy I, Ohulchanskyy TY, Goswami LN, Bonoiu AC, Bergey EJ, *et al.* Covalently dye-linked, surface-controlled, and bioconjugated organically modified silica nanoparticles as targeted probes for optical imaging. *ACS Nano.* 2008;2(3):449–56.

8 Uchiyama S, Tsuji T, Ikado K, Yoshida A, Kawamoto K, Hayashi T, *et al.* A cationic fluorescent polymeric thermometer for the ratiometric sensing of intracellular temperature. *Analyst.* 2015;140(13):4498–506.

9 Zhegalova NG, Aydt A, Wang ST, Berezin MY, editors. *Molecular thermometers for potential applications in thermal ablation procedures. Proc. SPIE 8596, Reporters, Markers, Dyes, Nanoparticles, and Molecular Probes for Biomedical Applications V.* International Society for Optics and Photonics; 2013:85960I.

10 Kucsko G, Maurer PC, Yao NY, Kubo M, Noh HJ, Lo PK, *et al.* Nanometre-scale thermometry in a living cell. *Nature.* 2013;500(7460):54–U71.

11 Wu Y, Alam MNA, Balasubramanian P, Ermakova A, Fischer S, Barth H, *et al.* Nanodiamond theranostic for light-controlled intracellular heating and nanoscale temperature sensing. *Nano Lett.* 2021;21(9):3780–8.

12 Sekiguchi T, Sotoma S, Harada Y. Fluorescent nanodiamonds as a robust temperature sensor inside a single cell. *Biophys Physicobiology.* 2018;15:229–34.

13 Neumann P, Jakobi I, Dolde F, Burk C, Reuter R, Waldherr G, *et al.* High-precision nanoscale temperature sensing using single defects in diamond. *Nano Lett.* 2013;13(6):2738–42.

14 Liu C-F, Leong W-H, Xia K, Feng X, Finkler A, Denisenko A, *et al.* Ultra-sensitive hybrid diamond nanothermometer. *Natl Sci Rev.* 2021;8(5):nwaa194.

15 Nishimura Y, Oshimi K, Umehara Y, Kumon Y, Miyaji K, Yukawa H, *et al.* Wide-field fluorescent nanodiamond spin measurements toward real-time large-area intracellular thermometry. *Sci Rep.* 2021;11(1):1–12.

16 Zhang S-C, Dong Y, Du B, Lin H-B, Li S, Zhu W, *et al.* A robust fiber-based quantum thermometer coupled with nitrogen-vacancy centers. *Rev Sci Instrum.* 2021;92(4):044904.

17 Singam SK, Nesladek M, Goovaerts E. Nitrogen-vacancy nanodiamond based local thermometry using frequency-jump modulation. *Nanotechnology.* 2019;31(10):105501.

18 Childress L, Dutt MG, Taylor J, Zibrov A, Jelezko F, Wrachtrup J, *et al.* Coherent dynamics of coupled electron and nuclear spin qubits in diamond. *Science.* 2006;314(5797):281–5.

19 Gruber A, Dräbenstedt A, Tietz C, Fleury L, Wrachtrup J, Borczyskowski Cv. Scanning confocal optical microscopy and magnetic resonance on single defect centers. *Science.* 1997;276(5321):2012.

20 Grinolds M, Warner M, De Greve K, Dovzhenko Y, Thiel L, Walsworth RL, *et al.* Subnanometre resolution in three-dimensional magnetic resonance imaging of individual dark spins. *Nat Nanotechnol.* 2014;9(4):279–84.

21 Staudacher T, Shi F, Pezzagna S, Meijer J, Du J, Meriles CA, *et al.* Nuclear magnetic resonance spectroscopy on a (5-nanometer) 3 sample volume. *Science.* 2013;339(6119):561–3.

22 Mamin H, Kim M, Sherwood M, Rettner C, Ohno K, Awschalom D, *et al.* Nanoscale nuclear magnetic resonance with a nitrogen-vacancy spin sensor. *Science*. 2013;339(6119):557–60.

23 Hsiao WW-W, Le T-N, Chang H-C. Applications of fluorescent nanodiamond in biology. In: Meyers, R.A.(Ed.), *Encyclopedia of Analytical Chemistry*. Hoboken, NJ, USA: John Wiley & Sons; 2022. pp. 1–43. https://doi.org/10.1002/9780470027318.a9776.

24 Acosta VM, Bauch E, Ledbetter MP, Waxman A, Bouchard LS, Budker D. Temperature dependence of the nitrogen-vacancy magnetic resonance in diamond. *Phys Rev Lett*. 2010;104(7): 070801.

25 Toyli DM, Christle DJ, Alkauskas A, Buckley BB, Van de Walle CG, Awschalom DD. Measurement and control of single nitrogen-vacancy center spins above 600 K. *Phys Rev X*. 2012;2(3): 031001.

26 Plakhotnik T, Gruber D. Luminescence of nitrogen-vacancy centers in nanodiamonds at temperatures between 300 and 700 K: Perspectives on nanothermometry. *Phys Chem Chem Phys*. 2010;12(33):9751–6.

27 Plakhotnik T, Doherty MW, Cole JH, Chapman R, Manson NB. All-optical thermometry and thermal properties of the optically detected spin resonances of the NV- center in nanodiamond. *Nano Lett*. 2014;14(9):4989–96.

28 Tzeng YK, Tsai PC, Liu HY, Chen OY, Hsu H, Yee FG, *et al.* Time-resolved luminescence nanothermometry with nitrogen-vacancy centers in nanodiamonds. *Nano Lett*. 2015;15(6):3945–52.

29 Fujiwara M, Dohms A, Suto K, Nishimura Y, Oshimi K, Teki Y, *et al.* Real-time estimation of the optically detected magnetic resonance shift in diamond quantum thermometry toward biological applications. *Phys Rev Res*. 2020;2(4):043415.

30 Hui YY, Chen OY, Azuma T, Chang B-M, Hsieh F-J, Chang H-C. All-optical thermometry with nitrogen-vacancy centers in nanodiamond-embedded polymer films. *J Phys Chem C*. 2019;123(24):15366–74.

31 Chen XD, Dong CH, Sun FW, Zou CL, Cui JM, Han ZF, *et al.* Temperature dependent energy level shifts of nitrogen-vacancy centers in diamond. *Appl Phys Lett*. 2011;99(16): 161903.

32 Fukami M, Yale C, Andrich P, Liu X, Heremans F, Nealey P, *et al.* All-optical cryogenic thermometry based on nitrogen-vacancy centers in nanodiamonds. *Phys Rev Appl*. 2019;12(1):014042.

33 Tetienne J-P, Lombard A, Simpson DA, Ritchie C, Lu J, Mulvaney P, *et al.* Scanning nanospin ensemble microscope for nanoscale magnetic and thermal imaging. *Nano Lett*. 2016;16(1):326–33.

34 Tsai P-C, Chen OY, Tzeng Y-K, Hui YY, Guo JY, Wu C-C, *et al.* Gold/diamond nanohybrids for quantum sensing applications. *EPJ Quantum Technol*. 2015;2(1):1–12.

35 Lanin A, Fedotov I, Ermakova YG, Sidorov-Biryukov D, Fedotov A, Hemmer P, *et al.* Fiber-optic electron-spin-resonance thermometry of single laser-activated neurons. *Opt Lett.* 2016;41(23):5563–6.

36 Simpson DA, Morrisroe E, McCoey JM, Lombard AH, Mendis DC, Treussart F, *et al.* Non-neurotoxic nanodiamond probes for intraneuronal temperature mapping. *ACS Nano.* 2017;11(12):12077–86.

37 Sotoma S, Zhong C, Kah JCY, Yamashita H, Plakhotnik T, Harada Y, *et al.* In situ measurements of intracellular thermal conductivity using heater-thermometer hybrid diamond nanosensors. *Sci Adv.* 2021;7(3):eabd7888.

38 Choi J, Zhou H, Landig R, Wu H-Y, Yu X, Von Stetina SE, *et al.* Probing and manipulating embryogenesis via nanoscale thermometry and temperature control. *Proc Natl Acad Sci USA.* 2020;117(26):14636–41.

39 Yukawa H, Fujiwara M, Kobayashi K, Kumon Y, Miyaji K, Nishimura Y, *et al.* A quantum thermometric sensing and analysis system using fluorescent nanodiamonds for the evaluation of living stem cell functions according to intracellular temperature. *Nanoscale Adv.* 2020;2(5):1859–68.

40 Alkahtani MH, Alghannam F, Jiang L, Almethen A, Rampersaud AA, Brick R, *et al.* Fluorescent nanodiamonds: Past, present, and future. *Nanophotonics.* 2018;7(8):1423–53.

41 Tsai P-C, Epperla CP, Huang J-S, Chen OY, Wu C-C, Chang H-C. Measuring nanoscale thermostability of cell membranes with single gold–diamond nanohybrids. *Angew Chem Int Ed.* 2017;56(11):3025–30.

42 Pedroza-Montero F, Santacruz-Gómez K, Acosta-Elías M, Silva-Campa E, Meza-Figueroa D, Soto-Puebla D, *et al.* Thermometric characterization of fluorescent nanodiamonds suitable for biomedical applications. *Appl Sci.* 2021;11(9):4065.

43 Wang N, Liu G-Q, Leong W-H, Zeng H, Feng X, Li S-H, *et al.* Magnetic criticality enhanced hybrid nanodiamond thermometer under ambient conditions. *Phys Rev X.* 2018;8(1):011042.

44 Zhang T, Liu G-Q, Leong W-H, Liu C-F, Kwok M-H, Ngai T, *et al.* Hybrid nanodiamond quantum sensors enabled by volume phase transitions of hydrogels. *Nat Commun.* 2018;9(1):1–8.

45 Mazuel F, Espinosa A, Luciani N, Reffay M, Le Borgne R, Motte L, *et al.* Massive intracellular biodegradation of iron oxide nanoparticles evidenced magnetically at single-endosome and tissue levels. *ACS Nano.* 2016;10(8):7627–38.

10

Nanodiamond-Enabled Drug Delivery

Stefanny Angela, Raymond Hsin, Steven Che-Wei Lu, Trong-Nghia Le, and Wesley Wei-Wen Hsiao**

* *Corresponding authors*

Drug delivery is a multidisciplinary area that includes chemistry and biology. A biomolecule carrier must possess many critical features for a suitable medication delivery mechanism. The most important requirement is that the molecule's loading capacity corresponds to the carrier's weight. It must also allow for sufficient bioactive chemical surface attachment, develop a working mechanism for targeted release, and be easily monitored utilizing noninvasive, highly sensitive techniques.

Nanodiamonds (NDs) satisfy all these requirements as follows: they have a large specific surface area, are biocompatible, have a simple bioconjugation process, and have distinctive optical features. ND is a suitable carrier because of its large loading capacity and ability to maintain and keep the natural therapeutic benefits of the attached substances. NDs' diverse surface capabilities also allow for innovative delivery strategies, such as regulated and long-term release. In some experiments, ND surfaces have been used to attach medications chemically. However, most current research has focused on physical adsorption approaches. NDs have been used to transport a variety of compounds, such as genes, small molecules, peptides, and proteins. This chapter discusses recent applications of NDs in drug delivery systems. We also discuss how optical imaging techniques, such as using radiation-damaged NDs as carriers, can be utilized to monitor or sense medicine distribution.

Nanodiamonds in Analytical and Biological Sciences: Principles and Applications, First Edition.
Edited by Wesley Wei-Wen Hsiao and Trong-Nghia Le.
© 2023 John Wiley & Sons Ltd. Published 2023 by John Wiley & Sons Ltd.

10.1 Introduction

Drug development is increasingly incorporating nanotechnology. Nanoparticle (NP)-based treatments can bypass biological boundaries, transport hydrophobic medicines and biological products, and target illness areas more precisely. Many NPs have been investigated as carriers in methods for delivering and monitoring drugs. Lipid-based vehicles (including micelles, solid lipid NPs, and liposomes), inorganic NPs (metal oxides and silica), polymer-carriers (dendrimers, hydrogels, polymersomes, and nanofibers), carbon structures (graphenes, fullerenes, NDs, and nanotubes), and metallic and semiconducting NPs (nanogold and quantum dots) are among some of the NPs that have been investigated (1, 2).

The unique physicochemical features of NPs, such as their enormous surface-area-to-mass ratios, quantum characteristics, and capacity to transport chemical and biological substances, make them good candidates for medicinal applications (3). Due to their vast functional surface areas, NPs can bind, adsorb, and transport molecules like medicines, proteins, and genes. Applied nanotechnology holds promise for enhancing our understanding of disease pathophysiology. It can also help provide more advanced diagnostic options and better therapeutics (4).

Nevertheless, due to their tiny size, NPs can be highly reactive. Interaction between NPs, cells' biomolecules, and the surrounding can occasionally set off a chain of unintended or undesirable biological consequences. The NPs' biocompatibility and targeting performance depend on their dynamic properties. Although NPs have several evident benefits, only a few NP-based medicines have been approved for therapeutic use. Scientists face several problems and roadblocks at every level of drug development (5, 6).

NDs are members of the nanocarbon family. As a result, they have significant advantages over other NPs for biological implementation. Detonation nanodiamonds (DNDs) and high-pressure, high-temperature (HPHT) NDs are some of the most commonly used NDs in clinical studies (7). Both are nontoxic and biocompatible (8). NDs inside living cells for an extended period have no effect on their function *in vitro* or *in vivo*. DNDs, with principal particle sizes ranging from 2 to 8 nm, are tiny enough to enter cell membranes noninvasively and cause no harm to treated cells (9). These properties, combined with their steady inert centers, reactive surfaces, and potential to develop hydrogels, mean that these NPs could effectively transport tiny chemical medicines (10).

Many scientists and practitioners in the field of nanomedicine have investigated the surface properties of NDs. Researchers have attached biomolecules (such as peptides, tiny molecules, genes, proteins, and short interfering ribonucleic acid [siRNA]) to NDs and transported the bioconjugated NPs directly to target cells (11, 12). Biochemical activities within the cells are then measured, monitored, and even altered using these biomolecules. These studies have found that ND has the potential to be used both as a tool for tracking (e.g., to detect precancerous alterations) and as a therapeutic tool (e.g., with the carried drugs). By combining numerous critical qualities required for potential treatments, including biological stability and compatibility, and manufacturing scalability, ND has emerged as a viable choice for biomedical activities (13–17).

Drug delivery is a multidisciplinary domain encompassing biochemistry and medicine. As a result, expanding ND research out of the lab and into more clinical settings necessitates a detailed knowledge of contemporary medication and gene delivery technologies and an understanding of how to employ NDs in such disciplines. Despite tremendous advancements, drug delivery tracking using NDs as detectors or biosensors is still in its early stages. This chapter covers the primary features of NDs that will almost certainly play a vital role in establishing enhanced, more personalized drug delivery and monitoring systems.

10.2 Drug Delivery Through Nanodiamonds

Three requirements must be met for a drug delivery system to be effective: an adequate loading ability in relation to the carrier's load, a robust active particle adhering to the surface, and a functioning targeted release mechanism. NDs satisfy all three requirements due to their large specific surface area, high biocompatibility, and ease of biomolecule functionalization. Drugs, genes, and protein molecules bind to ND particles as a result of reactions among functional groups on the NDs' interfaces and the particles of interest (18, 19). Surface groups are chosen for optimum adsorption, and the drug can be conjugated to ND surfaces. Surface groups all meditate on van der Waals forces, electrostatic interactions, and drug molecules with hydrogen bonds. In a solution, most functional groups on an ND are charged; the dissociation degree is determined by the surface group's pKa and the pH of the solution. The creation of carriers for the distribution of tiny biomolecules, focusing on chemotherapeutic drugs, has been the focus of ND drug delivery research in the literature. With a few exceptional cases, this work has focused on the noncovalent attachment of active chemicals to molecules' surfaces, allowing for more straightforward carrier generation.

10.2.1 Delivery of Small Molecules

Ho *et al.* have put much work into NDs' research and how they might be used in drug delivery (20–31). A noncovalent bonding approach has been established for doxorubicin (DOX) functionalization of NDs (2–8 nm) (22). They developed a strategy based on ionic interactions between DOX's positively charged amino group and the negatively charged ND surface groups. The ND-DOX compounds effectively cause DNA fragmentation and cellular mortality, and the DOX-loaded ND hydrogels effectively transport medicine to living cells. The pace of ND-DOX migration into live cells was studied using transmission electron microscopy (TEM) of ND-DOX particles and confocal imaging of NDs with fluorescein isothiocyanate (FITC)-linked poly-L-lysine physically adsorbed on their surface. Using both procedures, they discovered that NDs were instantly attached to the cell interface and progressively absorbed into the cytoplasm. The inherent biocompatibility of NDs has been demonstrated using cytotoxicity testing, reverse transcription polymerase chain reaction (RT-PCR), and DNA fragmentation assays. These cellular and genetic bioassays show that the ND-DOX conjugates cause cell death. NDs are a therapeutically useful drug delivery vehicle for systemic and localized drug administration. This research introduced a comprehensive ND-multicomponent drug delivery method for paclitaxel (PTX) targeting, imaging, and therapy enhancement (31). The heterofunctionalization of NDs was used to create this system by adding fluorescently tagged anti–epidermal growth factor receptor (EGFR) monoclonal antibodies, and PTX-DNA conjugates were applied on the ND's surface (Figure 10.1).

The anticancer medication cis-dichlorodiammineplatinum (II) (CDDP) was effectively loaded into NDs (32). CDDP was loaded in the same way as DOX, with carboxylic groups on the surface and ionic interactions with CDDP's amino groups. Scientists created a pH-responsive release feature for the system to reduce harmful side effects. They discovered that CDDP-ND particles provide a small CDDP concentration in the blood (pH 7.4) but release a considerably larger quantity into acidic lysosomes (pH 6) throughout their circulation (Figure 10.2). ND was also covalently coupled to paclitaxel, which was found to drastically diminish the feasibility of A549 human lung cancer cells (33). In A549 cells, both mitotic arrest and apoptosis are caused by the paclitaxel complex. Furthermore, in xenograft severe combined immunodeficiency disease (SCID) mice, ND-paclitaxel was found to suppress carcinogenesis and the development of lung cancer cells. The anticancer properties of the adsorbed chemotherapeutic agent were preserved, resulting in apoptosis, mitotic inhibition, and anti-tumorigenesis in human lung carcinoma cells.

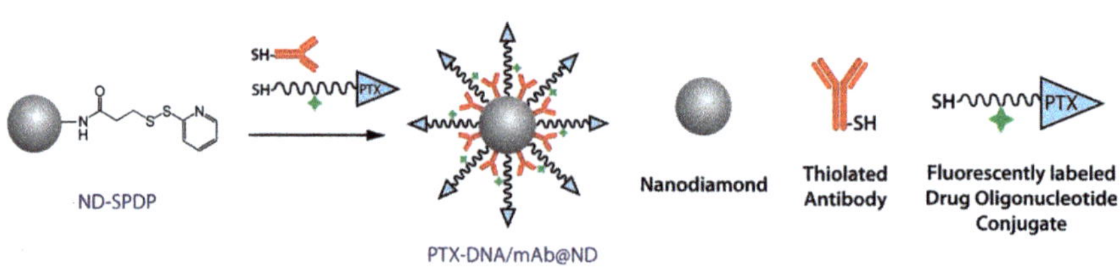

Figure 10.1 The production of PTX-DNA/mAb@ND. *Source:* (31) Zhang et al 2011 / John Wiley & Sons.

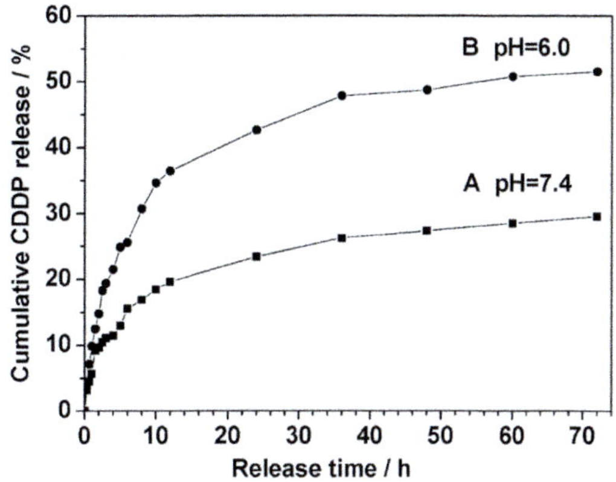

Figure 10.2 CDDP release profile derived from ND-CDDP composites. *Source:* (32) Guan et al 2010 / John Wiley & Sons.

In another investigation, Li *et al.* looked into ND's promise as a delivery system for anticancer drugs (34). Their research found that 10-hydroxycamptothecin (HCPT) can bind to NDs by increasing HCPT's solubility with NaOH, resulting in greater ND internal diffusion. In a low-pH medium, NDs released HCPT for an extended period. Later studies found that HCPT-ND complexes had a substantially stronger chemotherapeutic impact than standalone HCPT, indicating that NDs are a viable multipurpose drug delivery system for cancer treatment.

A unique ND nanocomposite capable of label-free bioimaging was demonstrated by Prabhakar *et al.* These researchers utilized a porous silica coating to increase the water dispersibility of NDs (35). Through literature studies and experiments, Adnan *et al.* showed the impact of pH on the amount of DOX deposited on ND's interface (36). DOX molecules are solely attracted to NDs at basic pH when 10% of the ND's surface area is unclaimed. The quantity of DOX that is bound rises as the environment's ionic strength and basicity rise. These trends are due to a greater electrostatic interaction (37).

Several researchers have also studied PEG-ND preparation for DOX administration and gradual release (38, 39), indicating that these intracellular delivery approaches may deliver DOX to cancer cells. For example, the carboxylic groups of oxidized NDs can be covalently bonded to DOX (40). In addition, the cell-penetrating transactivator of transcription (TAT) peptide, which acts as the HIV TAT protein, is attached to DOX-ND, thus preventing premature release and improving intracellular delivery (Figure 10.3).

The fundamental constraint in the systemic delivery of some medicines is their poor water solubility. To solve this issue, Chen *et al.* explored the water solubilization of water-insoluble drugs when they were coupled with NDs (20). Purvalanol A and 4-hydroxytamoxifen (4–OHT) were chosen as representative medications because they have low water solubility, they are the most widely used pharmaceuticals for

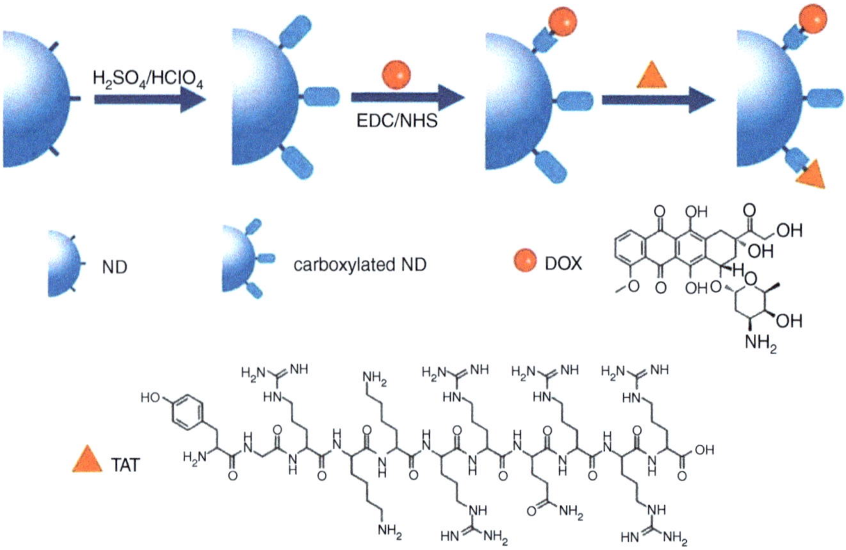

Figure 10.3 The production of TAT-ND-DOX. *Source:* (40) Li XX et al 2011 / Royal Society of Chemistry.

treating breast and liver cancer. These findings suggest that water-insoluble chemicals could be used in treatment-related situations using NDs.

Chemotherapy resistance is the greatest obstacle in cancer treatment. This resistance frequently results in tumor progression and cross-resistance to other chemotherapy medicines. Advancements in nanotechnology have developed scalable and biocompatible nanotherapies to overcome drug resistance and improve anticancer treatment effectiveness. Chow *et al.*, for example, have recently demonstrated that ND-DOX overcomes drug efflux and promotes apoptosis in mouse liver and mammary cancer cell models (41). In addition to halting tumor growth, ND-DOX has much lower toxicity *in vivo* than regular DOX therapy. However, DOX is ejected from tumor cells on a regular basis by a variety of proteins, which is a critical stumbling block to its efficiency. NaOH treatment reversibly binds DOX to ND, releasing the medication slowly in vivo and *in vitro*. Furthermore, compared to DOX alone, the novel ND-DOX system shows enhanced drug retention in tumor cells and increased safety and effectiveness. These findings point to ND-conjugated chemotherapeutics as a possible avenue for overcoming chemoresistance and improving chemotherapy effectiveness and biocompatibility.

Other prospects for NDs exist, including cancer research, where they might identify critical extracellular receptors and signaling cascades. Researchers have tried to design drug delivery systems that improve chemotherapy's efficacy and lower toxicity in patients with aggressive tumors. Breast cancer treatment often includes medications that target specific receptors, including Her2 inhibitory antibodies and particular estrogen receptor enhancers. Nevertheless, the cancer cells must possess the right receptors for these hormone receptor-specific drugs to benefit an individual. Triple-negative breast cancer lacks the progesterone and estrogen receptor, or Her2 protein, all of which are routinely specific in breast cancer therapy (42, 43). To combat this especially difficult kind of malignancy, Moore *et al.* created ND-lipid hybrid materials that were antibody functionalized. In mice with MDAMB-231 (triple-negative breast cancer) cells implanted, the EGFR was targeted by these antibodies, allowing for the delivery of the epirubicin chemotherapeutic simultaneously. Their findings revealed significant increases in treatment effectiveness and reduced chemotherapeutic tolerance (26). A conjugation of NDs with non-covalently bound epirubicin was coated with a lipid double layer, including a modest quantity of biotinylated lipids in the ND–lipid hybrid particles. A streptavidin anchor was used to bind the lipids to biotinylated EGFR-targeting antibodies. The ND–lipid hybrid particles' biocompatibility, scalability, and stability were excellent, indicating their bright future in chemotherapy administration.

Medication-eluting contact lenses allow for quick and precise drug administration to the eye. Researchers studying glaucoma, a condition that leads to the loss of vision in millions of individuals worldwide, have been using drug-eluting contact lenses for a long time as a therapy option. Although when used with eye drops, these drug-soaked lenses allow for local therapy, their value is limited since they rely on burst release rather than continuous therapeutic administration. Furthermore, most soft contact lenses require moist storage, contributing to premature drug elution. To address these shortcomings, Kim *et al.* created an ND-embedded contact lens that can deliver timolol maleate, an antioxidative glaucoma medicine, via lysozyme-triggered release, avoiding the need for recurrent dosing (30). ND nanogel-embedded lenses retain their optical clarity, water content, and oxygen permeability. Despite this, the mechanical strength of these lenses improved, indicating their clinical and commercial promise.

Glioblastoma is one of the most prevalent and deadliest malignant brain tumors. However, no effective therapy for this tumor type has been developed. This is largely due to difficulties related to breaching the blood-brain barrier (BBB) and retaining standard medicines. To help combat this illness, researchers have turned to convection-enhanced delivery approaches involving the injection of NDs adsorbed with DOX (28). In convection-enhanced delivery, drugs are administered using catheters implanted stereotactically into the tumor mass, around the tumor, or in the resection cavity. In both the glioma cell line and normal rat parenchyma, conjugation with NDs effectively increased DOX uptake and retention (Figure 10.4). Moreover, conjugation with NDs localized toxicity and greatly increased the lethal effectiveness of DOX on brain tumors via convection-enhanced transport.

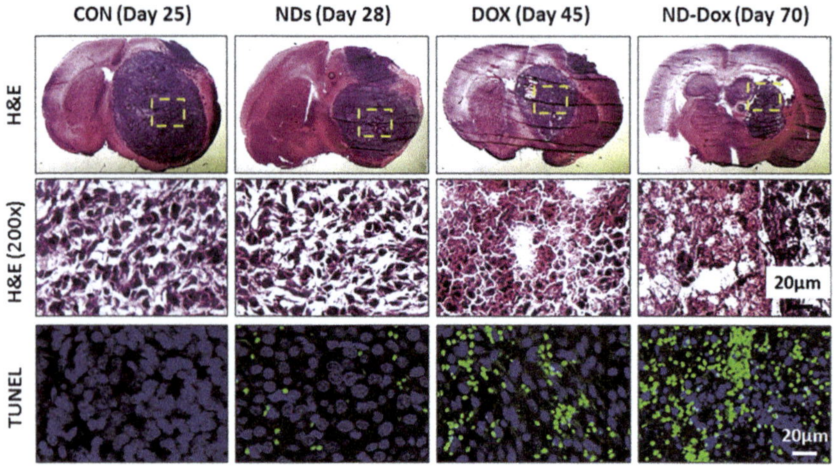

Figure 10.4 Through convection-enhanced administration, ND-DOX was able to improve the efficiency of glioblastoma therapeutics significantly. *Source:* Reproduced from ref. (28) / Elsevier.

Efavirenz (EFV) is a very effective non-nucleoside reverse transcriptase inhibitor (NNRTI) commonly used for the treatment of HIV-1 as part of combination antiretroviral therapy (cART). Due to its poor pharmacokinetics, low bioavailability, and eventual drug resistance, therapeutic delivery agents must be modified to make EFV more focused and efficacious. Of the various nanocarbon materials studied for use as drug carriers, ND holds the most promise due to its biocompatibility, inertness, and ease of accessibility. *In vitro*, researchers investigated the anti-HIV-1 performance of ND-EFV by administering the composition to HIV-1-infected primary human macrophages (44). Initially, human peripheral blood mononuclear cells (PBMC) were isolated and grown to differentiate them from macrophages. After differentiation, cells were infected with HIV, and free drug (FD) and ND-EFV were given to HIV-1-infected macrophages separately. The presence of HIV-1 p24 in the supernatant of drug-treated cells revealed a significant difference in viral suppression between unformulated EFV and ND-EFV. Over a seven-day period, the effect of ND-EFV on HIV-1 replication was both practical and long-lasting. On the other hand, the unformulated EFV was only able to suppress viral replication up to day five, after which p24 levels reverted to untreated groups. Positive controls included infected and untreated macrophages. As demonstrated in Figure 10.5, in comparison to unformulated EFV, ND-EFV was far more successful in preventing HIV-1 replication. It was also shown that employing ND-EFV for targeted drug delivery to the central nervous system (CNS) has potential prospects. Despite this, the current investigation found no

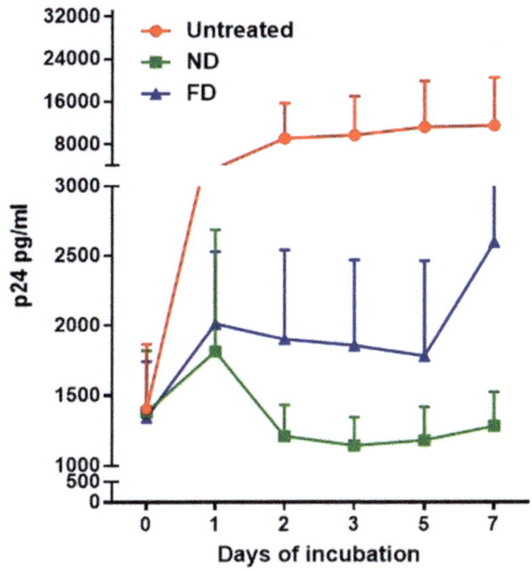

Figure 10.5 Therapeutic ND-EFV's effectiveness in HIV-1-infected macrophages. Three distinct circumstances were used to track HIV replication. Positive control cells were HIV-infected cells that were not treated (red line). Unformulated 40 g/ml EFV was used to treat HIV-infected cells as a control (in blue). The cells treated with 40 g/ml ND-EFV were used as a control (green line). The p24 level was measured at various time intervals in each of the three treatment groups to determine the impact of therapy on HIV replication. In this situation, day 0 was defined as 24 hours after infection. Statistical significance was determined using p values ($p < 0.0001$). *Source:* (44) Roy et al 2018 / Springer Nature.

evidence of ND-EFV being delivered directly to the CNS. As a result, this formulation will be employed in the future for direct anti-HIV-1 medicine delivery to the CNS, combined with a particular targeting agent (ligand, antibody, etc.).

10.2.2 Delivery of Peptides and Proteins

In addition to carrying small compounds, ND can also help release therapeutic proteins. To do this, charged molecules like proteins are deposited on an ND. Anionic end groups ($-COO^-$) and protonated amino groups ($-NH_3^+$) on ND's surface create favorable circumstances for a charge–charge interaction with ND. NDs form hydrogen bonds with these biomolecules, increasing adsorption. A study by Shimkunas *et al.* investigated whether NDs could be used as protein delivery vehicles (45). They examined insulin as a possible stimulator of wound repair and vascularization in severe burns and perhaps other disorders. Researchers created a highly effective insulin noncovalently conjugated on the ND's interface through physical adsorption. Insulin was released when the ND–insulin combination was introduced to a basic environment. Using imaging techniques and adsorption/desorption experiments, they demonstrated NDs' ability to bind and release insulin effectively. After desorption, the protein's function was preserved in cytotoxicity and RT-PCR studies. However, adsorbed peptides and proteins ceased to function. This finding suggests that ND could play an essential role in insulin delivery.

ND's use does not hamper protein activity in the administration of drugs. NDs have been employed in the administration of transforming growth factor beta (TGF-beta) antibody, which, in addition to insulin therapy, can be used as an anti-scarring agent (46). Incubation in a serum-containing medium causes the antibodies to liberate, while ND-antibody compounds are stable in water. The retention of protein activity following release was confirmed by ELISA. ND-conjugated protein molecules' biological activity has also been investigated by Nguyen *et al.* (47). They discovered that lysozyme retains its hydrolytic activity after adsorption onto NDs (100 nm). At pH 5, the relative hydrolytic activity of ND-adsorbed lysozyme is 60%. Furthermore, Liu *et al.* discovered that alpha-bungarotoxin, a neurotoxin derived from *Bungarus multicinctus*, maintains its bioactivity after ND by inhibiting the membrane proteinα-7-nicotinic acetylcholine (α-7-nAchR) (48). Bovine serum albumin (BSA) was another protein successfully loaded on NDs (49).

Taking this work a step further, Moore *et al.* developed new NP solutions that might be employed as injectable alternatives for BMP-2 and modest doses of basic-fibroblast (b-FGF) growth aspect in oral surgery (50). These researchers examined the efficacy of ND suspensions for the simultaneous and targeted delivery of BMP-2 and b-FGF and their capacity to accelerate localized bone formation. BMP-2 and b-FGF can be quickly loaded onto ND clusters because of their strong adsorption capabilities via physisorption. Using Fourier transform infrared spectroscopy to confirm the loading and enzyme-linked immunosorbent assay (ELISA) to quantify the uptake of b-FGF and BMP-2 onto the ND's interface, Moore *et al.* effectively established that NDs could simultaneously deliver two functional proteins (BMP-2 and b-FGF) essential for bone repair *in vivo*. The findings of these investigations all point to one conclusion: ND has excellent potential as a protein delivery vehicle because of its capacity to maintain the bioactivity of coupled proteins or peptides.

Nuclear-targeting treatment is thought to be a viable illness-treatment technique. However, developing a delivery system that is both safe and nucleus-permeable has proven to be a challenging endeavor. In aqueous solutions, Leung *et al.* utilized 30-nm NDs loaded with dual-function cationic peptides, including the HIV TAT protein and a nuclear localization signal (NLS) peptide, to create a nuclear-targeted delivery platform (51). Confocal fluorescence investigations were used to evaluate the intracellular localization of modified NDs. Nitrogen-vacancy (NV) centers in NDs generate red light when activated by visible light sources, such as a 633-nm laser, with a maximal fluorescence emission at 700 nm. In this study, NDs demonstrated effective cellular uptake after being treated with TAT-NLS-NDs coated with Cy3-labeled DNA4625 for 24 hours. In addition, since the ND signals (red) overlapped with nucleus staining, Hoechst (blue) in HeLa, and MCF-7 cells, most of them were found in the nucleus. Researchers also observed a Cy3 fluorophore signal inside the nuclei that completely overlapped with the ND and Hoechst fluorescence traces (Figure 10.6a and 10.6b). Indeed, living cells might transfer the laden cargos to the nucleus using this TAT-NLS-ND platform. However, following cellular absorption in HeLa and MCF-7 cells, the bulk of TAT-NDs were unevenly scattered in the cytoplasm and did not demonstrate nuclear-targeting capabilities (Figure 10.6c–10.6e). MCF-7 and HeLa cells rarely took up the uncoated carboxyl-NDs (Figure 10.6f and 10.6g). These findings show the TAT-NLS-ND systems' ability to transport mi RNA, siRNA, plasmid DNA, and other negatively charged cargos to the nucleus. However, due to NDs' wide size distribution (polydispersityindex1), the number of NDs entering the nucleus is minimal. Thus, only TAT-NLS-NDs of smaller sizes can pass through nucleus membrane pores without agglomeration and enter the nucleus. TAT-NLS-ND has been confirmed to be an effective, safe, and simple carrier, allowing for higher cellular absorption than carboxyl-ND. Furthermore, this compound has nucleus-targeting capabilities in a variety of cell types, using fluorescence colocalization assays, flow cytometry, and 3-[4,5-dimethylthiazol-2-yl]-2,5-diphenyltetrazolium bromide (MTT) investigations (51).

To improve the intracellular delivery of therapeutic peptides, Gu *et al.* used ND to create a stimulus-responsive platform for drug delivery (52). This intracellular ND–peptide drug delivery platform was used to deliver SALL4 inhibitory peptides against hepatocellular carcinoma (HCC) caused by the SALL4 oncogene and found that ND–peptide complex formations enhanced the stability of peptides as well as cellular and tumor retention. Compared with conventional therapeutic peptides, the ND--peptide combination demonstrated better therapeutic efficacy against SALL4-driven HCC due to improved stability and retention (Figure 10.7). While the ND–peptide complex's natural homing makes it well-suited for HCC, more research is needed to determine if its efficacy is generalizable to other cancers. One study using typical chemotherapy medications in breast cancer demonstrated that this was feasible and that incorporating an active tumor-targeting motif improved results even more (26, 53). When taken together, this research shows that intracellular ND-specific peptide drug delivery devices can potentially increase the clinical translation of therapeutic peptides to their successful adoption in cancer therapy.

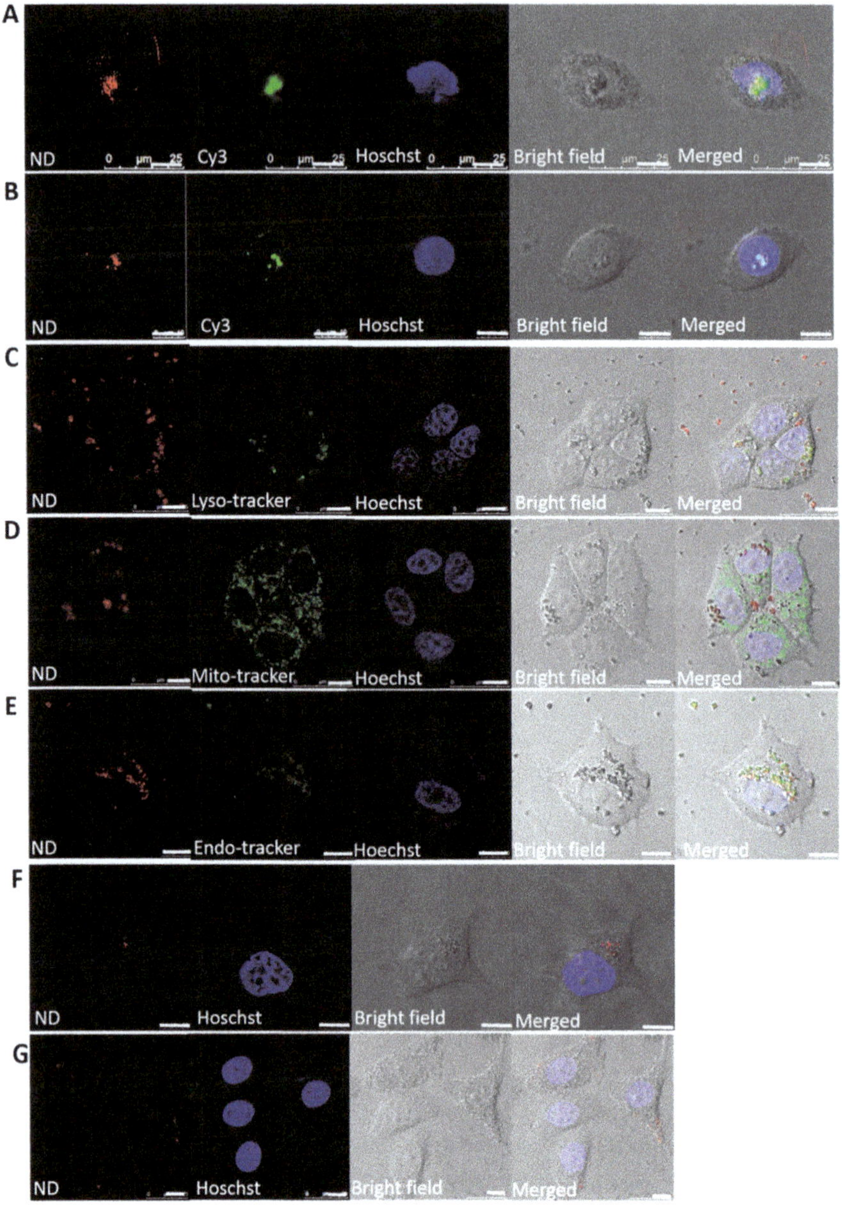

Figure 10.6 DNA4625-Cy3-TAT-NLS-NDs' confocal fluorescence images in (a) MCF-7 and (b) HeLa cells stained with Hoechst. DNA4625-TAT-NDs in MCF-7 cells stained with (c) lyso-tracker, (d) mitoracker, and (e) endotracker in confocal fluorescence pictures. Carboxyl-NDs' confocal fluorescence pictures in (f) MCF-7 and (g) HeLa cells. The scale bar is 10 meters. *Source:* Reproduced from ref. (51) / American Chemical Society.

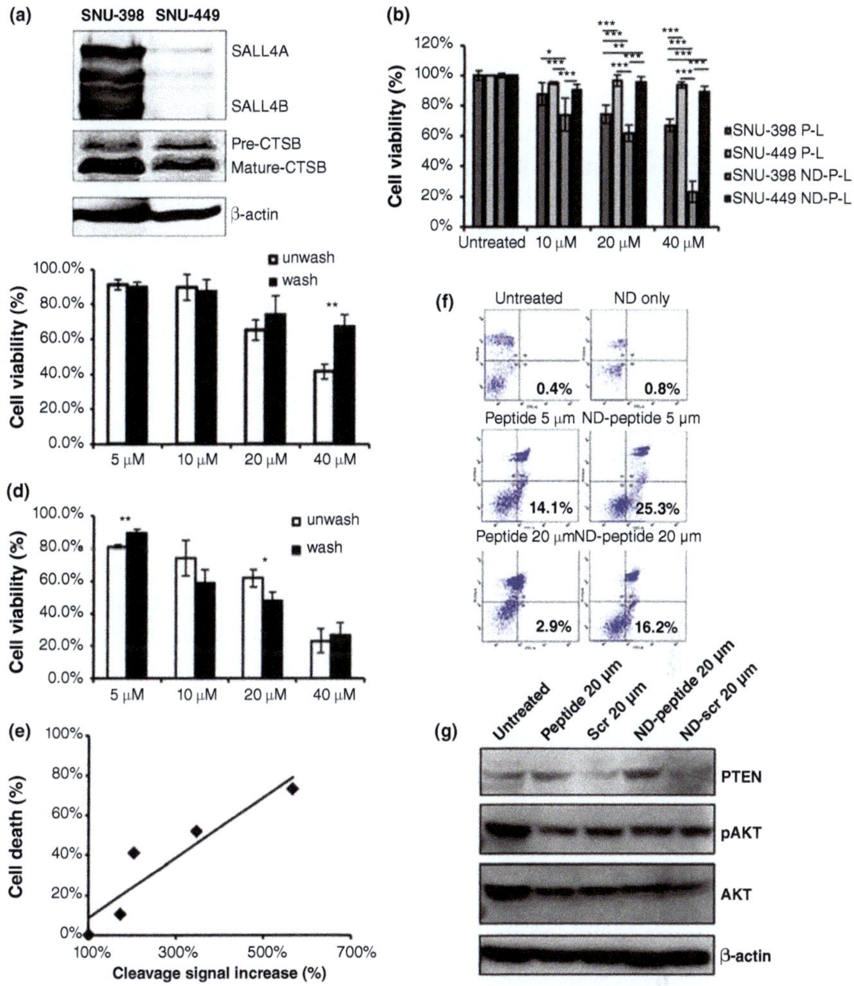

Figure 10.7 SALL4-driven HCC cells are selectively killed by the ND–peptide complex. (a) Expression of SALL4 and Cathepsin B (CTSB) in SNU-398 and SNU-449 cells were analyzed using Western blots. (b) After the treatment of peptide (P-L) and ND–peptide (ND-P-L), cell viability in SNU-398 and SNU-449 cells was assessed. (c) SNU-398 cell viability following peptide treatment with and without rinse step. (d) SNU-398 cell viability following ND–peptide treatment with and without a rinse step. (e) SNU-398 cell death and ND–peptide cleavage signal have a linear relationship. (f) Apoptosis detection with Annexin V/PI in SNU-398 cells after various doses of peptide and ND–peptide therapy. The numbers indicate the proportion of cells that undergo early apoptosis. (g) Western blot examination of PTEN, AKT, and phosphor-AKT in SNU-398 cells following treatment with peptide (either with CTSB cleavage linker or scrambled linker [scr]) and ND–peptide (either with CTSB cleavage linker or scrambled linker [ND-scr]). The mean and standard deviation (n = 3) were used to depict the data. The experiments for Western blot analysis were done three times. The unpaired two-tailed Student's t-test was used for statistical analysis. *$p < 0.05$; **$p < 0.01$; ***$p < 0.005$. *Source:* Reproduced from ref. (52) / John Wiley & Sons.

10.2.3 Gene Delivery

Gene therapy is the leading possibility for developing approaches and solutions in healing genetic and nongenetic disorders. However, before gene therapy can be used on humans, the development of secure and reliable carriers of genetic material is required. A wide range of NDs has been considered for transporting genetic material (54–59). Most of these suggested NDs work by encapsulating nucleic acids into NPs that cells can absorb.

When selecting a carrier of genetic material, researchers must examine several essential parameters. Outside the cells, the polymer's capacity to shield its constituents from destruction is necessary for efficient gene delivery. Polymers should also promote cellular absorption and nucleic acid intracellular release. Transfecting cells with a DNA plasmid encased in the polymer, which might be used to substitute a defective gene into the target cell genome, was the old technique of gene therapy. RNA interference (RNAi) is gaining popularity as a novel diagnostic approach for silencing damaged genes. In this technique, target cells are infected with complementary siRNA. Nonetheless, polymer-assisted siRNA distribution has the same drawbacks as previous DNA delivery methods. This chapter looks at a few different ND gene and siRNA delivery techniques. After outlining some of the significant challenges associated with gene delivery, we present recent achievements in ND research and show how ND-based drug delivery approaches can help overcome some of these barriers.

The fundamental aim of gene therapy is to transfer exogenous genetic elements into cells in such a method that they replace the defective gene or enable new activities. There are two types of gene delivery systems: viral and nonviral. Research has demonstrated unmatched transgene transfection efficacy. However, concerns with gene carrying capacity, NP storage, formulation, and remaining viral elements that might cause insertional cytotoxicity, mutagenesis, tumorigenicity, and immunogenicity have prevented it from becoming an all-purpose vector. As a result of these issues, the field of nonviral vectors has emerged. These vectors comprise biocompatible materials created and manufactured using novel synthetic techniques. Nonviral vectors are relatively simple to make. They are also much less mutagenic and carcinogenic and can accommodate any gene size (54).

NPs have become increasingly popular as a gene therapy nanocarrier. NDs have various characteristics that distinguish them from other NPs. ND's intrinsic solubility in water, biocompatibility, flexibility, high surface-area-to-volume ratio, and various functional sites for simple bioconjugation with multiple medicines and specific compounds make this NP a viable gene delivery vehicle (60).

Researchers all over the world are attempting to integrate gene delivery with nanomedicine. The difficulty of merging safety and effectiveness into a single platform piqued this new interest, and the ND appears to be up to the task (55, 56).

Conventional methods are either very effective but hazardous or ineffective yet safe. ND circumvents many of these challenges. Not only is it practical, but it has few adverse biological side effects (57).

NDs successfully covalently and noncovalently fabricated with a typical polyethyleneimine-800 (PEI800) function as effective plasmid DNA transport carriers (56). The cytotoxicity of the ND-PEI800 compound is minimal, and its transfection effectiveness appears comparable with that of PEI25k, which has a high molecular weight. Cross-linked ND-PEI800 attaches to plasmid DNA via electrostatic interactions and shields it from degradation. Unlike PEI800, ND-PEI800 increases transfection efficiency by 70-fold while maintaining plasmid DNA biocompatibility. ND-PEI800 exhibits 400- and 800-fold increases in transfection effectiveness relative to an amine- and carboxyl-terminated NDs, respectively. The following list of increasing transfection efficiencies: naked DNA < ND < ND-NH_2 < PEI800 < ND-PEI800. The increased transfection efficiency of ND-PEI800 over ND-NH_2 is owing to proton sponge impact, leading to endosome instability and bursting.

Martin *et al.* investigated the Fenton handling of DNDs and found that following functionalization, these nanomaterials can traverse the cell membrane and hit the nucleus (58). The Fenton approach removes any graphitic material from the surface of ND while also derivatizing it with additional hydroxyl groups (–OH), aiding in biomolecule conjugation. Due to their dense- OH groups, Fenton-treated NDs have a bioconjugation advantage over other NPs. They are also more soluble in water. Using fluorescently tagged Fenton-treated NDs, these researchers showed that the ND complex could successfully infiltrate HeLa cells, with some reaching the cell nucleus. As we can see, ND functionalized with bulkier charged groups can load nucleic acid satisfactorily. This laboratory research demonstrates ND's potential as a gene delivery device, bringing promise to anyone hoping to use this nanomaterial in gene therapy.

The issue of ND loading is relevant to RNAi. As a therapeutic, RNAi can reversibly shut down any gene. To exploit the clinical promise of RNAi, siRNA, a short (usually 20–27 bp) dsRNA particle that can be found in a cell endogenously or delivered from exogenous supplies (e.g., synthetic siRNAs) by transfection, must be transported to the action site in cells of targeted tissues using delivery agents. However, to fully grasp the therapeutic possibilities of siRNA, we must first develop a safe and effective delivery mechanism. Since unprocessed siRNA is unstable in circulation, immunogenic, and difficult to penetrate through cell membranes, before transporting it to its target, it must first be chemically modified and have delivery components added before researchers can use it. Apart from that, siRNA will be subjected to negative moderation. To solve siRNA's in vivo delivery issues, researchers have looked at various materials, including peptides, polymers, antibodies, lipids, aptamers, and small compounds (61). Recently, the surface modification of ND combined with drug loading has been added to the set. This is because researchers hope that ND's unique properties, such as its biocompatibility and ability to transport large quantities of medicine, can mitigate some of the challenges associated with siRNA delivery.

Chen *et al.*'s research confirmed ND's siRNA carrier capability (55). By coating NDs with PEI800 and treating them with siRNA, ND-PEI-siRNA compounds can be produced (Figure 10.8). This compound was created by the contrarily charged ND, PEI, and siRNA, which interact electrostatically. In cells, the second-biggest hit performance was achieved by knocking down GFP at a ratio of 1:3 siRNA:ND-PEI (the first is Lipofectamine, the gold-standard transfection reagent). When transfection is performed in a medium containing 10% serum, ND-PEI surpasses Lipofectamine.

The biocompatibility of the ND-PEI-siRNA complex has been further confirmed by toxicity studies using an MTT assay. Chen *et al.* effectively blended ND and PEI to create a noncytotoxicity vector yet had the same transfection effectiveness as high-molecular-weight PEI25k. This finding demonstrates that ND can be used as a platform for DNA delivery and add an effective vector for RNA delivery.

Research by Xu *et al.* further highlighted NDs' therapeutic potential by using this nanomaterial as a platform for gene delivery in tumor parts (62). The ability of NDs with an appropriate size and surface potential to load siRNA was demonstrated as a nanoscale pharmaceutical carrier using electrostatic interactions. Simultaneous release of siRNA from ND-siRNA compounds was shown as the existence of protein-aided siRNA release. Endosomal leakage and excellent cellular absorption and persistence were also observed, demonstrating ND-siRNA compounds' capability for efficient internal transport and siRNA-mediated gene

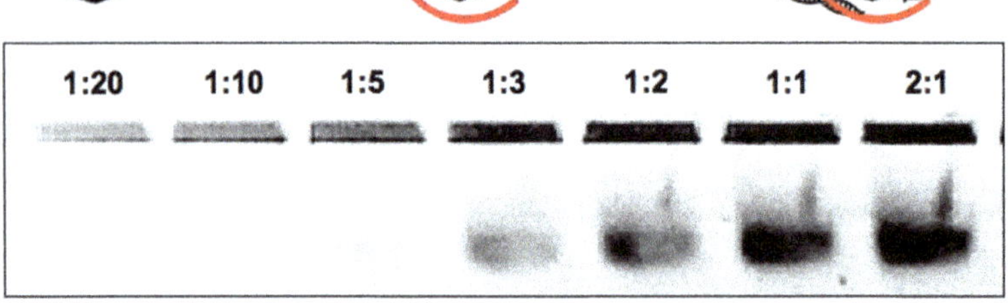

Figure 10.8 The synthesis of polyethylenimine 800 (PEI800)-ND. For varied loading ratios, gel electrophoresis demonstrated the degree of siRNA loading onto ND-PEI800 complexes. *Source:* Reproduced from ref. (55) / American Chemical Society.

regulation. Further research has revealed the ability of ND-siRNA compounds to penetrate HCC tumor spheroids, showing their ability to be used in therapeutic applications. ND-siRNAs' effect on solid tumor formations was studied using 3-D HCC spheroids. The penetrating potential of ND-siRNA compounds was examined in different HCC spheroid types, SNU398 and LM3 (Figure 10.9). HCC spheroids were incubated in culture media with lipo-Cy3 or ND-Cy3 siRNA growth factors. Hoechst was used to stain the spheroids and then photographed using a confocal microscope at each time point. To better illustrate the fluorescence spectra of Cy3, the channel was presented in LUT fire (Figure 10.9b and 10.9e). Fiji analyzer then used the radial profile plot plugin to illustrate the radial distribution of Cy3 fluorescence intensity in the images. According to the findings, Cy3 siRNA increasingly penetrates the spheroids over time. When ND-siRNA was compared with lipo-siRNA, a higher intensity of the Cy3 response was seen inside the HCC spheroids, indicating a higher penetration capacity. The ND-siRNA group had a considerably lower Cy3 signal, and complexes were mainly stuck in the outer cells after 24–48 hours. Still, the lipo-siRNA exhibited a much lower Cy3 signal as complexes were attached primarily on the outside cells (62).

G9a, also known as euchromatic histone-lysine N-methyltransferase 2, is a histone methyltransferase primarily in control for dimethylating histone H3 lysine 9 (H3K9). While G9a has the potential to be a successful therapeutic target for a variety of malignancies, including HCC, the need for more clinically usable inhibitors has hampered its acceptance in the clinic. Gu *et al.* used an ND-based delivery strategy to solve the problems associated with current G9a inhibitors, such as agglomeration and dispersion in aqueous solutions. Using a physical adsorption approach, they could generate complexes of the ND-UNC0646 type with high loading efficiencies. Dynamic light scattering (DLS) findings revealed that ND-UNC0646 might be used as a nanoscale drug delivery device because of its size and surface potential. The effective adsorption of UNC0646 onto ND carriers was further validated by Fourier transform infrared spectroscopy (FTIR) spectra. TEM was used to examine NDs and ND particle UNC0646's shape and surface lattice structure. Furthermore, the pH dependence of UNC0646 release from ND-UNC0646 was demonstrated; release under physiological pH circumstances was relatively stable and sustained; however, the release may be initiated under acidic pH, making it appropriate for pH-sensitive intracellular release. On the other hand, proteins were able to boost UNC0646 release through a competitive mechanism. Furthermore, in a closed environment, ND UNC0646 demonstrated comparable *in vitro* cell-killing efficacy to UNC0646, whereas ND-UNC0646 showed better-maintained efficacy in the long run. Additionally, ND-UNC0646 was more efficient in lowering H3K9 methylation by inhibiting methyltransferase activity and significantly impacted invasion suppression. This ND–inhibitor method for HCC therapy is not restricted to G9a inhibitors. This platform for

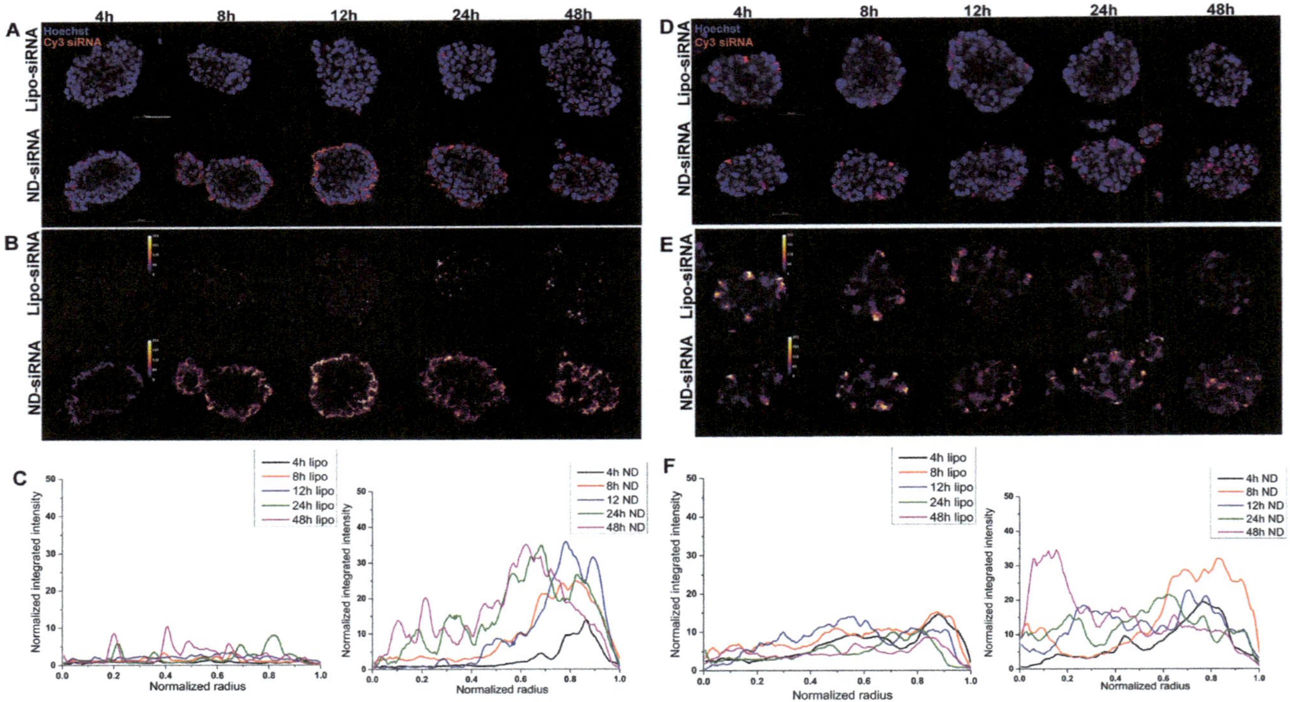

Figure 10.9 In comparison to the lipo-siRNA group, ND-siRNA complexes had a higher penetration capacity in SNU398 (a) and LM3 (d) 3-D spheroids. The red signal represents the Cy3-labeled siRNA, while the blue signal represents the 4′,6-diamidino-2-phenylindole (DAPI) staining. The scale bars are 50 meters. Using the Fiji software, the intensity of Cy3 siRNA fluorescence in SNU398 spheroids (b) and LM3 spheroids (e) was adjusted and presented in LUT fire. Estimation of Cy3 intensity in SNU398 spheroids (c) and LM3 spheroids (f) using the radial profile plot plugin in (f), as illustrated in (a) and (d). *Source:* Reproduced from ref. (62) / Royal Society of Chemistry.

ND-mediated delivery could be used to load various inhibitors of small molecules to improve the water solubility while maintaining activity, hence expanding their biomedical uses. Notably, the preparation procedure for these ND–inhibitor complexes is suitable for scale-up production, suggesting a viable route to commercialization. This work adds to the growing evidence that NDs can be effectively used in many biological applications. In particular, it shows how NDs can improve small-molecule drugs' therapeutic and safety profiles *in vivo* that are usually inappropriate due to their hydrophobic or insoluble nature. As a result, NDs can be used as a strategy for developing new therapy alternatives for earlier untreatable targets.

10.3 Fluorescent Nanodiamond-Based Theranostics Platform

Fluorescence imaging is a typical technique for studying single-molecule biological interactions and dynamics. Researchers use organic dyes as external fluorescent labels to observe the medication delivery mechanisms of NDs (14, 31, 58). However, photobleaching of the dye molecules has made long-term tracking impossible. Chang *et al.* were the first to use NDs' large arrays of crystallographic flaws (such as color centers of NV) to counter the effects of photobleaching (63–67). Fluorescent nanodiamonds are NDs that include color centers (FNDs). Furthermore, after adequate surface modification using contrast agents such as Gd(III) (68, 69) and ^{18}F (70) or ^{188}Re (71) that provide a qualitative distribution in different organs, NDs can be used as a light source and can monitor drug delivery *in vivo* using magnetic resonance imaging (MRI) and positron emission tomography (PET).

Li *et al.* investigated the use of FND to deliver DOX intracellularly. The authors used physical adsorption to deposit DOX on FNDs that emit far-red light (72). Initially, FND surfaces were treated with concentrated H_2SO_4/HNO_3 to provide negatively charged carboxyl functionalities, which allowed the adsorption of positively charged DOX. FND-DOX compounds with a diameter of about 140 nm can be successfully internalized via clathrin-dependent endocytosis trafficking into HeLa cells with rates equivalent to nonadsorbed DOX. According to confocal fluorescence imaging, although FND-DOX particles that have been absorbed are dispersed in the cytoplasm, DOX that has been separated from FND-DOX can move and enter the nucleus. As a DOX intracellular transporter, more recent investigations using FNDs coupled with transferrin (TfR) by human hepatoma (HepG2) cells revealed extremely selective TfR-mediated NP bioconjugate absorption (73). This means that FNDs could be used in cancer therapy in specific cancer bioimaging and precise drug delivery.

FND–poly(ethyleneimine) (PEI) complexes have also been employed in other research to treat specific disease types, such as Ewing's sarcoma, a kind of pediatric bone cancer (59). Using FNDs to transport siRNA suppressed the expression of the EWS-Fli1 oncogene, which boosts the ability of cancer cells to proliferate. Fluorescence readings confirmed the presence of FNDs and siRNA in cells. Colocalization tests and phase contrast imaging were used to verify the findings. Researchers can, thus, utilize these techniques to track drug delivery and release in cells in unprecedented detail.

The effect of ferritin metalloprotein on the spin characteristics of single NV^- centers in NDs was investigated explicitly by Ermakova *et al.* (74). Using magnetic noise as a contrast method, these researchers discovered that due to the proteins' inner paramagnetic iron atoms, a few ferritin molecules stuck to the surface of a single 30-nm ND (Figure 10.10). Despite the technical difficulties of the magnetic sensing investigations, this research suggests that FND is an ideal material for creating nanoprobes that can detect single molecules of biomolecules such as proteins or DNA.

Since it can be activated by near-infrared light and has a strong emission peak, the upconversion of NaYF4:Yb,Er nanoparticles (UCNP) exhibits good luminous characteristics. By combining NDs and UCNPs, a new nanoplatforms for drug delivery and bioimaging in cells has been developed (75). The UNCP-ND nanocomposite was made by adding an amino group to the UCNPs and a carboxyl group to the NDs, followed by a condensation reaction. Under 980 nm laser

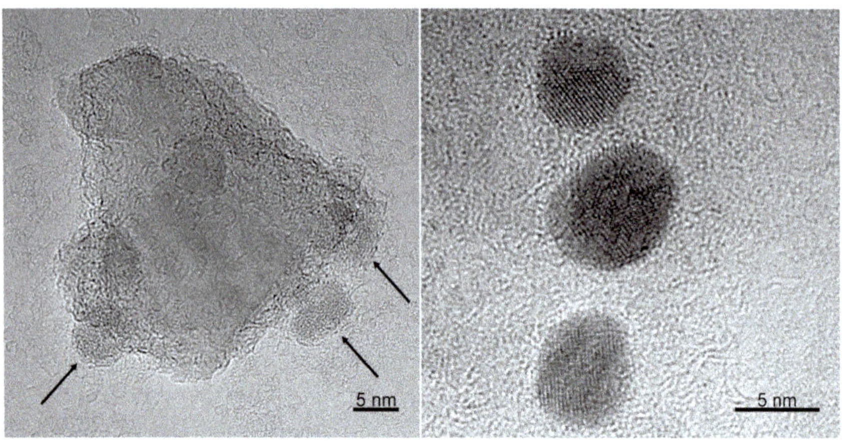

Figure 10.10 High-resolution TEM pictures of a single ferritin-coated ND (left). Black arrows indicate the locations of the metalloprotein molecules on the surface. A picture of ferritin with the iron core (right). *Source:* Reproduced from ref. (74) / American Chemical Society.

excitation, the nanocomposite generates a brilliant green emission, and cell labeling results indicate their intriguing ability for bioimaging. The composite also has good drug retention and pH-responsive drug release capabilities, making it a viable and regulated anticancer drug delivery vehicle for tumor locations. Moreover, DOX-loaded UCNP-ND had higher therapeutic efficacy in HeLa cells than DOX monomers, implying that DOX distribution via UCNP-ND could be an excellent way to avoid DOX side effects while increasing its efficacy. The UCNP-ND emission spectrum was studied under 980-nm excitation to confirm the inheritance of this capacity. The transition between the levels of energy of 4H11/2, 4S3/2, and 4F9/2 to 4I15/2 of Er^{3+} can be attributed to the emission peaks of UCNPs at 521, 539, and 659 nm, as illustrated in Figure 10.11a. UCNP@SiO_2 and UCNP-NDs have the same emission peaks as UCNPs, whereas NDs have no emission peaks. The inset in Figure 10.11a is an acceptable digital photograph taken indoors (left) with a 980-nm laser (right). The strong green light emitted by UCNP-ND can be seen. Preprepared UCNP-ND retains the luminous features of UCNP, offering up prospects for bioimaging applications. Cell imaging on HeLa cells was used to

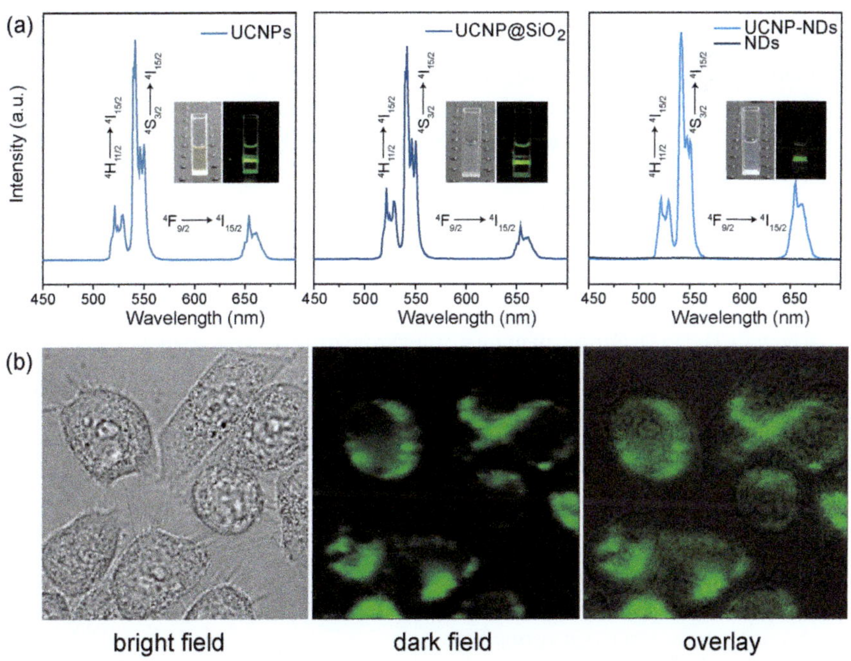

Figure 10.11 (a) UCNPs, UCNP@SiO_2, UCNP-NDs, and NDs PL spectra and pictures (excitation with a 980 nm laser). *Source:* (75) Zhanga et al 2019 / with permission of Elsevier. (b) Fluorescence pictures of HeLa cells after 10 hours of incubation with UCNP-NDs in bright-field (white light), UV channel (980 nm), and overlay micrographs. *Source:* Reproduced from ref. (75) / American Association for the Advancement of Science (AAAS).

assess the capacity of the generated composite UCNP-NDs for *in vitro* imaging. After co-incubation with UCNP-NDs, cell micrographs were obtained using a confocal fluorescence microscope. Under bright-field mode, the group of cells can be seen clearly in Figure 10.11b. In dark field mode, a brilliant green fluorescence can be observed in the relevant cell cluster area. The fluorescent areas and cell cluster positions could be clearly seen in the overlay plane, and they were very similar, indicating that UCNP-NDs had correctly labeled HeLa cells.

To summarize, NDs have the potential to effectively deliver active agents for a wide range of disorders treatment, such as cancer. However, relying too strongly on a single substance can inhibit major clinical improvements. Only after resolving the constraints of ND-based treatments in clinical contexts can real ND-based drug delivery systems be developed. We believe that in clinical science, NDs bound with tiny compounds, proteins, and genes have a favorable future. ND-based delivery and monitoring systems have made significant progress in generating potent, long-lasting, and high targetability carriers. After removing some of the barriers to this technique, ND platforms will stimulate the development of rationally designed, novel nanomedicines. The effectiveness of NDs in clinical trials, as well as whether they are more successful than other NP-based systems for drug, gene delivery, and monitoring systems, is yet to be seen. Nonetheless, research on ND has demonstrated the device's versatility: The ability to equip NDs with a range of surface capabilities and monitor them using a variety of imaging modalities is a feature of ND that keeps this nanomaterial on scientists' minds.

References

1 Chow EKH, Ho D. Cancer nanomedicine: From drug delivery to imaging. *Sci Transl Med*. 2013;5(216):216rv4.
2 Koo OM, Rubinstein I, Onyuksel H. Role of nanotechnology in targeted drug delivery and imaging: A concise review. *Nanomedicine-Uk*. 2005;1(3):193–212.
3 Niemeyer CM. Nanoparticles, proteins, and nucleic acids: Biotechnology meets materials science. *Angew Chem Int Ed*. 2001;40(22):4128–58.
4 De Jong WH, Borm PJA. Drug delivery and nanoparticles: Applications and hazards. *Int J Nanomedicine*. 2008;3(2):133–49.
5 Bae YH, Park K. Targeted drug delivery to tumors: Myths, reality and possibility. *J Control Release*. 2011;153(3):198–205.
6 Desai N. Challenges in development of nanoparticle-based therapeutics. *AAPS J*. 2012;14(2):282–95.
7 Mochalin VN, Shenderova O, Ho D, Gogotsi Y. The properties and applications of nanodiamonds. *Nat Nanotechnol*. 2012;7(1):11–23.

8 Zhu Y, Li J, Li WX, Zhang Y, Yang XF, Chen N, *et al.* The biocompatibility of nanodiamonds and their application in drug delivery systems. *Theranostics.* 2012;2(3):302–12.

9 Man HB, Ho D. Diamond as a nanomedical agent for versatile applications in drug delivery, imaging, and sensing. *Phys Status Solidi A.* 2012;209(9):1609–18.

10 Man HB, Ho D. Nanodiamonds as platforms for biology and medicine. *J Lab Autom.* 2013;18(1):12–18.

11 Kaur R, Badea I. Nanodiamonds as novel nanomaterials for biomedical applications: Drug delivery and imaging systems. *Int J Nanomedicine.* 2013;8:203–20.

12 Moosa B, Fhayli K, Li S, Julfakyan K, Ezzeddine A, Khashab NM. Applications of nanodiamonds in drug delivery and catalysis. *J Nanosci Nanotechno.* 2014;14(1):332–43.

13 Vaijayanthimala V, Lee DK, Kim SV, Yen A, Tsai N, Ho D, *et al.* Nanodiamond-mediated drug delivery and imaging: Challenges and opportunities. *Expert Opin Drug Del.* 2014:735–49.

14 Lam R, Ho D. Nanodiamonds as vehicles for systemic and localized drug delivery. *Expert Opin Drug Del.* 2009;6(9):883–95.

15 Xing Y, Dai LM. Nanodiamonds for nanomedicine. *Nanomedicine-Uk.* 2009;4(2):207–18.

16 Vaijayanthimala V, Chang HC. Functionalized fluorescent nanodiamonds for biomedical applications. *Nanomedicine-Uk.* 2009;4(1):47–55.

17 Hsiao WW-W, Le T-N, Chang H-C. Applications of fluorescent nanodiamond in biology. In: Meyers, R.A. (Ed.), *Encyclopedia of Analytical Chemistry.* Hoboken, New Jersey, U.S.: John Wiley & Sons; 2022. pp. 1–43.

18 Mochalin VN, Pentecost A, Li X-M, Neitzel I, Nelson M, Wei C, *et al.* Adsorption of drugs on nanodiamond: toward the development of a drug delivery platform. *Mol Pharm.* 2013;10(10):3728–35.

19 Paci JT, Man HB, Saha B, Ho D, Schatz GC. Understanding the surfaces of nanodiamonds. *J Phys Chem C.* 2013;117(33):17256–67.

20 Chen M, Pierstorff ED, Lam R, Li SY, Huang H, Osawa E, *et al.* Nanodiamond-mediated delivery of water-insoluble therapeutics. *Acs Nano.* 2009;3(7):2016–22.

21 Ho DA. Beyond the sparkle: The impact of nanodiamonds as biolabeling and therapeutic agents. *Acs Nano.* 2009;3(12):3825–9.

22 Huang H, Pierstorff E, Osawa E, Ho D. Active nanodiamond hydrogels for chemotherapeutic delivery. *Nano Lett.* 2007;7(11):3305–14.

23 Huang HJ, Pierstorff E, Osawa E, Ho D. Protein-mediated assembly of nanodiamond hydrogels into a biocompatible and biofunctional multilayer nanofilm. *ACS Nano.* 2008;2(2):203–12.

24 Lam R, Chen M, Pierstorff E, Huang H, Osawa EJ, Ho D. Nanodiamond-embedded microfilm devices for localized chemotherapeutic elution. *Acs Nano.* 2008;2(10):2095–102.

25 Man HB, Kim H, Kim HJ, Robinson E, Liu WK, Chow EK, *et al.* Synthesis of nanodiamond-daunorubicin conjugates to overcome multidrug chemoresistance in leukemia. *Nanomedicine-Uk*. 2014;10(2):359–69.

26 Moore L, Chow EK-H, Osawa E, Bishop JM, Ho D. Diamond-lipid hybrids enhance chemotherapeutic tolerance and mediate tumor regression. *Adv Mater*. 2013;25(26):3532–41.

27 Pierstorff E, Lam R, Ho D. Nanoscale architectural tuning of parylene patch devices to control therapeutic release rates. *Nanotechnology*. 2008;19(44):445104.

28 Xi G, Robinson E, Mania-Farnell B, Vanin EF, Shim KW, Takao T, *et al.* Convection-enhanced delivery of nanodiamond drug delivery platforms for intracranial tumor treatment. *Nanomedicine-Uk*. 2014;10(2):381–91.

29 Moore L, Grobárová V, Shen H, Man HB, Míčová J, Ledvina M, *et al.* Comprehensive interrogation of the cellular response to fluorescent, detonation and functionalized nanodiamonds. *Nanoscale*. 2014;6(20):11712–21.

30 Kim H-J, Zhang K, Moore L, Ho D. Diamond nanogel-embedded contact lenses mediate lysozyme-dependent therapeutic release. *Acs Nano*. 2014;8(3):2998–3005.

31 Zhang X-Q, Lam R, Xu X, Chow EK, Kim H-J, Ho D. Multimodal nanodiamond drug delivery carriers for selective targeting, imaging, and enhanced chemotherapeutic efficacy. *Adv Mater*. 2011;23(41):4770-5.

32 Guan B, Zou F, Zhi JF. Nanodiamond as the pH-responsive vehicle for an anticancer drug. *Small*. 2010;6(14):1514–19.

33 Liu KK, Zheng WW, Wang CC, Chiu YC, Cheng CL, Lo YS, *et al.* Covalent linkage of nanodiamond-paclitaxel for drug delivery and cancer therapy. *Nanotechnology*. 2010;21(31):315106.

34 Li J, Zhu Y, Li W, Zhang X, Peng Y, Huang Q. Nanodiamonds as intracellular transporters of chemotherapeutic drug. *Biomaterials*. 2010;31(32):8410–8.

35 Prabhakar N, Näreoja T, von Haartman E, Karaman DŞ, Jiang H, Koho S, *et al.* Core–shell designs of photoluminescent nanodiamonds with porous silica coatings for bioimaging and drug delivery II: Application. *Nanoscale*. 2013;5(9):3713–22.

36 Adnan A, Lam R, Chen H, Lee J, Schaffer DJ, Barnard AS, *et al.* Atomistic simulation and measurement of pH dependent cancer therapeutic interactions with nanodiamond carrier. *Mol Pharm*. 2011;8(2):368–74.

37 Yan JJ, Guo Y, Altawashi A, Moosa B, Lecommandoux S, Khashab NM. Experimental and theoretical evaluation of nanodiamonds as pH triggered drug carriers. *New J Chem*. 2012;36(7):1479–84.

38 Wang DX, Tong YL, Li YQ, Tian ZM, Cao RX, Yang BS. PEGylated nanodiamond for chemotherapeutic drug delivery. *Diam Relat Mater*. 2013;36:26–34.

39 Zhang XY, Wang SQ, Fu CK, Feng L, Ji Y, Tao L, *et al.* PolyPEGylated nanodiamond for intracellular delivery of a chemotherapeutic drug. *Polym Chem*. 2012;3(10):2716–9.

40 Li XX, Shao JQ, Qin Y, Shao C, Zheng TT, Ye L. TAT-conjugated nanodiamond for the enhanced delivery of doxorubicin. *J Mater Chem.* 2011;21(22):7966–73.

41 Chow EK, Zhang XQ, Chen M, Lam R, Robinson E, Huang HJ, *et al.* Nanodiamond therapeutic delivery agents mediate enhanced chemoresistant tumor treatment. *Sci Transl Med.* 2011;3(73):73ra21.

42 Rakha EA, Ellis IO. An overview of assessment of prognostic and predictive factors in breast cancer needle core biopsy specimens. *J Clin Pathol.* 2007;60(12):1300–6.

43 Rakha EA, El-Sayed ME, Green AR, Lee AH, Robertson JF, Ellis IO. Prognostic markers in triple-negative breast cancer. *Cancer.* 2007;109(1):25–32.

44 Roy U, Drozd V, Durygin A, Rodriguez J, Barber P, Atluri V, *et al.* Characterization of nanodiamond-based anti-HIV drug delivery to the brain. *Sci Rep.* 2018;8(1):1603.

45 Shimkunas RA, Robinson E, Lam R, Lu S, Xu XY, Zhang XQ, *et al.* Nanodiamond-insulin complexes as pH-dependent protein delivery vehicles. *Biomaterials.* 2009;30(29):5720–8.

46 Smith AH, Robinson EM, Zhang XQ, Chow EK, Lin Y, Osawa E, *et al.* Triggered release of therapeutic antibodies from nanodiamond complexes. *Nanoscale.* 2011;3(7):2844–8.

47 Nguyen TTB, Chang HC, Wu VWK. Adsorption and hydrolytic activity of lysozyme on diamond nanocrystallites. *Diam Relat Mater.* 2007;16(4–7):872–6.

48 Liu KK, Chen F, Chen PY, Lee TJF, Cheng CL, Chang CC, *et al.* Alpha-bungarotoxin binding to target cell in a developing visual system by carboxylated nanodiamond. *Nanotechnology.* 2008;19(20):205102.

49 Wang HD, Niu CH, Yang Q, Badea I. Study on protein conformation and adsorption behaviors in nanodiamond particle-protein complexes. *Nanotechnology.* 2011;22(14):145703.

50 Moore L, Gatica M, Kim H, Osawa E, Ho D. Multi-protein delivery by nanodiamonds promotes bone formation. *J Dent Res.* 2013;92(11):976–81.

51 Leung HM, Chan MS, Liu LS, Wong SW, Lo TW, Lau C-H, *et al.* Dual-function, cationic, peptide-coated nanodiamond systems: facilitating nuclear-targeting delivery for enhanced gene therapy applications. *ACS Sustain Chem Eng.* 2018;6(8):9671–81.

52 Gu M, Wang X, Toh TB, Hooi L, Tenen DG, Chow EK-H. Nanodiamond-based platform for intracellular-specific delivery of therapeutic peptides against hepatocellular carcinoma. *Adv Ther.* 2018;1(8):1800110.

53 Chow Edward K, Zhang X-Q, Chen M, Lam R, Robinson E, Huang H, *et al.* Nanodiamond therapeutic delivery agents mediate enhanced chemoresistant tumor treatment. *Sci Transl Med.* 2011;3(73):73ra21–73ra21.

54 Jin S, Leach JC, Ye K. Nanoparticle-mediated gene delivery. *Methods Mol Biol.* 2009;544:547–57.

55 Chen M, Zhang XQ, Man HB, Lam R, Chow EK, Ho DA. Nanodiamond vectors functionalized with polyethylenimine for siRNA delivery. *J Phys Chem Lett*. 2010;1(21):3167–71.

56 Zhang XQ, Chen M, Lam R, Xu XY, Osawa E, Ho D. Polymer-functionalized nanodiamond platforms as vehicles for gene delivery. *Acs Nano*. 2009;3(9):2609–16.

57 Purtov KV, Petunin AI, Burov AE, Puzyr AP, Bondar VS. Nanodiamonds as carriers for address delivery of biologically active substances. *Nanoscale Res Lett*. 2010;5(3):631–6.

58 Martin R, Alvaro M, Herance JR, Garcia H. Fenton-treated functionalized diamond nanoparticles as gene delivery system. *Acs Nano*. 2010;4(1):65–74.

59 Alhaddad A, Adam MP, Botsoa J, Dantelle G, Perruchas S, Gacoin T, *et al*. Nanodiamond as a vector for siRNA delivery to Ewing sarcoma cells. *Small*. 2011;7(21):3087–95.

60 Zhang BL, Li YQ, Fang CY, Chang CC, Chen CS, Chen YY, *et al*. Receptor-mediated cellular uptake of folate-conjugated fluorescent nanodiamonds: A combined ensemble and single-particle study. *Small*. 2009;5(23):2716–21.

61 Kanasty R, Dorkin JR, Vegas A, Anderson D. Delivery materials for siRNA therapeutics. *Nat Mater*. 2013;12(11):967–77.

62 Xu J, Gu M, Hooi L, Toh TB, Thng DKH, Lim JJ, *et al*. Enhanced penetrative siRNA delivery by a nanodiamond drug delivery platform against hepatocellular carcinoma 3D models. *Nanoscale*. 2021;13(38):16131–45.

63 Yu S-J, Kang M-W, Chang H-C, Chen K-M, Yu Y-C. Bright fluorescent nanodiamonds: No photobleaching and low cytotoxicity. *J Am Chem Soc*. 2005;127(50):17604–5.

64 Fu CC, Lee HY, Chen K, Lim TS, Wu HY, Lin PK, *et al*. Characterization and application of single fluorescent nanodiamonds as cellular biomarkers. *Proc Natl Acad Sci USA*. 2007;104(3):727–32.

65 Chang Y-R, Lee H-Y, Chen K, Chang C-C, Tsai D-S, Fu C-C, *et al*. Mass production and dynamic imaging of fluorescent nanodiamonds. *Nat Nanotechnol*. 2008;3(5):284–8.

66 Mohan N, Chen C-S, Hsieh H-H, Wu Y-C, Chang H-C. In vivo imaging and toxicity assessments of fluorescent nanodiamonds in Caenorhabditis elegans. *Nano Lett*. 2010;10(9):3692–9.

67 Wee TL, Mau YW, Fang CY, Hsu HL, Han CC, Chang HC. Preparation and characterization of green fluorescent nanodiamonds for biological applications. *Diam Relat Mater*. 2009;18(2–3):567–73.

68 Manus LM, Mastarone DJ, Waters EA, Zhang XQ, Schultz-Sikma EA, MacRenaris KW, *et al*. Gd(III)-Nanodiamond conjugates for MRI contrast enhancement. *Nano Lett*. 2010;10(2):484–9.

69 Nakamura T, Ohana T, Yabuno H, Kasai R, Suzuki T, Hasebe T. Simple fabrication of Gd(III)-DTPA-Nanodiamond particles by chemical modification for use as magnetic resonance imaging (MRI) contrast agent. *Appl Phys Express.* 2013;6(1):015001.

70 Rojas S, Gispert JD, Martin R, Abad S, Menchon C, Pareto D, *et al.* Biodistribution of amino-functionalized diamond nanoparticles. In vivo studies based on F-18 radionuclide emission. *Acs Nano.* 2011;5(7):5552–9.

71 Zhang X, Yin J, Kang C, Li J, Zhu Y, Li W, *et al.* Biodistribution and toxicity of nanodiamonds in mice after intratracheal instillation. *Toxicol Lett.* 2010;198(2):237–43.

72 Li YQ, Zhou XP, Wang DX, Yang BS, Yang P. Nanodiamond mediated delivery of chemotherapeutic drugs. *J Mater Chem.* 2011;21(41):16406–12.

73 Wang D, Li Y, Tian Z, Cao R, Yang B. Transferrin-conjugated nanodiamond as an intracellular transporter of chemotherapeutic drug and targeting therapy for cancer cells. *Ther Deliv.* 2014;5(5):511–24.

74 Ermakova A, Pramanik G, Cai JM, Algara-Siller G, Kaiser U, Weil T, *et al.* Detection of a few metallo-protein molecules using color centers in nanodiamonds. *Nano Lett.* 2013;13(7):3305–9.

75 Zhang K, Zhao Q, Qin S, Fu Y, Liu R, Zhi J, *et al.* Nanodiamonds conjugated upconversion nanoparticles for bio-imaging and drug delivery. *J Colloid Interface Sci.* 2019;537:316–24.

11

Nanodiamond for Mass-Spectrometry-Based Analysis of Peptides, Proteins, and Proteomes

Dinh Minh Pham, Ryu Endo, Chris Chen-Hua Chang, Thi-Hong-Hanh Le, and Wesley Wei-Wen Hsiao**

* *Corresponding authors*

The mass of an object is normally measured by a balance, and the finest analytical balance available today in a laboratory can measure at an accuracy of 0.01 mg. However, mass measurement is not as straightforward for a "tiny" molecule. In these instances, mass spectrometry (MS) is used to "weigh" molecules. The mass range for current mass spectrometers fluctuates from several hundreds of kilodaltons (kDa) to sub-daltons (Da). One atomic mass unit (amu), the standard unit used for indicating the mass on an atomic or molecular scale is equal to $1.660538921 \times 10^{-27}$ kg, or one-twelfth of the mass of carbon-12 (^{12}C).

In 1886, the German physicist Eugen Goldstein made one of the first observations of rays in gas discharges under low pressure. These observations led to the development of MS. Nobel Prizes have been awarded to researchers who contributed to this technique. Today, mass spectrometers can be found in almost every research institution and in fields including chemistry, physics, and biology. The mass spectrometer is also an important analytical instrument in several industries.

As we all know, the basic dogmas of molecular biology are DNA, RNA, and proteins. Proteins are far more complex than DNA and RNA. With modern MS techniques, we can probe proteins, gaining information on their molecular weight, structure, function, posttranslational modifications (PTMs), and protein–protein interactions. MS has been the core component of proteomics studies, allowing for high-throughput sequencing and characterization of complicated protein mixtures and proteomes. MS instruments' precision, sensitivity, and versatility have enabled novel applications in biological research, biopharmaceutical characterization, and diagnostic detection. With its many ionizations and measurement (e.g., electrospray ionization [ESI], matrix-assisted laser desorption/ionization [MALDI], Fourier transform mass spectrometry [FT-MS], ion trap, and quadrupole) options, MS can analyze proteins and peptides ranging in mass from hundreds daltons to hundreds of kilodaltons and in attomole to nanomole quantities. This chapter provides foundational information on protein analysis using MS and summarize the applications of nanodiamonds (NDs) in MS-based analyses of peptides, proteins, and proteomes.

Nanodiamonds in Analytical and Biological Sciences: Principles and Applications, First Edition.
Edited by Wesley Wei-Wen Hsiao and Trong-Nghia Le.
© 2023 John Wiley & Sons Ltd. Published 2023 by John Wiley & Sons Ltd.

11.1 Principles of Analyzing Macro Biomolecules with Mass Spectrometry

Joseph Thomson, who earned the Nobel Prize in 1906 for his investigations into the conduction of electricity by gases, was the first scientist to demonstrate and describe the ability to separate molecules based on different sizes and charges in a mass spectrometer. While the roots of molecule separation can be traced back to the early twentieth century, scientists could not analyze large macromolecules until more than 70 years later. This ability emerged when the chemical ionization (CI) of molecules in a gas was followed by the physical separation of ions in a vacuum. The process that transforms macromolecules into ions in the gas phase is called *desorption technology*. One early key advance, published in 1966 by Munson and Field (1), was the first use of CI, which enabled the ionization of thermolabile biomolecules. Plasma desorption (1976) (2) and fast atom bombardment (1981) (3) followed CI as other important milestones in the ionization of macro biomolecules, especially protein molecules. Although these techniques achieved moderate success, they were unreliable for proteins with higher molecular masses (>10 kDa). Researchers in the late twentieth century also faced challenges in converting MS into a powerful detector for liquid separation and liquid chromatography (LC). The invention of electrospray (ES) and soft laser desorption (SLD) technologies solved both these issues (the effective ionization of large proteins and their coupling with LC). Scientists soon demonstrated that the ES technique was so mild that viral particles remain viable even after an ESI process (4). Proteins ionized by ESI carry multiple charges and are distributed in a series of low m/z peaks. In contrast, the SLD method provided easy access to singly charged ionizations of intact biomolecules in complex matrices, allowing for the direct observation of big protein molecules or protein complexes in a single shot. John Fenn and Koichi Tanaka shared half of the Nobel Prize in Chemistry in 2002 for introducing the ESI and SLD methods, and advancing MS as a viable tool for the analysis of biomolecules and a wide range of molecular classes. Over the past two decades, these methods have formed the basis for a revolution in the field of MS-based protein characterization.

11.2 The Post-Era of Genomics

Proteomics is generally defined as the large-scale study of protein functions and structures. Protein research is not new, but proteomics is a relatively new field of research that has emerged from genomics. An estimated 20 000–25 000 protein coding genes out of more than 3.2 billion base pairs compose the most advanced being on the planet – the human animal. However, the human proteome is much more complicated because of gene splicing, epigenetics, RNA processing, and the PTMs of proteins. A proteome is defined as the entire set of proteins produced or modified by an organism or a system at a particular time and condition. Many new questions have emerged in the wake of the human genome project, which was completed in 2000. Among them are: What is the relationship between a genome and its proteome? What happens if a gene is missing or damaged? How do proteins in a cell work together? What is the relationship between proteomes and diseases?

After more than two decades of development, proteomics has become an important research field. Several biological processes, such as cell signaling, protein–protein interactions, protein function, protein PTMs, and disease biomarkers, have been investigated and addressed using proteomics approaches. Proteomics can employ three distinct methodologies: top-down, middle-up, and bottom-up. In general, top-down proteomics studies proteomes at the protein level, whereby intact proteins are separated and directly analyzed by mass spectrometers. Although the analysis of large intact proteins in a high-throughput manner is challenging, the top-down approach has advantages for studying proteoforms (the same proteins with different PTMs or even conformations) (5). Bottom-up proteomics, on the other hand, offers information about proteins by analyzing peptides that come from the proteolytic cleavages of proteins. The typical mass range of peptides is under 5000 Da in a routine workflow of bottom-up approach. This proteomics approach is extremely robust in identifying proteins in complicated samples or proteomes. With advances in MS, separation technology for peptides, and bioinformatics, several projects have claimed to successfully obtain the complete proteomes of some model organisms, quantify the copy number of proteins in cells, and comprehensively analyze different types of protein PTMs (6). The last approach, the middle-down approach, deals with peptides in mid-range sizes. However, this approach is not widely used.

11.3 Bioinformatics for the Coming of Big-Data Biology

Bioinformatics is an indispensable component of genomics and proteomics studies. It is an interdisciplinary research field that develops methods and software tools for interpreting and extracting useful information from biological data. For biologists, the size and complexity of data collected now or in the future will increase far-ahead of the famous Moore's law. Gordon E. Moore, the co-founder of Intel and Fairchild Semiconductor, observed that the number of transistors per integrated circuit doubled approximately every two years. However, advancements in computer chip design are neglecting the speed of information generated by mankind. "Some say that about 90% of all the data in the world today has been created in the past few years" – BBC's 2014 report (www.bbc.com/news/business-26383058). With advances in instrumentation, data storage technology, and interdisciplinary studies, biologists need to be prepared to deal with scientific data that are not only big but also complicated.

11.4 Nanodiamonds for the Mass-Spectrometry-Based Analysis of Peptides, Proteins, and Proteomics

Nanotechnology, which manipulates matters at the nanometer scale (1 nm = 10^{-9} m), is emerging as a new and powerful field of research. Imagine the usefulness of tiny nanoparticles (or nanomachines) that can enter deep inside your body to find and fix damaged cells (e.g., cancer cells). Until recently, such a technology has been a fiction. Currently, nanotechnology has been mobilized to produce new materials, devices, applications in medicine, electronics, and even consumer products.

The first nanodiamonds (NDs) were produced via detonation by the Soviet Union in the 1960s. Interestingly, these particles were virtually unknown to the rest of the world until the 1990s. It is only since the end of the Cold War, and after a number of scientific breakthroughs, that the nanoparticle has attracted significant attention from the international scientific community (7). Nanoscale diamonds are nontoxic, have highly specific areas, and possess excellent optical/mechanical properties. Researchers can now create defect centers in NDs, making them amenable to biological imaging, magnetic sensing, and temperature measurement at nanoscale space resolutions. The defect color centers, such as NV^-, buried inside the crystal lattice of a diamond nanoparticle are unique quantum systems that can be manipulated at ambient temperatures. In this chapter, we discuss the application of NDs in the MS-based analysis of peptides, proteins, and proteomes. In addition, we elucidate and demonstrate the unique characteristics of this nanoparticle that make it special and useful in this area of inquiry.

11.5 Origin of Interactions Between Nanodiamonds and Proteins or Peptides

11.5.1 Surface Properties of Carboxylated/Oxidized Nanodiamonds

The core materials used in most MS-based analysis of peptides, proteins, and proteomes with NDs are synthetic diamond powders made by the high-pressure, high-temperature (HPHT) approach. These ND particles, which are commercially acquired with a nominal size of 100 nm (due to the relation between density and ease-of-handling with conventional centrifuges), must be surface-functionalized to be used within MS applications. As a result, surface oxidation with strong oxidative acids under microwave heating is often employed (8). Sulfuric acid and nitric acid with the volumetric ratio of 3:1 is the most commonly used protocol for this purpose. However, special caution must be taken (see ref. (8) for the detailed requirements of hazard prevention). After being treated, a number of oxygen-containing groups can be found on the surface of ND (9). Figure 11.1 presents a typical Fourier transform infrared spectroscopy (FTIR) absorption spectrum of dried acid-oxidized ND particles. The most dominant component observed in the spectrum is the C=O stretching vibrations of the carbonyl and/or carboxyl moieties represented by the absorption band at 1824 cm^{-1}. The broad absorption band at ~3500 cm^{-1} can be attributed to the O–H stretching vibrations of adsorbed water and/or surface carboxylic groups. On the basis of conductometric titration, it is estimated that the oxidative-acid-treated ND contributes only ~7% of the –COOH moiety on its surface. Infrared spectroscopy has also compared the surface characteristics of acid-oxidized NDs before and after

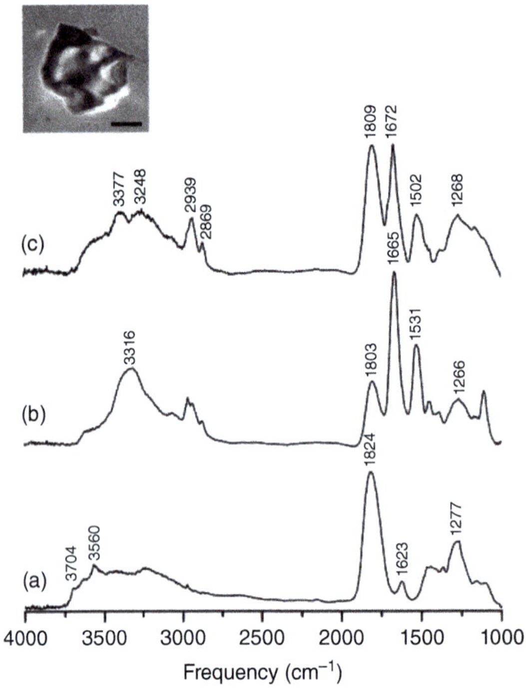

Figure 11.1 Infrared spectra of 100-nm NDs after (a) treatment with strong oxidative acids, (b) physisorption of horse cytochrome *c*, and (c) coating with PLs. Polynomial fits for baseline subtraction were applied to each spectrum. Inset on top: Scanning electron microscopy image of a single 100-nm ND particle. The scale bar is 40 nm. *Source:* (9) Huang et al 2004 / American Chemical Society.

protein coating and poly L-lysine (PLs) coating (9). As seen in Figure 11.1c, the FTIR spectrum of NDs coated noncovalently with PLs showed marked changes. The absorption bands at 2939 and 2869 cm^{-1} now represent the C–H stretching vibrations and the most prominent features at 1665 and 1531 cm^{-1} to the vibrations (amide I and II, respectively) of the amide bonds in the physically adsorbed PLs. According to the results, the PLs are more likely to connect to the ND in an extended form, adorning the surfaces of the particles with amino groups. For the NDs immobilized with horse heart cytochrome *c*, some prominent surface characteristics are preserved, indicating the noncovalent interactions between the protein and ND particles.

11.5.2 Interactions of Proteins and Peptides with Carboxylated/Oxidized Nanodiamonds

Carrying a variety of oxygen-containing groups after acid treatment, the acid-oxidized NDs have negatively charged surfaces (8). The zeta potential of a typical 100-nm ND after acid washing is approximately −45 mV. Carboxylated/oxidized NDs, therefore, can conjugate proteins or peptides or any other biomolecules by covalent bonding through their carboxyl/or carboxyl-derived groups present on the surface or by electrostatic interaction. However, NDs are carbon-based materials and inherently hydrophobic. For these reasons, acid-oxidized NDs show an exceptionally high affinity for water-soluble proteins in different buffers (10) and membrane proteins solubilized in detergent micelles or chaotropic solutions (8) through a special physical absorption mechanism. Figure 11.2 illustrates the adsorption isotherms of some tested water-soluble proteins (cytochrome *c*, myoglobin, and bovine serum albumin) on the surfaces of ND. A common feature of these proteins is that they all present the same adsorption behavior: adsorption saturation is reached extremely quickly, even at protein concentrations as low as 10 mM. For a particular protein, the maximum absorption peaks as the pH of the protein buffer reaches the protein pI. Experimental data show that a ratio of ~50:1 (ND:protein) (w/w) is sufficient to completely capture water-soluble proteins from a solution (11). This ratio can be reduced to around 30:1 (w/w) for hydrophobic membrane proteins in detergent solutions (8).

The interaction between carboxylated/oxidized NDs with membrane proteins is also very strong and pH-dependent (8). However, since membrane proteins are originally embedded or anchored to cellular lipid bilayers, they must first be solubilized to form a homogenous soluble protein solution. Detergents, salts, and chaotropic reagents are commonly used to extract these proteins from cell membranes and effectively solubilize them. Pham *et al.* demonstrated that carboxylated ND is able to selectively capture membrane proteins from detergents and other solubilizing chemicals (8). As shown in Figure 11.3, ND proved its high-affinity interaction with membrane proteins, dispersed in various surfactant solutions or even in a mixture of detergents and chaotropic reagents. The study also

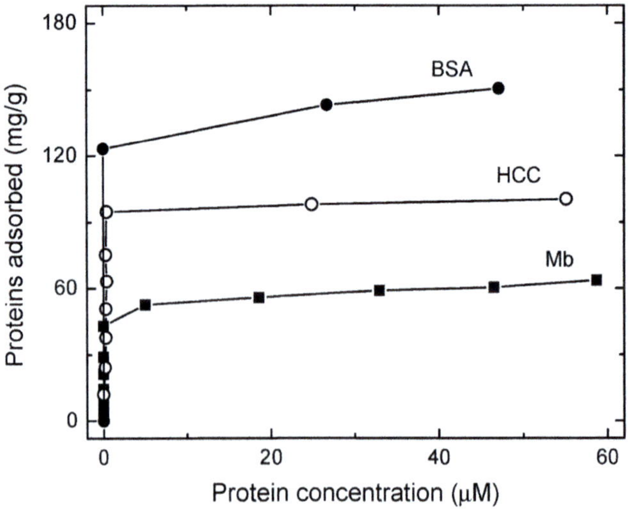

Figure 11.2 Adsorption isotherms of horse heart cytochrome *c* (○), horse heart myoglobin (■), and bovine serum albumin (●) on 100-nm diamonds at pH 10.5, 6.9, and 4.7, respectively. The typical error involved in the measurements is (10%). *Source:* (10) Kong et al 2005 / American Chemical Society.

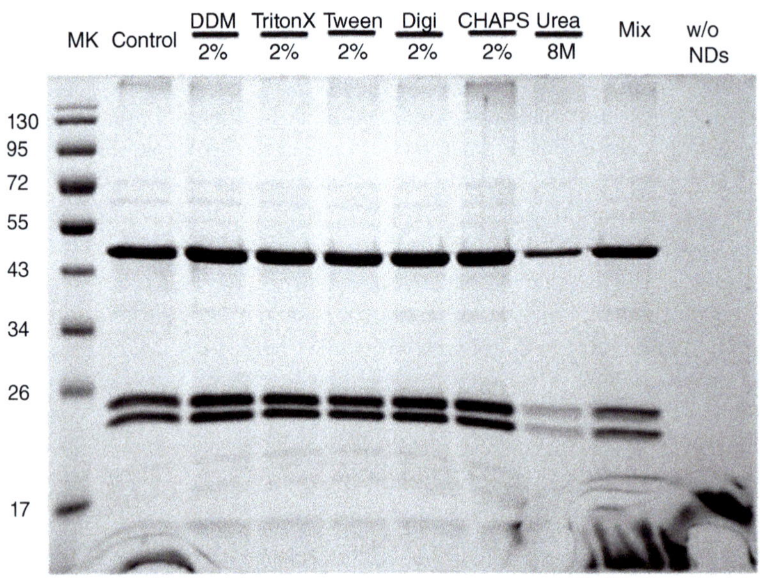

Figure 11.3 SDS-PAGE analysis of a model multisubunit membrane protein complex, particulate methane monooxygenase (pMMO), extracted from solutions containing different detergents or a mixture of detergents (mix) using NDs. This experiment used 200 μg of NDs to extract 6 μg of the pMMO proteins solubilized in 200 μl of the detergent solutions. Control experiment (control, the first well): directly loading 6 μg protein into a well for SDS-page analysis. The last well (w/o NDs) represents the sample without the addition of NDs to the pMMO solutions containing the mixture of detergents. *Source:* Reproduced from ref. (8) / American Chemical Society.

showed that interactions between membrane proteins and the surface of ND are due to the interplay of hydrophobicity and ionic forces.

The effect of ionic strength of the buffers used to isolate protein is clear when NDs are mixed with a complicated mixture of globular water-soluble proteins or a detergent micelle-solubilized membrane protein sample. In the first experiment, with water-soluble proteins, Pham and colleagues found NDs switched quickly and clearly from strong interactions with certain types of proteins at pH 3 to other proteins at pH 7 and then at pH 11 (as demonstrated in Figure 11.4b). Another interesting observation from the experiment is that, in contrast to pH effect, detergent (Triton X-100, 2% w/w) does not affect the interactions. The membrane proteins solubilizing in Triton X100 micelles are observed with a slight effect of pH to their interaction with NDs. Figure 11.4a shows that membrane proteins in the detergent phase (D-phase).

The ND–peptide interaction is weaker than that with proteins due to the smaller size of peptides. Experimental data have shown that peptides derived from the digestion of some water-soluble proteins, such as cytochrome *c*, dependently interact with the pH of surface-oxidized ND particles (13). Whenever the solution is 1 or 2 pH units more acidic than the pI value of a peptide, that peptide is

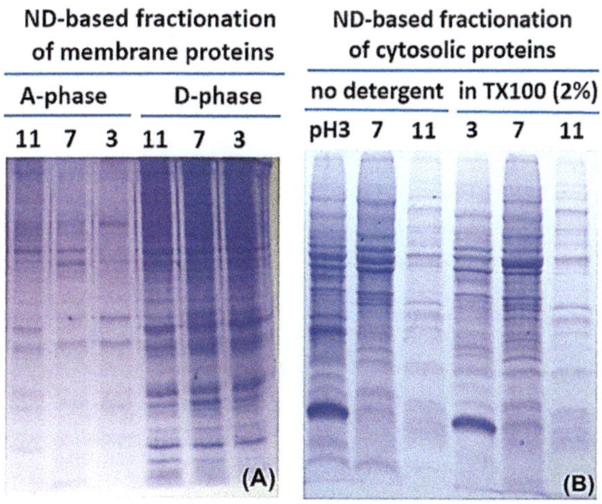

Figure 11.4 The effects of pH and detergent on the interaction between carboxylated/oxidized NDs with water-soluble proteins or detergent micelles solubilized membrane proteins. The input samples used were membranous pellet and cytosolic supernatant harvested from ultracentrifugation of mouse myeloma cell lysate. (a) Membrane proteins were first enriched from the membranous pellet with TX-100 two-phase separation. The proteins in the detergent phase (D-phase) were fractionated by sequential pH-tuned extractions with NDs. (b) The cytosolic supernatant or water-soluble proteins were prepared in two different buffers, with and without the addition of detergents, to test the influence of the surfactant and pH on protein extraction by NDs. *Source:* Reproduced from ref. (12) / MDPI/CC BY 4.0.

absorbed onto ND particles. In other words, peptides leave the ND's surface when the buffered pH increases from acidic to alkaline (Figure 11.5). This is a clear difference between how proteins and peptides interact with the surface of an ND. Proteins seemingly love NDs to the extent that they do not leave a ND easily, even under extreme alkaline conditions (11).

However, peptides and proteins share some common characteristics. For peptides with sizes of more than several tens of amino acids, adsorption to the ND surface reaches a maximum when the solution's pH is adjusted to the pI of those peptides. Understanding the similarities and differences between the interactions of ND with peptides and proteins has greatly contributed to the development of an ND-based platform for "shotgun" proteomics (8). However, an important aspect of the peptide–ND interaction that remains elusive is the interplay of hydrophobicity and ionic force. Nevertheless, because ND is a carbon-based material, hydrophobicity might also play a significant role.

11.5.3 Other Surface-Functionalized Nanodiamonds

Turning away from surface-oxidized NDs, which have been the most extensively used substances in MS-based analyses so far, researchers are increasingly

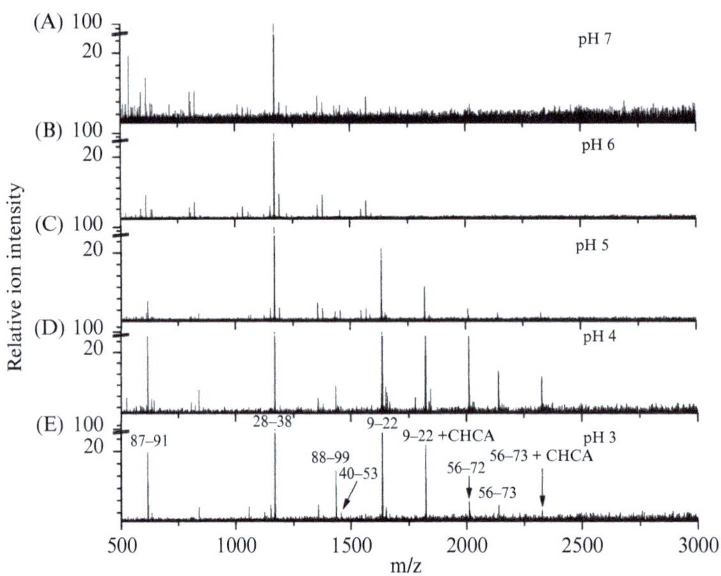

Figure 11.5 AP–MALDI mass spectra of cytochrome *c* trypsin digest on ND extraction from different pH-buffered solutions: pH 7 (a), pH 6 (b), pH 5 (c), pH 4 (d), and pH 3 (e). Each spectrum was obtained by concentrating 200 μl of a 5-nM solution with NDs. *Source:* (13) Sabu et al 2007/ with permission of Elsevier.

studying the functionalization of NDs for other applications. NDs can be readily surface-modified with a wide range of functional groups, such as carboxylate, hydroxyl, and amine (14). These groups can be extended to conjugate covalently or noncovalently with biomolecules. For applications in MS analysis, NDs combined with bioactive ligands or polymers are often used as high-affinity probes for the selective extraction and enrichment of molecules. For instance, after being coated with PL, ND particles can be applied to extract and quantitatively analyze DNA molecules with MALDI-MS (15). Other successful examples include the characterization of protein PTMs (14, 16), such as the enrichment and MS analysis of phosphorylated peptides (17, 18) and glycosylated proteins (19–21). Another exciting aspect of the application of surface-functionalized ND to the MS-based analysis of proteins is the immobilization of proteolytic enzyme to the surface of NDs. An enhanced proteolysis of trypsin can be achieved by covalently immobilizing the enzyme to the surface of a 3–10 nm carboxylated detonation nanodiamond (dND) (22). Another study has found that the enzymatic activity of lysozyme, which is noncovalently immobilized to acid-washed NDs' surfaces, depends on nanoparticle size (23).

11.6 SPEED Platform for MS-Based Study of Cellular Cytoplasmic Proteins and Human Fluids

The remarkable high-affinity capture of proteins by acid-oxidized NDs has led to the development of a platform called SPEED, an abbreviation for "solid-phase extraction and elution on diamond." The platform facilitates protein purification and isolation for sodium dodecyl sulfate polyacrylamide gel electrophoresis (SDS-PAGE), and it is especially suited for the MS-based analysis of proteins. The platform is able to extract and enrich diluted proteins from solutions containing high concentrations of salts, chaotropic reagents, and detergents, including sodium dodecyl sulfate (SDS). Proteins bound to the ND's surface can be eluted from the nanoparticles via electrophoresis buffers containing SDS (>1%) for facile analysis by SDS-PAGE. After a contaminant clean-up, the ND–protein complexes can also be directly deposited on a MALDI plate for MALDI-MS analysis. Moreover, chemical processing (i.e., disulfide bond reduction, cysteine alkylation, and proteolytic digestion) can be conducted efficiently on the particles to give proteolytic digests. This allows for their further application in proteomics. Remarkably, acid-oxidized ND is the only nanoparticle that has proved useful for practical applications, such as in membrane proteomic studies, so far (24).

Most proteins in the cytosol of cells and in human fluids are water-soluble and hydrophilic. Today, the MS analysis of this protein class is routinely easy. However, in many cases, such as in biomarker discovery and the analysis of complicated or highly contaminated samples, the need to concentrate low-concentration proteins and to cleanse samples before submission to an MS-based analysis is critical in every experimental design. Samples treated with NDs are one of the most expedient, effective, and sensitive approaches for the MS-based analysis of peptides, proteins, and proteomes. The versatility of NDs also allows for a wide range of MS-based studies, from the matrix-assisted laser desorption/ionization time-of-flight mass spectrometry (MALDI-TOF-MS) of individual proteins to the high-throughput and large-scale shotgun proteomics of proteomes. Figure 11.6 illustrates the "SPEED platform" with ND as its core component for the preparation of peptide, protein, and proteome samples for multiple applications of MS. This protocol was initially developed to analyze cellular cytosolic proteins and human fluids with MS (11).

The SPEED platform presents a wide range of applications for NDs in the preparation of samples for analysis at the peptide, protein, or even proteome level. ND is extremely efficient at capturing and enriching proteins from highly diluted and contaminated solutions. The method is simple and fast: Acid-oxidized NDs are added directly to protein solutions and mixed for 1 minute. To collect the proteins in an ND-bound form, a 3-minute centrifugation of 13 000 × g at room temperature is enough. Due to their strong noncovalent interactions with NDs, proteins can sustain extensive washes, during which undesired contaminants are

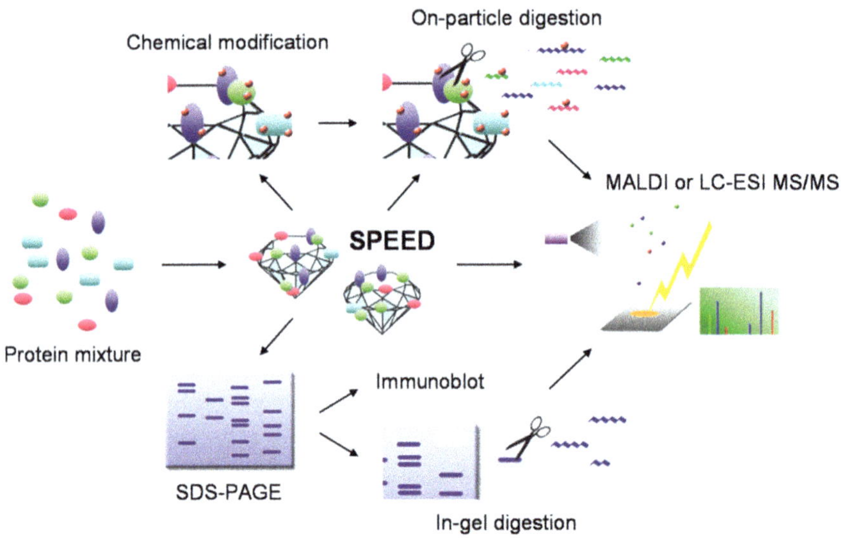

Figure 11.6 Scope of typical applications of the SPEED platform for proteome analysis. Abbreviations: ESI: electrospray ionization; LC: liquid chromatography; MALDI: matrix-assisted laser desorption/ionization; MS: mass spectrometry; SDS: sodium dodecyl sulfate; and PAGE: polyacrylamide gel electrophoresis. *Source:* (16) Wu C-C et al 2010 / John Wiley & Sons.

eliminated. For water-soluble proteins, an ND amount with a ratio of roughly 50:1 (ND:protein, w/w) is sufficient to capture all proteins in a solution. Kong *et al.* have reported that a conjunction of sample preconcentrations with NDs and laser desorption/ionization with UV matrixes introduces a new method for the time-of-flight mass spectrometric analysis of proteins in highly diluted and impure solutions (10). In a 1-ml solution, proteins with a high molecular weight, such as cytochrome *c*, myoglobin, and albumin, can be detected at concentrations as low as 100 pM. For the MS-based analysis of proteins in general, and MALDI-MS analysis in particular, the presence of contaminants including salts and surfactants often preclude the direct analysis of biological samples. With the help of NDs, researchers can now easily eliminate such interferences. NDs have been tested in the analysis of human blood serum, which is rich in salts and metabolites. MALDI-TOF-MS on proteins captured and cleansed by NDs shows a 10-fold increase in sensitivity and considerably richer spectral characteristics than those obtained using the Zip Tip method (Figure 11.7) (10).

The next step after protein enrichment and contaminant removal is the processing of the molecules for further characterization. Proteins bound to the surfaces of NDs can be eluted and separated by SDS-PAGE gels. They can also be chemically modified (e.g., by reduction and alkylation) and digested into peptides. Chen *et al.* demonstrated that the proteome analysis of one urine sample can be finished in 8 hours with the SPEED platform (11). In comparison with conventional techniques like protein depletion via affinity column, acid precipitation, filtration, dialysis, and 2-D SDS-PAGE followed by in-gel digestion, the SPEED platform is faster and more easily deployed.

Perona Martinez *et al.* developed a very simple approach based on the selective binding of serum proteins to the surface of NDs. By adding salt and NDs, high-abundance proteins could be significantly depleted, and low-abundance proteins could be accessed for analysis. After aggregation, the samples were centrifuged (13 200 rpm for 21 minutes) to remove supernatant called “soft corona.” The pellets were resuspended in distilled water and centrifuged once more to cleanse the particles. The supernatant was then removed, leaving just the firmly bound proteins behind in the pellet. The pellet was then freeze-dried (Figure 11.8). Remarkably, this work could identify proteins at a concentration of pg/ml following only a simple depletion step. Moreover, some biomarkers revealed in this study were undetectable with the use of a cutting-edge ultrahigh-performance LC column (which depletes the 64 most abundant serum proteins) (25).

In a recent study, Khuong *et al.* developed a novel approach of human proteome shotgun proteomics by analyzing human serum (26). This work proposes an innovative strategy for using NDs in conjunction with Triton X-114 protein extraction to fractionate crude serum into six pH-tuned fractions, simplifying the whole proteome and making protein profiling highly efficient. In this research, a total of 663 proteins were identified and evenly distributed throughout the fractions.

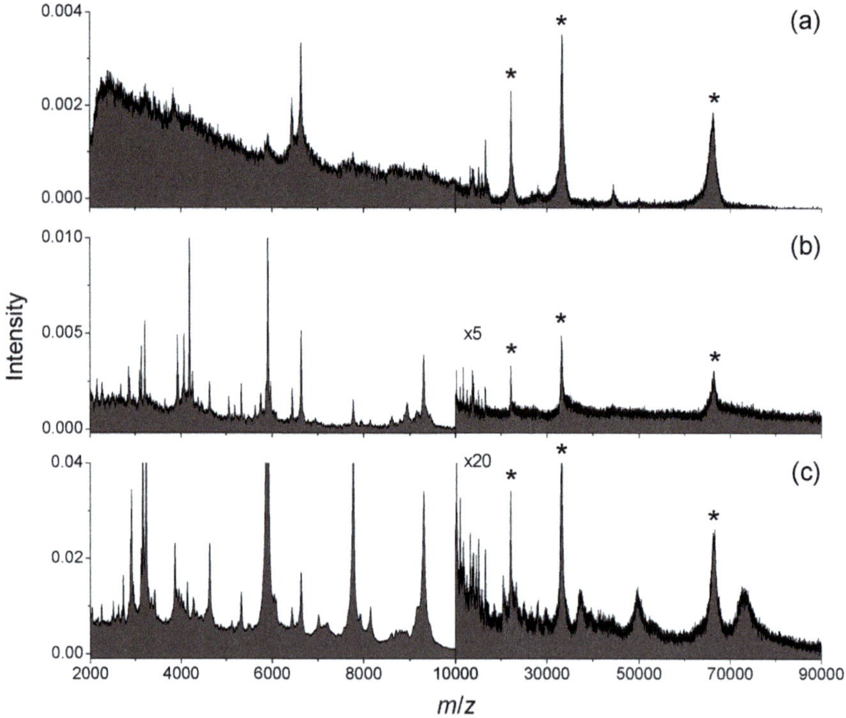

Figure 11.7 MALDI-TOF mass spectra of human blood serum acquired after 50-fold dilution with the α-Cyano-4-hydroxycinnamic acid (4HCCA) matrix solution (a), after pretreatment with the Zip Tip (b), and after 100-fold dilution with deionized water and then treated with diamond nanoparticles (c). The relative quantities of the serum samples consumed in obtaining these spectra are roughly 1:600:50. Note the changes of both x and y scales at m/z 10 000. Albumin peaks are denoted by asterisks (*). *Source:* (10) Kong et al 2005 / American Chemical Society.

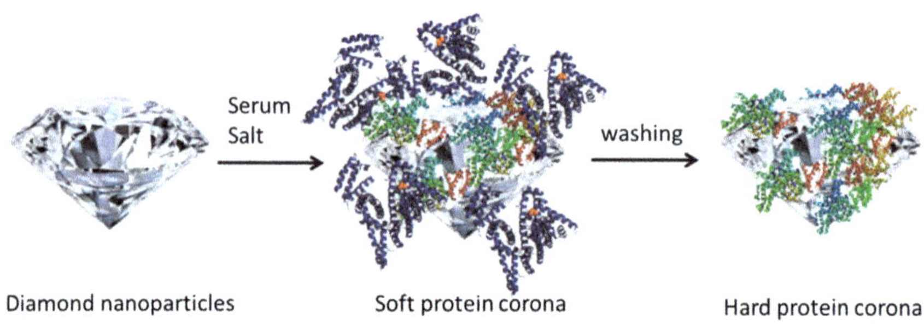

Figure 11.8 Schematic representation of the experiment: First, NDs and salts are mixed with the serum samples. Certain proteins (mostly proteins whose biological function is binding to negatively charged molecules) adhere to the diamonds. Analyzing proteins on the diamond particles reveals that high-abundance proteins were successfully depleted. At this point, loosely binding proteins, the so-called soft corona, still adhered. These proteins can be removed by an additional washing step, which further depletes some proteins. *Source:* Reproduced from ref. (25) / American Chemical Society.

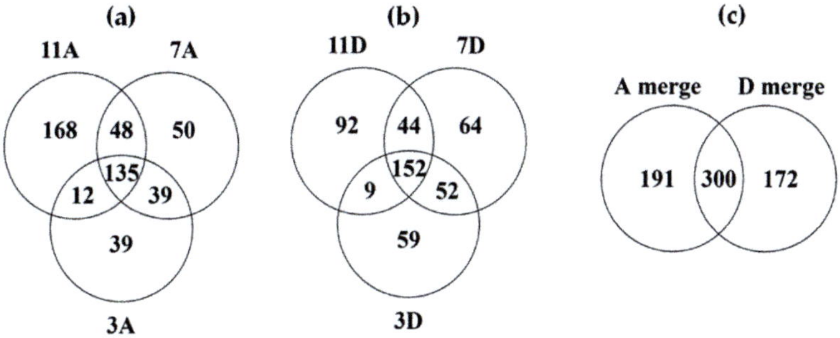

Figure 11.9 Overlapping Venn diagram. The diagram shows the overlapping results of each of the three fractions (pH11, pH7, and pH3) and graphically describes the data in Table 11.1. Diagrams (a) and (b) show the three fractions' overlapping results of the aqueous phase and detergent-rich phase, respectively. Diagram (c) shows the overall overlap of AP and DP, merged with the data of the three fractions from each phase. The graph was created by online Venny 2.0 (BioinforGP, Spanish National Biotechnology Center). *Source:* (26) Khuong et al 2022 / American Chemical Society.

Table 11.1 Number of proteins identified by Bioinformatic Analysis. The number of proteins is summarized in each fraction of pH values. Duplicated proteins are removed to give the total protein figures. *Source:* Ref. (26) / American Chemical Society.

Fractions of AP	Number of proteins	Fraction of DP	Number of proteins
pH 11	363	pH 11	297
pH 7	272	pH 7	312
pH 3	225	pH 3	272
AP merge	491	DP merge	472
Total number of proteins: 663			

Thirty-nine FDA-approved biomarkers were also discovered – a remarkable achievement (Figure 11.9). By fractionating crude serum using NDs in conjunction with denaturing detergent, this study lays the groundwork for a comprehensive, straightforward, and novel technique for human serum protein profiling.

11.7 SPEED Platform for MS-Based Study of Membrane Proteins

Membrane proteins, including integral membranes and membrane-associated proteins, play important roles in many cellular processes. They can be used as conduits for transmitting molecular and electrical signals within cells as well as between cells and their wider surroundings. Membrane proteins are the targets of 60–70% of modern medicinal drugs. Therefore, the identification and characterization of these proteins is crucial in many fields of research. Researchers estimate that about 25% of all genes in most genomes encode membrane proteins (27). Despite their importance, membrane proteins have eluded characterization due to their hydrophobic nature, low abundance in cells, and high level of dynamics in their expression levels in cell membranes.

There are several factors that challenge the application of MS-based analyses of membrane proteins, including its extreme hydrophobicity, the excessiveness of chemicals used to prepare membrane protein samples, and the difficulty of preparing samples. For example, to prepare membrane protein isolate for MS analysis, detergents and denaturing reagents are necessary to solubilize and purify these intrinsically hydrophobic proteins. Unfortunately, the presence of these reagents interferes with the downstream analysis of proteins and peptides due to ion suppression, inadequate digestion, and mass assignment errors. The elimination or reduction of these interfering compounds is, thus, an important step in membrane protein sample preparation. However, preparing a good sample for MS-based analysis of membrane proteins is a really time-consuming and labor-intensive process. Furthermore, processing procedures to reduce the amount of these chemicals in membrane protein isolate contributes to significant sample loss. In this section, we discuss how NDs can be utilized to facilitate the sample preparation of membrane proteins for MS analysis (Figure 11.10).

In the literature, we found various MS-based strategies for enhancing and simplifying the characterization of membrane proteins. Typically, detergents or

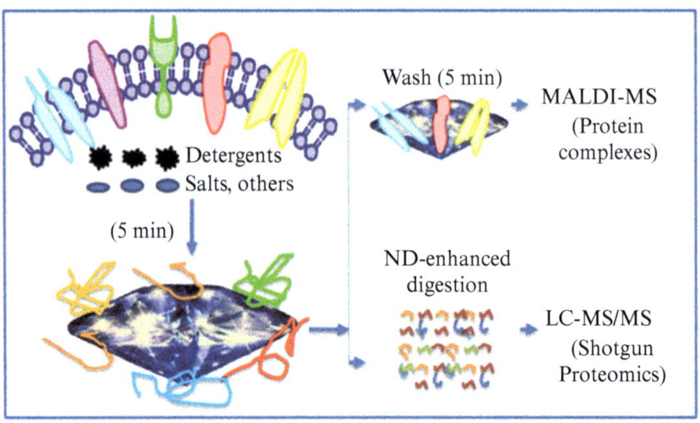

Figure 11.10 Flowchart of extraction, solubilization, and sample preparation of membrane proteins for MS-based studies using ND technology. *Source:* (8) Pham et al 2013 / American Chemical Society.

detergent-like reagents in the presence or absence of organic solvents – and sometimes denaturants – are utilized in the analysis of membrane proteins via MALDI-MS. However, even when "MS-compatible detergents" are used, deleterious effects are frequently reported. These negative effects include the suppression of signals and the undesired loss or gain of protein modifications. Thus, it is critical to reduce or eliminate detergents and other interfering contaminants prior to MS analysis. As demonstrated in Figure 11.11, ND treatment has greatly improved the MALDI-TOF analysis of a model membrane protein named particulate methane monooxygenase (pMMO). When performing MS of the protein in the absence of ND assistance, we could not see a clear signal of the membrane protein, especially when the protein is solubilized in urea or ionic detergents. But its MS signal reached high levels when the membrane was extracted onto NDs' surface for detergent cleaning and protein enrichment. These results indicate that membrane proteins prepared onto the surface of NDs are still in their subunit forms. In other words, NDs do not cut or cleave the amino acid chain. However, the native globular structure of the protein is somehow disrupted by the strong interaction with NDs' surface. The question of whether this disruption is severe and happens to other proteins remains open (8).

In "bottom-up" membrane protein analysis, a method whereby proteins are analyzed by fragmenting them into smaller pieces, scholars employ two major strategies to convert proteins into peptides: "in-gel" and "in-solution" digestion. Both of these commonly used methods have advantages and disadvantages.

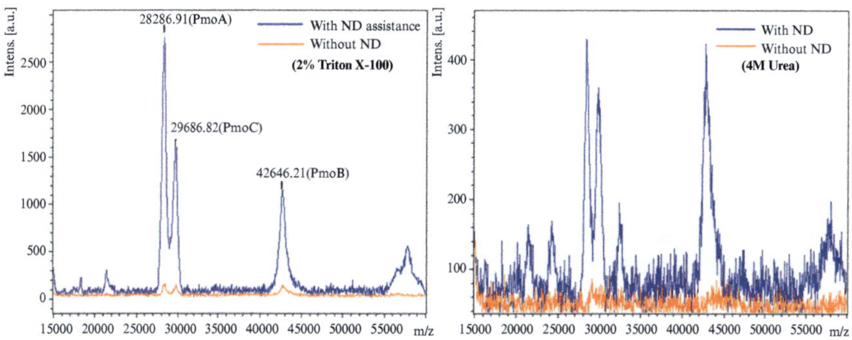

Figure 11.11 Demonstration of ND-assisted MALDI-MS for the analysis of the pMMO complex solubilized in mild detergent 2% Triton X-100 (left) and in 4 M urea (right). For experiments conducted without the aid of the nanoparticle (labeled "without ND"), 0.5-μl protein detergent micelles at a concentration of 50 nM and 0.5 μl matrix are directly deposited on the MALDI plate for analysis. The blue signals (labeled as "with ND") are obtained by MALDI-TOF analysis of the pMMO complex extracted onto the surface of diamond nanoparticles. *Source:* (8) Pham et al 2013 / American Chemical Society.

For example, the advantage of in-gel digestion is its resistance to impurities. However, the gel inhibits the access of proteolytic enzymes and the extraction of hydrophobic peptides. This approach is also labor-intensive as well as time-consuming. Although in-solution digestion, particularly filter-based in-solution digestion, is more easily automatable and reduces sample handling, research on this technique has revealed sample loss, partial digestion, and poor repeatability. Pham *et al.* have introduced an ND-based platform – an extension of SPEED for soluble proteins – for the preparation of membrane digestion (12). This ND method is a relatively new approach. However, it is inexpensive, easily deployed, and effective for membrane protein cleaning-up and proteolytic digestion. This research has shown that enzymatic digestions of membrane proteins are substantially facilitated because the proteins extracted on the NDs are exposed to the surface of the nanoparticles, rather than SDS gels or detergent solutions. The utility of NDs for analyses of complicated mixtures of membrane proteins and effective SDS removal has been demonstrated in shotgun proteomics analyses of the membrane-enriched fraction of *Escherichia coli* (TB1 strain) (8). A single 2-h 1D-LC-MS/MS run of tryptic peptides prepared via the on-surface ND digestion of a membrane fraction of *E. coli* (solubilized in 2% SDS) leads to the identification of more than 400 proteins, 70% of which can be classified as membrane proteins. For comparison, the best protein coverage obtained by the same 1-D LC-MS/MS analysis came from the RapiGest-assisted in-solution digestion with a number of identified proteins similar to what the ND-based method has obtained.

ND-based platforms have also shown prominent advantages when combined with detergent two-phase separations for the enrichment and fractionation of membrane proteins. As a first step in solubilizing and enriching the membrane proteins of mammalian cells, researchers have developed and optimized two-phase separation conditions using Triton X-100 or Triton X-114 as detergents (28). Membrane proteins solubilized in detergent micelles are fractionated and enriched by sequential pH-tuned extraction with NDs. Thanks to their high affinity for NDs, the adsorbed membrane proteins can sustain extensive washes to remove salts, detergents, and other impurities, thereby facilitating proteolytic digestion and downstream analysis. Combined with ultracentrifugation, this new platform can achieve a purity of approximately 70% membrane proteins, out of more than 2000 proteins identified in a shotgun proteomic assay for the whole membrane proteome of cultured mouse myeloma cells. This new platform is not only simple, fast, and cheap, but also effective in the sample preparation of membrane proteins for both GE and shotgun membrane proteomics (Figure 11.12).

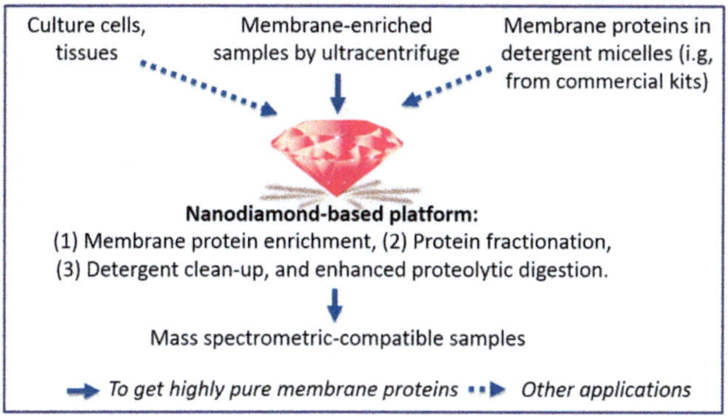

Figure 11.12 A general ND-based platform for the enrichment, fractionation, and sample preparation of membrane proteins. *Source:* This figure is adapted from ref. (12).

11.8 Surface-Functionalized Nanodiamonds for MS-Based Study of Protein Posttranslational Modifications

Protein PTMs refer to the covalent associations of functional molecules to proteins during or after protein biosynthesis. It is an interesting and effective way to increase the functional diversity of proteomes. Researchers have discovered that these modifications are key to many cellular biological processes and influence almost all aspects of normal cell biology and pathogenesis. Understanding the function of proteins through PTM profiling and characterization has been critical to the study of cell biology, disease treatment, and healthcare prevention. Together with mass spectrometric analysis, ND particles have actively been developed to analyze PTMs. ND-based approaches can improve and facilitate identifying and quantitatively analyzing many important PTMs, such as phosphorylated proteins and protein glycosylation.

11.8.1 Nanodiamonds for the Selective Enrichment and MS-based Analysis of Phosphopeptides

Protein phosphorylation is a reversible and complicated process. It is the common PTM of proteins, providing a mechanism for the rapid and flexible control of protein function and molecular signaling in cells. Due to evolution and the abundance of phosphate ions on earth, phosphorylation of proteins has emerged as one of the most prominent types of PTMs in all species (29).

To effectively analyze phosphorylation sites in proteins, the first and most crucial step is to enrich and purify phosphopeptides from complicated cellular lysates. Nowadays, commercial kits often use Fe^{3+}-IMAC and TiO_2 beads for the isolation and enrichment of phosphorylated peptides/proteins. However, Chang *et al.* used polyarginine (PA)-coated ND particles as a specific probe for the same purpose and were able to selectively extract and enrich multiphosphorylated peptides from complex samples (17). NDs and PA were conjugated by covalent bonding between the carboxylic moieties in NDs with the amine group of arginine via a 1-ethyl-3-(3-dimethylaminopropyl)-carbodiimide (EDC)-mediated coupling reaction. Since arginine has the guanidine moiety that is a typical recognition site in protein–protein interactions involving phosphate or sulfate moieties, the PA-ND conjugates showed an exceptionally high affinity for multiphosphorylated peptides due to multiple arginine–phosphate interactions. The application of PA-coated diamond NPs as high-affinity probes was found to selectively extract and concentrate multiphosphorylated peptides from the tryptic digestion products of α-casein, β-casein, and nonfat milk. The advantage of using these nanoparticle probes is that the extracted phosphopeptides can be directly examined by MALDI-TOF-MS without the need to remove the solid supports. This technique greatly simplifies analytical procedures. More recently, PA-coated ND has been coupled with TiO_2-coated ND for stepwise enrichment and fractionation of both multiply and singly phosphorylated peptides prior to on-probe MALDI-MS analysis (18). The method's viability was demonstrated utilizing synthetic peptides with varying numbers of phosphate groups, tryptic digests of α-casein, β-casein, and complicated protein mixtures. The approach's excellent specificity was demonstrated by its efficient enrichment and fractionation of phosphopeptides from β-casein and bovine serum albumin digest at a molar ratio as low as 1:1000, outperforming both commercial Fe^{3+}-IMAC and TiO_2 isolation kits (Figure 11.13). Thus, this method serves as a simple and powerful alternative for the fractionation and identification of multiple and singly phosphorylated peptides by MALDI-MS, allowing for more information to be deduced from limited starting materials.

11.9 Nanodiamonds for Enrichment and MS-Based Analysis of Glycopeptides

Protein glycosylation is the covalent attachment of "sugar" molecules to a polypeptide chain of proteins during or after protein biosynthesis. Glycosylated proteins play many important roles in various biological functions. Like protein phosphorylation, the glycosylation of proteins serves as a control of molecular

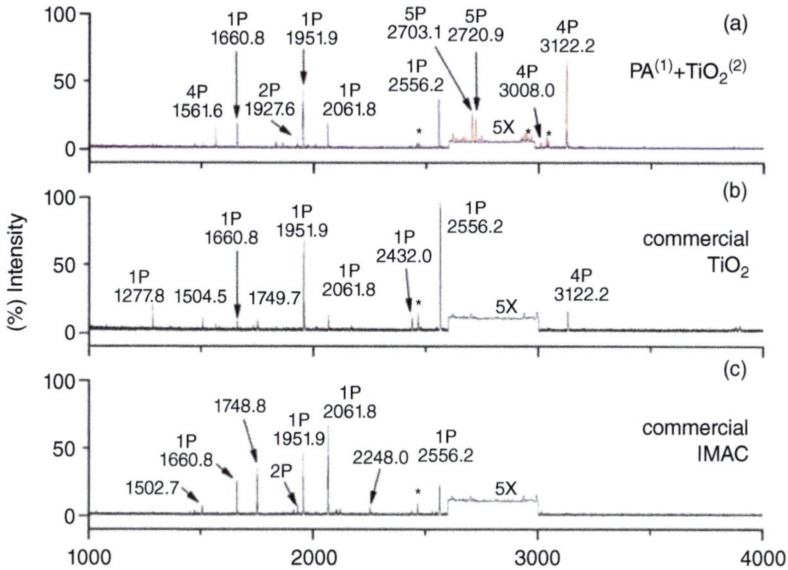

Figure 11.13 MALDI-TOF mass spectra of a peptide mixture (total 1 pmol), originating from tryptic digests of two phosphoproteins (a-casein, and b-casein) and five nonphosphoproteins (BSA, carbonic anhydrase, cytochrome *c*, myoglobin, and transferrin). The mass spectra were obtained by using (a) PA-coated NDs and TiO_2-coated NDs in series, (b) commercial TiO_2 beads, and (c) commercial Fe^{3+}-IMAC beads, respectively. *Source:* (18) Shiau et al 2011 / Royal Society of Chemistry.

signaling inside cells. More importantly, glycosylated proteins have been found to be the most prominent PTM on the surface of both mammalian cells and viruses. Researchers have revealed that the "glycan" molecules that are covalently attached to cell surfaces or membrane proteins are crucial for interactions between cells and their surrounding environment. In addition, glycosylation has profound effects on cell–cell interactions, the immune response, cell growth, and pathogenesis. The discovery and identification of glycopeptides and glycosylated proteomes has, therefore, become an important area of basic and translational research.

MS-based analysis is currently the only method applicable for the high-throughput characterization of protein glycosylation. It is, so far, the key technology to unraveling the biological function of glycosylated proteins in a context of complicated cellular networks. However, protein glycosylation is labile, which results in poor reproducibility and complications in analysis. MS-based characterizations of glycopeptides/proteins also suffer from poor ionization efficiency with respect to unmodified counterparts, the inherent heterogeneity of each glycosylation site, and the complex nature of glycan structure. Moreover, many glycosylated proteins, including signaling components, have low abundance in cellular lysates, further complicating their analysis. ND-particle-based methods have, therefore, emerged as important tools in characterizing important protein PTMs.

To enrich and purify glycosylated peptides/proteins from complicated samples, ND particles functionalized with boronic acid have been developed (19, 20). In earlier work, Yeap *et al.* investigated the specific capture of glycoproteins by dND conjugated with alkyl spacer chain and aminophenyl boronic acid (APBA) (dND-spacer-APBA) (19). Using a mixture of some model proteins (three glycoproteins (ovalbumin, fetuin, and RNase) and BSA [nonglycoprotein]), authors demonstrated that dND-spacer-APBA can selectively extract glycoproteins from nonglycosylated ones. After glycoprotein capture, complexes are ready to be analyzed directly by MALDI-MS. This study also pointed out that the spacer between ND surface and boronic acid moiety is critical to eliminating the nonselective binding of proteins. The role of the alkyl spacer chain applied in this work was to create an exclusion shell that inhibits nonspecific binding with nonglycated proteins and minimizes steric hindrance among the bound glycoproteins. Although the nonspecific binding of proteins was not totally eliminated, this study showed that the insertion of an alkyl spacer chain resulted in a much lower degree of nonspecific binding in comparison with the direct functionalization of carboxylated ND. Figure 11.14(I) illustrates the reaction scheme to prepare the dND-spacer-APBA.

In order to further reduce nonspecific binding between nonglycoproteins with boronic-acid-functionalized NDs, Xu *et al.* introduced a new spacer design linking boronic acid with the carboxylic groups on the surface of NDs (20). As seen in Figure 11.14(II), PL was first covalently attached to the ND via EDC/NHS-mediated coupling following the conjugation of boronic acid moieties to the amine group of PL assisted by PEG-diglycolic acid (PEG). Compared to the preparation of the dND-spacer-APBA, although there are additional chemical reactions, the introduction of PL as the first spacer greatly amplifies the number of boronic acid moieties attaching to one nanoparticle. As a result, the new functionalized ND (dND-p-APBA) has good dispersibility in aqueous solution and is promising for glycoproteome analysis. The dND-p-APBA was highly specific and efficient in capturing glycopeptides from tryptic digests of horseradish peroxidase (HRP), bovine fetuin, horse heart myoglobin (30), b-casein, and mouse liver tissue. The glycopeptides captured by the beads could then be used for the direct analysis with MALDI-MS or they could be eluted from the beads for characterization by LC-MSMS analysis. In comparison with commercial boronic acid-functionalized magnetic nanoparticles, the dND-p-APBA showed higher specific enrichment efficiency for glycopeptides.

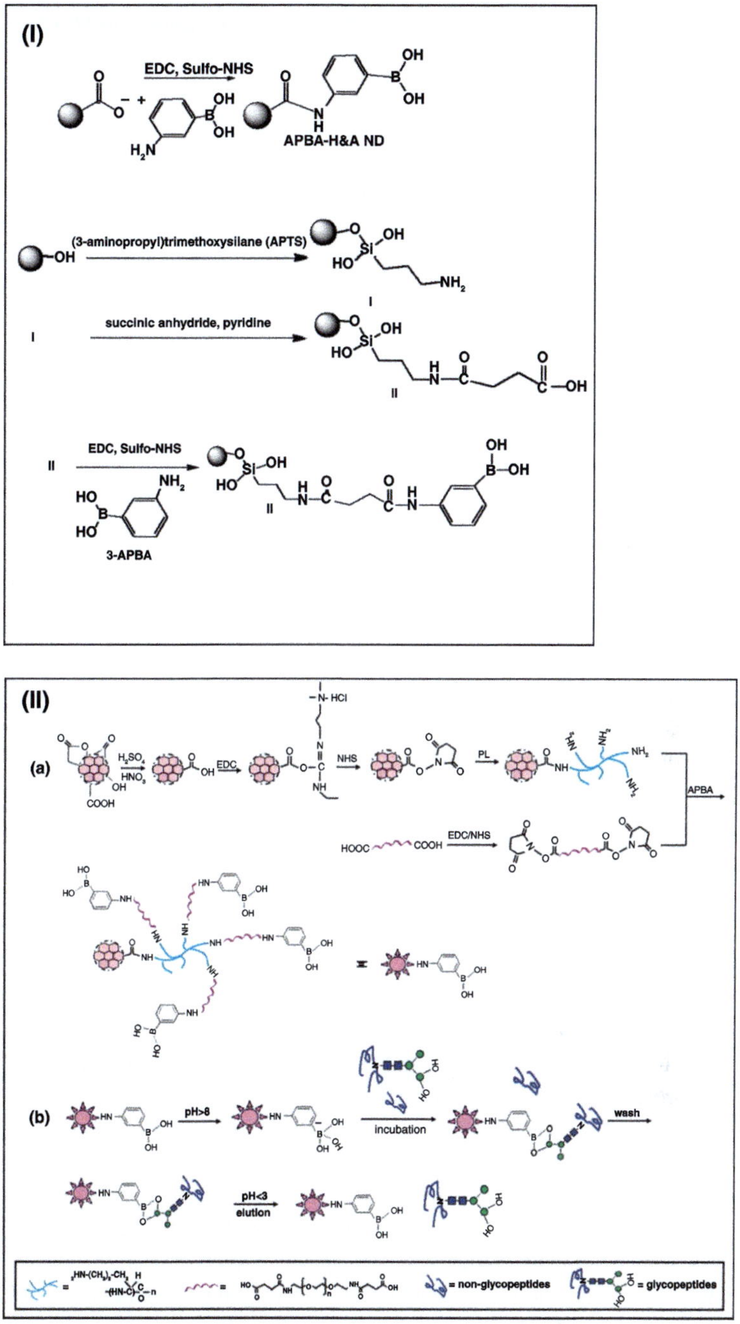

Figure 11.14 Functionalization of ND with boronic acid for the specific enrichment and purification of glycopeptides and glycoproteins. (I) "First generation." *Source:* (19) Yeap et al 2008 / American Chemical Society. (II) "Second generation." *Source:* (20) Xu GB et al 2013 / Royal Society of Chemistry. The term "first generation" and "second generation" merely indicates the different time points of the two boronic acid-functionalized ND conjugates.

Acknowledgments

This work was supported by the National Foundation for Science and Technology Development (NAFOSTED), Vietnam, under Grant No. 106.02-2018.38.

References

1 Munson MSB, Field FH. Chemical ionization mass spectrometry. I. General introduction. *J Am Chem Soc*. 1966;88(12):2621–30.

2 Macfarlane RD, Torgerson DF. Californium-252 plasma desorption mass spectroscopy. *Science*. 1976;191(4230):920–5.

3 Barber M, Bordoli RS, Sedgwick RD, Tyler AN. Fast atom bombardment of solids (F.A.B.): A new ion source for mass spectrometry. *J Chem Soc, Chem Commun*. 1981(7):325–7.

4 Bothner B, Dong XF, Bibbs L, Johnson JE, Siuzdak G. Evidence of viral capsid dynamics using limited proteolysis and mass spectrometry*. *J Biol Chem*. 1998;273(2):673–6.

5 Whitelegge J. Intact protein mass spectrometry and top-down proteomics. *Expert Rev Proteom*. 2013;10(2):127–9.

6 Mann M, Kulak Nils A, Nagaraj N, Cox J. The coming age of complete, accurate, and ubiquitous proteomes. *Mol Cell*. 2013;49(4):583–90.

7 Mochalin VN, Shenderova O, Ho D, Gogotsi Y. The properties and applications of nanodiamonds. *Nat Nanotechnol*. 2012;7(1):11–23.

8 Pham MD, Yu SS-F, Han C-C, Chan SI. Improved mass spectrometric analysis of membrane proteins based on rapid and versatile sample preparation on nanodiamond particles. *Anal Chem*. 2013;85(14):6748–55.

9 Huang LCL, Chang H-C. Adsorption and immobilization of cytochrome c on nanodiamonds. *Langmuir* 2004;20(14):5879–84.

10 Kong X, Huang L, Hsu C-M, Chen W-H, Han C-C, Chang H-C. High-affinity capture of proteins by diamond nanoparticles for mass spectrometric analysis. *Anal Chem*. 2005;77(1):259–65.

11 Chen WH, Lee SC, Sabu S, Fang HC, Chung SC, Han CC, *et al*. Solid-phase extraction and elution on diamond (SPEED): A fast and general platform for proteome analysis with mass spectrometry. *Anal Chem*. 2006;78(12):4228–34.

12 Pham MD, Wen T-C, Li H-C, Hsieh P-H, Chen Y-R, Chang H-C, *et al*. Streamlined membrane proteome preparation for shotgun proteomics analysis with Triton X-100 cloud point extraction and nanodiamond solid phase extraction. *Materials*. 2016;9(5):385.

13 Sabu S, Yang F-C, Wang Y-S, Chen W-H, Chou M-I, Chang H-C, *et al*. Peptide analysis: Solid phase extraction–elution on diamond combined with atmospheric

pressure matrix-assisted laser desorption/ionization–Fourier transform ion cyclotron resonance mass spectrometry. *Anal Biochem*. 2007;367(2):190–200.

14 Hartmann M, Betz P, Sun Y, Gorb SN, Lindhorst TK, Krueger A. Saccharide-modified nanodiamond conjugates for the efficient detection and removal of pathogenic bacteria. *Chem Eur J*. 2012;18(21):6485–92.

15 Kong XL, Huang LCL, Liau SCV, Han CC, Chang HC. Polylysine-coated diamond nanocrystals for MALDI-TOF mass analysis of DNA oligonucleotides. *Anal Chem*. 2005;77(13):4273–7.

16 Wu C-C, Han C-C, Chang H-C. Applications of surface-functionalized diamond nanoparticles for mass-spectrometry-based proteomics. *J Chin Chem Soc*. 2010;57(3B):583–94.

17 Chang C-K, Wu C-C, Wang Y-S, Chang H-C. Selective extraction and enrichment of multiphosphorylated peptides using polyarginine-coated diamond nanoparticles. *Anal Chem*. 2008;80(10):3791–7.

18 Shiau KJ, Hung S-U, Lee H-W, Wu C-C. Nanodiamond-based two-step sampling of multiply and singly phosphorylated peptides for MALDI-TOF mass spectrometry analysis. *Analyst*. 2011;136(9):1922–7.

19 Yeap WS, Tan YY, Loh KP. Using detonation nanodiamond for the specific capture of glycoproteins. *Anal Chem*. 2008;80(12):4659–65.

20 Xu GB, Zhang W, Wei LM, Lu HJ, Yang PY. Boronic acid-functionalized detonation nanodiamond for specific enrichment of glycopeptides in glycoproteome analysis. *Analyst*. 2013;138(6):1876–85.

21 Hsieh C-C, Guo JY, Hung S-U, Chen R, Nie Z, Chang H-C, *et al*. Quantitative analysis of oligosaccharides derived from sulfated glycosaminoglycans by nanodiamond-based affinity purification and matrix-assisted laser desorption/ionization mass spectrometry. *Anal Chem*. 2013;85(9):4342–9.

22 Wei LM, Zhang W, Lu HJ, Yang PY. Immobilization of enzyme on detonation nanodiamond for highly efficient proteolysis. *Talanta*. 2010;80(3):1298–304.

23 Perevedentseva E, Cai PJ, Chiu YC, Cheng CL. Characterizing protein activities on the lysozyme and nanodiamond complex prepared for bio applications. *Langmuir*. 2011;27(3):1085–91.

24 Alfonso-Garrido J, Garcia-Calvo E, Luque-Garcia JL. Sample preparation strategies for improving the identification of membrane proteins by mass spectrometry. *Anal Bioanal Chem*. 2015;407(17):4893–905.

25 Perona Martinez F, Nagl A, Guluzade S, Schirhagl R. Nanodiamond for sample preparation in proteomics. *Anal Chem*. 2019;91(15):9800–5.

26 Khuong HL, Chen C-H, Lin J-L, Le T-N, Pham TH, Le TBT, *et al*. Nanodiamond solid-phase extraction and Triton X-114 cloud point separation for robust fractionation and shotgun proteomics analysis of the human serum proteome. *J Proteome Res*. 2022;21(1):67–76.

27 Krogh A, Larsson B, von Heijne G, Sonnhammer ELL. Predicting transmembrane protein topology with a hidden Markov model: Application to complete genomes *J Mol Biol*. 2001;305(3):567–80.

28 Arnold T, Linke D. Phase separation in the isolation and purification of membrane proteins. *Biotechniques*. 2007;43(4):427–40.

29 Hunter T. Why nature chose phosphate to modify proteins. *Philos Trans R Soc Lond, B, Biol Sci*. 2012;367(1602):2513–6.

30 Kim Y, Kim HS, Yang Y, Baek K, Choi N, Park M-H. Quantitative analysis of hydrogen peroxide using ratiometric fluorescent probe-doped silica nanoparticles. *Toxicol Environ Health Sci*. 2017;9(2):108–15.

12

Emerging Roles of Artificial Intelligence in Nanodiamond Sensing

Shahzad Ahmad Qureshi, Haroon Aman, Yu-Chen Lin, and Wesley Wei-Wen Hsiao**

* *Corresponding authors*

Artificial intelligence (AI) can enhance the accuracy, data acquisition speed, and analysis for various types of nanosensing platforms and modalities. It can also predict theoretical models of efficient sensors with optimized physical and chemical properties for clinical purposes. Regardless of their type and size, nanosensor responses can be accurately predicted prior to any actual experimentation. The implementation of AI can speed up the drug discovery process and its clinical testing. In real-time medical screening applications (magnetic resonance imaging [MRI] and computed tomography [CT] scan), the nanosensor signals include unexpected fluctuations mixed with time-varying noisy features. These unpredictable events can cause abnormalities in the sensor behavior, including systematic and random errors in measurements. If these measurements are being applied to patients, then the reliability of treatment can be compromised. Therefore, in medical treatment, real-time monitoring with noise cancellation and filtering is indispensable. Large healthcare data repositories available worldwide can also be accessed to train and model the physical system for the required feature extraction. Some popular algorithms used to implement the necessary conditions of signal monitoring and processing are: artificial neural network (ANN), support vector machine (SVM), random forest (RF), and *k*-nearest neighbors (*k*-NN). Efforts to develop AI-based biosensors with high sensitivity and specificity are underway. These sensors can also train, detect, and classify biological signals arising from disease biomarkers in minimal time. There are also attempts to develop novel biosensing platforms to accurately classify different biomolecules (including proteins, nucleic acids, and enzymes). Large-scale data generated by online genomic and proteomic studies can be utilized for drug discovery. Thus, AI-developed biosensors have the potential to revolutionize medical diagnostics and personalized medicine for reliable disease detection. In conclusion, AI-based algorithms can be used to develop smart nanosensors and therapeutic tools with

Nanodiamonds in Analytical and Biological Sciences: Principles and Applications, First Edition.
Edited by Wesley Wei-Wen Hsiao and Trong-Nghia Le.
© 2023 John Wiley & Sons Ltd. Published 2023 by John Wiley & Sons Ltd.

optimized chemical, optical, and electrical properties to generate high-quality signals in minimal time for a wide range of diseases. A schematic of an AI-integrated biosensor is illustrated in Figure 12.1.

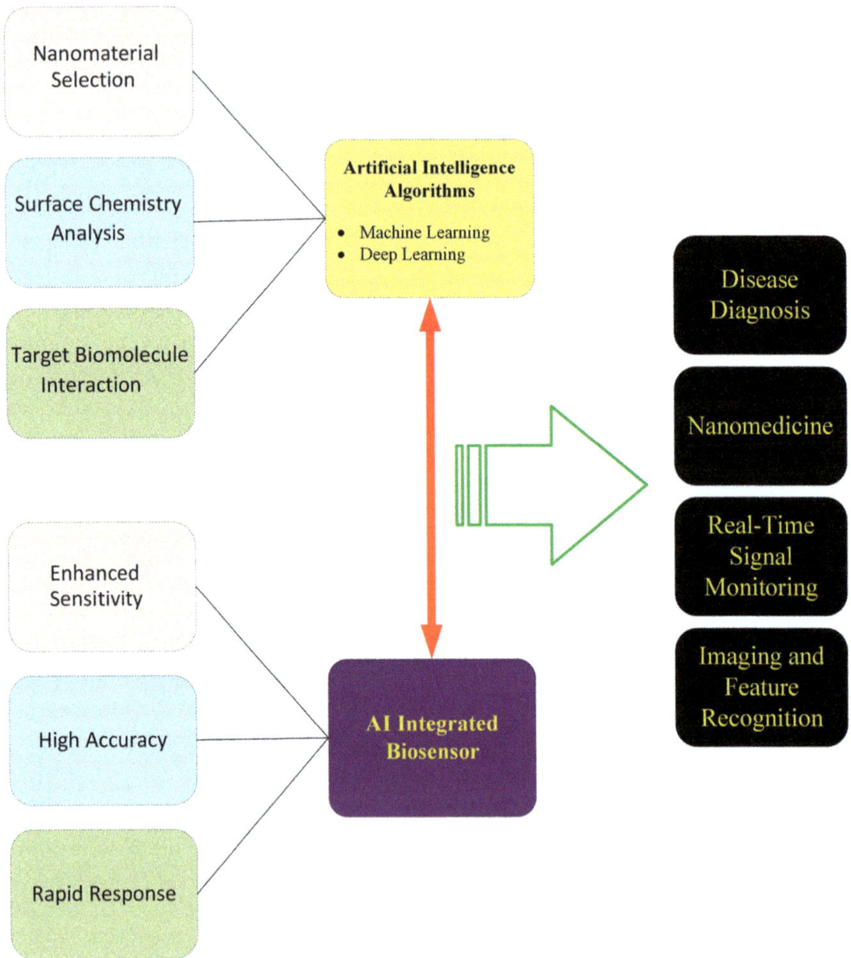

Figure 12.1 A block representation of an AI-based biosensor. The training data are provided regarding libraries consisting of physiochemical properties related to target molecules and nanomaterials. AI algorithms develop models using input data and predict the ingredients necessary for the synthesis of highly sensitive, accurate, and efficient nanosensors. The consequences are shown on the right side as applications in disease diagnosis, nanomedicine, live signal monitoring, and analysis.

In the forthcoming sections, we compile some recently reported applications of machine learning (ML) and deep learning (DL) that have been implemented to enhance the scope of diamond nanosensing in physics and biology.

12.1 Machine Learning Algorithms for Nanodiamond Sensing

ML, a subset of AI, involves the development of algorithms enabling computers to learn and make predictions. The process involves providing the ML algorithm with training data and model development to identify patterns and relationships in the data. The optimized model can then be used to make predictions or decisions about new test data that it has not seen before (1). Nitrogen-vacancy (NV) centers in FNDs are highly sensitive nanosensors that allow the detection of very weak electromagnetic signals, enabling nanoscale exploration in various applications (2). However, the data generated by these nanosensors can be complex, and analyzing them can be challenging and time-consuming. ML algorithms can help us to improve fluorescent nanodiamond (FND) sensing in this situation by analyzing large volumes of data in a very short time:

12.1.1 Support Vector Machine

In SVM, the mapping of supervised learning data ($\underline{x}^t$, y^t), as paired data vector and label, forms X as the dataset with feature space S so that the discrimination boundary (DB) is maximized across the least-confidence points using "m" features (3). The support vectors near the hyperplane depend on maximizing the margin. The hyperplane function, for the ith class is: $g_i(\underline{x}) = \boldsymbol{w}_i'\underline{x} + b\ ; \boldsymbol{w} \in S, b \in \mathbb{R}$, where $\underline{x}$ is training vector, w_i' is a vector that is orthonormal to DB, and b represents its bias. The optimum DB can be derived by maximizing the margin as depicted by minimize $\frac{1}{2} \| \boldsymbol{w} \|^2$, subject to

$$wx^t + b \begin{cases} \geq 1 & for\ y^t = 1, \\ otherwise & y^t = -1, \end{cases} \qquad \forall t\ x^t \in \mathbb{R}^m, y^t \in \{+1, -1\}. \tag{1}$$

12.1.2 *k*-Nearest Neighbors

The k-NN is built on the idea that the query q will have a higher number of nearest neighbors for the class it belongs to. The distance l is considered between q and k number of nearest neighbors ($\underline{x}^t$) that belong to class y^t. q is considered to lie in the class y^i having a larger number of nearest neighbors among k-NNs (4). The Minkowski distance, given by $l_{mk}(q, \underline{x}^t) = \left(\sum_{t=1}^{k} \left|q - \underline{x}^t\right|^r\right)^{\frac{1}{r}}$, is used to measure the distance between two instances in the feature space. The value of r estimates the effect of q for distance to the most far-off located nearest neighbor.

12.1.3 Random Forest

A decision tree is a sequence of queries about the data that eventually leads to a prediction. The RF algorithm is based on the multiple decision trees that act as base learners during the training phase. It uses bagging and bootstrapping procedures that help to boost performance. An unpruned tree is grown from each sample, and the node split is determined by selecting the appropriate predictor x^p at each node p from randomly chosen predictors. The class label u is assigned to an instance based on the majority of the labels v^i of the ith class predicted by the base learners (5): $u = \arg_i^{\max} P(v^i \mid p)$.

12.1.4 Artificial Neural Network

The ANN is a classifier that is inspired by the structure and function of biological "neurons" found in the human brain (6). The interconnected neurons or nodes form the network organized into layers and are designed to learn from input data through a process known as training. During training, the weights are adjusted in the network between its neurons to minimize a specified error function, such as the difference between the network's predicted and actual outputs. This allows the ANN to gradually improve its performance on a given task, such as classification or regression.

12.1.5 Convolutional Neural Networks

ANN is a shallow learning algorithm with one hidden layer. In contrast, DL algorithms have a multihidden layer structure. One of DL's types is convolutional neural network (CNN), commonly used for classification purposes. For 2-D input data, the arrangement of layers or architecture takes intensity-normalized, resized, and randomly permuted images from the input layer and processes them through convolution layers to get feature maps, followed by the selective activation of neurons (rectified linear unit [ReLU]) and downsampling (Pooling) to get useful information (6). A dropout layer is used before the fully connected layers to avoid overfitting. The softmax layer extracts training and test features, predicting output probabilities. The classification layer detects brain tumor class. In diamond nanosensing, the CNN can analyze FND signals and identify patterns or anomalies. A schematic diagram of ML and DL frameworks is illustrated in Figure 12.2a, whereas a residual CNN architecture is shown in Figure 12.2b (7). Some of the important components of CNN are given as follows.

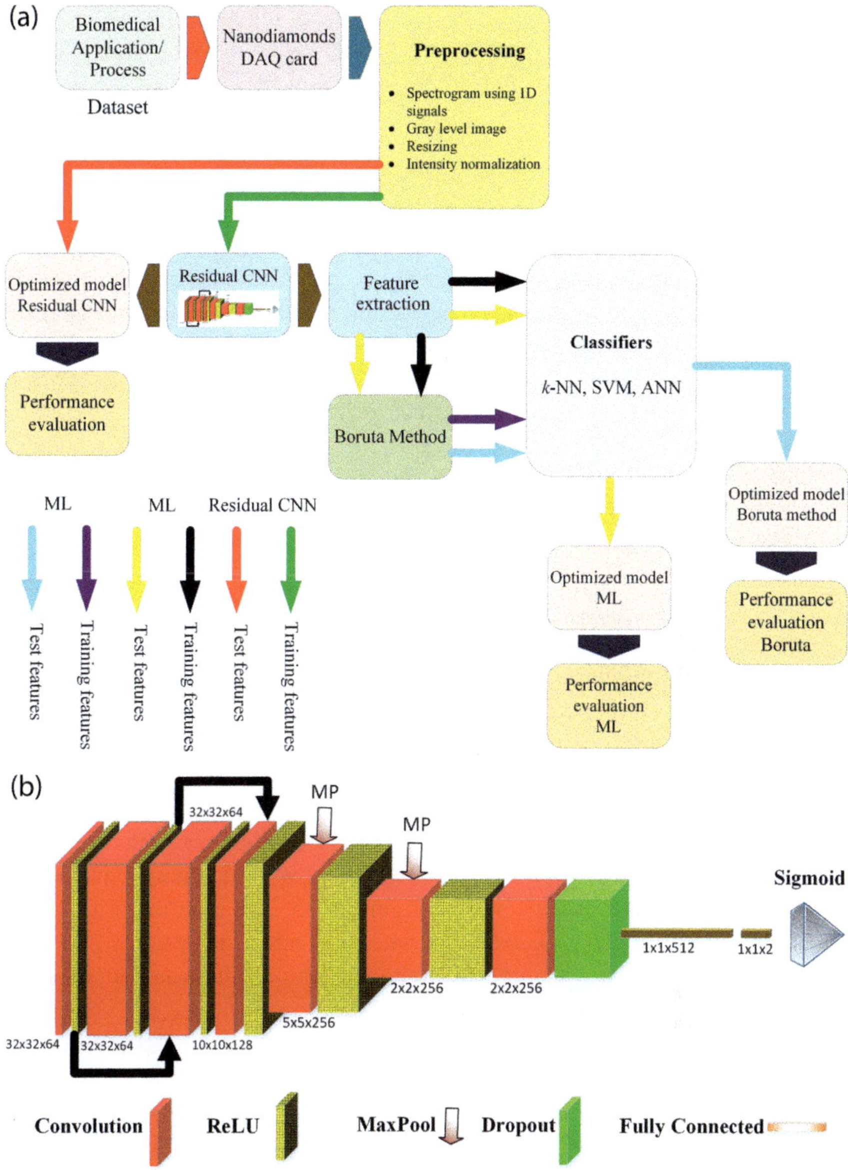

Figure 12.2 A schematic representation of ML- and DL-based frameworks. (a) ML and residual CNN-based system using Boruta feature selection algorithm to isolate discriminative features from the dynamic features extracted by DL. (b) Residual CNN architecture using two times downsampling via MaxPooling.

a) ***Input Layer***
The input layer is used to ensure the congruency of image size as required by the architecture and normalization of image data. The image matrix is fed to the layer-wise sequential setup of CNN that reduces loss by going through different mini-batches.

b) ***Convolution Layers***
The human brain captures important information or patterns in images. The CNN works similarly, collecting/extracting information by passing filters over an image in strides and storing convolution results in a matrix, namely a feature map. To capture the distinctive details present in the form of intensity variations in an image, the convolution filters have a receptive field, an area in an image to be convolved with a kernel. The term filter refers to a 2-D array of multiple kernels, constituting parameters, stacked together. For a particular convolution layer, the number of elements or kernels in a filter stack, determining the depth of the outcome or number of feature maps. The convolution process is given by

$$I'[m,n] = (I * h)[m,n] = \sum_{o}\sum_{p} h[o,p]\, I[m-o, n-p], \tag{2}$$

where I and I' represent the original image and feature map, respectively, and h is one of the kernels from the filter stack. A schematic diagram of the convolution process is illustrated in Figure 12.3a.

c) ***Activation Layer***
The ReLU is commonly used as the activation function in CNNs. Compared with other activation units, it is a nonsaturated function that reduces training time (8). It converts all negative values to zero (nonlinear behavior) while behaving linearly for all positive values, as real-world learning requires non-negative values. It is illustrated in Figure 12.3b. The equation of the ReLU model is given by $x_{out} = max(0, x_{in})$, where the output intensity, x_{out}, is equal to the input intensity, x_{in}, when it is a positive value; otherwise, output is considered as a zero value.

d) ***Maximum Pooling Layer***
The basic objective of a maximum pooling (MaxPool) layer is to reduce the data size to be processed in successive layers, modifying the pixel (matrix coefficient) values and making it insensitive to slight variation. Preferably, it is carried out by sliding a 2×2 sized window over a feature map and selecting the maximum value out of the four values in the matrix. The stride of the MaxPool layer is a tunable parameter. The principle of the MaxPool layer is illustrated in Figure 12.3c. At first glance, it seems as if the information is being lost. However, it is capturing useful information out of the noise in data (9–11).

e) ***Dropout Layer***
The basic objectives of the dropout layer are to improve generalization (by reducing overfitting) and to increase the computational speed. The

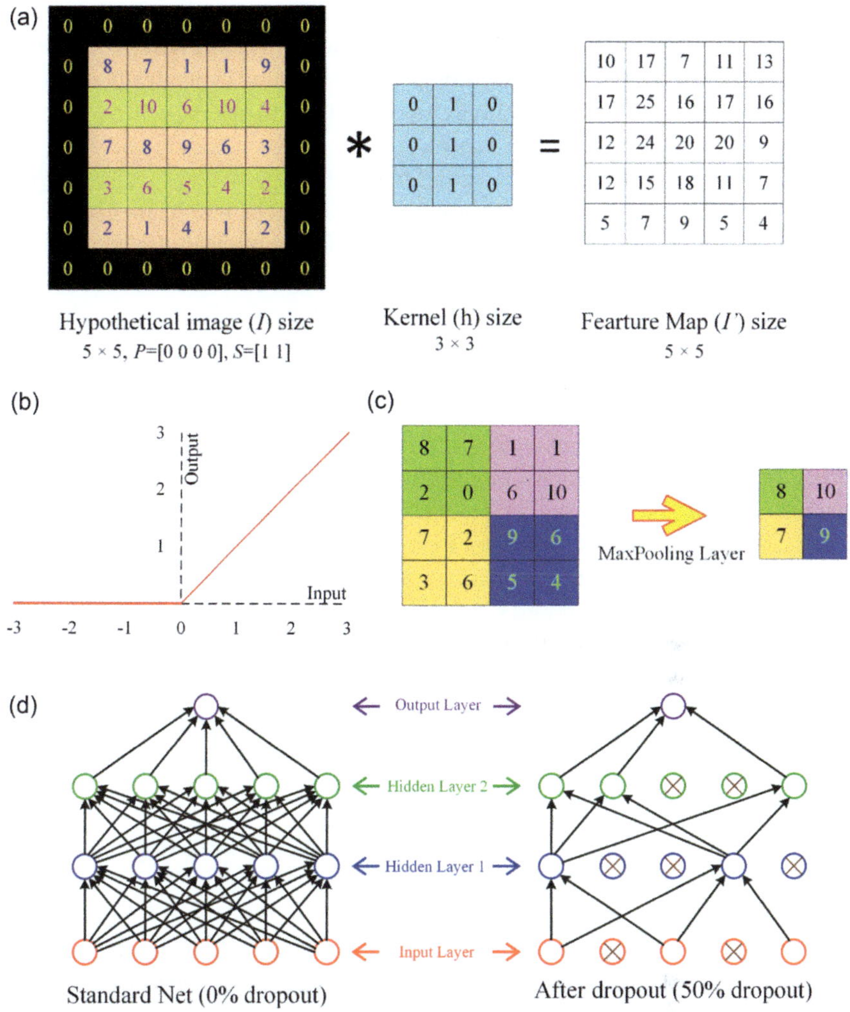

Figure 12.3 The schematic diagrams of various important layers of CNN. (a) The convolution process for a hypothetical image. (b) ReLU activation function for the proposed approach. (c) Operation of MaxPooling layer; input size = 4 × 4, stride $[S_x, S_y]$ = [2, 2], and output size = 2 × 2 (from the same colored boxes on the left, forming a 2 × 2 matrix, only the maximum value is selected going to a single small box of the corresponding color on the right, thereby reducing the size to one-half). (d) Function of dropout layer illustrated with a hypothetical network: without dropout layer applied, and with 50% units dropped out randomly (both inward and outward connections).

dropout refers to the random removal of hidden and visible activation units, thereby making them temporarily unavailable with all their inward and outward connections (12). The dropout layer is illustrated in Figure 12.3d.

f) ***Full Connected Layer***
This layer connects neurons in the preceding layer to neurons in the following layer. It helps classify images during training by using dynamically extracted features from forthcoming softmax and classification layers.

g) ***Softmax and Classification Layers***
Softmax layers are used for squashing, using the sigmoid function, the non-normalized output of a fully connected layer for multiclass categorization to a probability distribution for the predicted classes in the range [0, 1]. The classification layer determines the cross-entropy loss (CEL) for binary classification and predicts classes for all images.

12.1.6 Principal Component Analysis

The principal component analysis (PCA) is a technique used to discard similar features reducing dimensions in the input vector to discriminative features, thereby making the training task less complex (13). Therefore, the segregated most important features of a dataset are used to visualize high-dimensional data mapped to a lower-dimensional feature space.

These are just a few examples of ML algorithms highly dependent on the complexity of the underlying input data distribution. The choice of algorithm is determined by the specific diamond nanosensing and the characteristics of the data being analyzed. From this point onward, discussions are limited to FND biosensing.

12.2 Machine Learning Revolutionizes Biosensing and Disease Diagnosis

ML algorithms can significantly enhance the scope and performance of florescent nanosensors toward disease diagnosis and rapid on-site prediction of infectious species through a variety of customized sensing platform (14). Disease-related biomarkers with lower physiological concentrations are mostly accompanied by noisy interferences, making them difficult to observe using conventional techniques. Therefore, the chemical and optical properties of these nanosensors must be optimized through programmable nanoparticles with surface engineering. Microscopic interactions between biomolecules and nanosensors can create complex signals, which are difficult to analyze manually (15). In this situation, ML can assist in rapid on-chip disease prediction involving feature extraction, analysis, and reconstruction through predictive models.

The creation of customized smart, highly sensitive nanosensors with specificity enough for detecting ultralow physiological concentrations of target biomolecules (2) requires a demanding application of ML algorithms. The control and optimization of optical properties through nanostructure surface design are essential for improving performance. However, due to extensive chemical parametric space, discovering novel nanostructures with different compositions and sizes is laborious and time-consuming. Bayesian optimization (BO) algorithms that are a specific category of ML adopt a sequential approach, in which the model is initially formulated using defined rules within a certain parametric range to accomplish the required objectives. The results are then improved through an iterative process (14). This produces a rigorous framework for analyzing experimental and simulated data using iterations, leveraging key takeaways for the next iteration in a closed-loop fashion. BO algorithms can efficiently find the optimized experimental conditions within the chemical parametric space, providing a more efficient approach than traditional search procedures. The design and fabrication of nanosensors using this method can enhance performance, improving their sensitivity and specificity for detecting biomarkers in biological systems.

BO can also be integrated with other ML algorithms, such as ANNs and CNN for an inverse sensor design from the previous results (16). Sequentially, ANNs can fine-tune the optimal synthesis parameters of nanoparticles by interpolating between measured results while ensuring a thorough examination of the chemical parametric space. This improves the stability of nanoparticles' optical and electrical properties for targeting biofluids. ML-based dimensionality reduction techniques (PCA) can be used to simplify and associate input features to principal components for the quick interpretation of complex data from biomarker-rich biological samples. The PCA has been tested over a database of Raman scattering spectral data for effective differentiation between biomarkers and background interferences. Furthermore, ML algorithms can also evaluate the relative significance of input features for target discrimination through a feature-based analysis in terms of performance measures. Overall, the analysis and interpretation of complex data from chemical and biological matrices can be accurately evaluated using authentic scientific principles and foundations.

Accurate predictive models in the context of an unknown disease allow us to translate raw signals from nanosensors into an understandable logical relationship that reflects an individual's disease status. This process involves the isolation and deconvolution of complex sensor signals emerging due to different biomarkers while weighting net signal fluctuations in response to a list of predefined biomolecules. The SVMs can accommodate linear as well as nonlinear relations using kernel functions. The above-narrated models are also applicable to biomarkers concerning emerging infections like COVID-19 by offering an intuitive strategy to predict future data through simulations.

In conclusion, ML algorithms combined with nanosensors offer a robust technique for early-stage disease diagnosis and prediction for possible outbreaks caused by unknown mutations in preexisting infections. By using the available physiochemical characteristics of the disease, vaccine and drug development processes can be expedited and involve fewer side effects. Using AI for integrated sensing platforms represents a paradigm shift in the pharmaceutical industry by aiding the development of rapid, efficient, and cost-effective medicines.

12.3 Improving Speed and Accuracy with Magnetic Field

The sensor response time and accuracy are important nanosensors' reliability characteristics. The enhancement in data acquisition rate can help in real-time biological studies (17). AI algorithms can enhance the processing of large amounts of data from sensors, identifying patterns and signatures for abnormalities and noise correction. A recent study (18) evaluated the application of ML algorithms to optically detected magnetic resonance (ODMR)-based magnetometry using a group of NV centers in FNDs. The model utilizes the Gaussian regression (GR) algorithm to improve accuracy in magnetic field sensing, extending the notion for wide-field imaging. It is estimated that, by using GR, the field can be measured with an accuracy of 1.8 μT using 50-nm FNDs. The GR model relies upon a nonparametric learning that is generally utilized for function evaluation to estimate the magnetic field using an ensemble of NV centers in FNDs. The exact field and the OMDR spectra are combined as a set of input variables (x_i) represented as a kernel function $k(x, x')$ to iteratively determine the output vector as (y_i). The exact magnetic field is determined by using a commercially available tesla meter. The experimental ODMR data, along with exactly measured fields, are used as input data for model training, whereas predicting the magnitude of the unknown magnetic field is carried out by the GP algorithm. Experiments were conducted using glass and silicon substrates over a wide range of field strengths (0–2286 μT). Accuracy can be significantly altered by an order of magnitude due to field orientation and the nature of the substrate. This is because of variations in thermal conductivity and the dielectric constant of the system. The AI model combines Spin Hamiltonian of the NV center, NV quantum yield, and the collection efficiency of the objective lens. This method has also been used to observe field distribution around the NV center due to a current-carrying wire. The results derived here are compatible with different sensing modalities for accurate measurements in dynamic systems. Dushenko *et al.* (19) formulated a sequential Bayesian interference approach combining statistical prediction of the magnetic fields by reconstruction of the ODMR spectra and reducing measurement time by an order of magnitude less than conventional schemes.

12.4 Fluorescent-Nanodiamond-Based Immunomagnetic Microscopy of Tumors

The histological analysis of tumors is an essential technique for observing the shape, size, morphology, and growth of cancerous cells in the tissues. FND magnetic microscopy offers promising results for the histological analysis of cancer cells and the quantification of cancer biomarkers (20). Conventional microscopy techniques, such as immunofluorescence microscopy and multispectral imaging, suffer from undesired cell autofluorescence. Therefore, large datasets of images can help minimize the noise. However, they also require storage and computational resources. MRI images require strong external magnetic fields, but their resolution is limited to nearly 100 microns. FNDs, on the other hand, have demonstrated magnetic microscopy with micron-scale resolution and strong signal stability at low noise levels. Additional benefits of FND-based tumor imaging include detecting and quantifying magnetically labeled biomarkers (21).

FND wide-field immunomagnetic microscopy of human lung cancer enhanced with DL algorithms has been implemented to investigate the presence of a magnetically labeled epidermal growth factor receptor (EGFR) (22). Much like in ODMR spectroscopy, in this technique, the EGFR protein was detected from the distribution of magnetic nanoparticles deduced from optical photoluminescence images. EGFR protein expression was then correlated with the density of magnetic nanoparticles dispersed in the tissue with a resolution of ~20 μm. The high sensitivity of NV magnetic sensing demonstrates the detection of a very low number of magnetic nanoparticles. This enabled researchers to identify cancer cells in breast and lung tissues more accurately than flow cytometry and immunohistochemistry. In this experiment, the distribution of magnetic spins was evaluated by using the DL algorithm, which primarily reconstructs the magnetic images and the density of magnetic nanoparticles from the optical images, with additional benefits of noise suppression and background correction. The quantification of magnetic species and cell segmentation was also conducted. The technique can have wider implications in different studies related to blood samples and bodily fluids, enabling rapid disease diagnosis and effective treatment.

12.5 Scaling the Nanomedicine in Clinical Research

Nanomedicine deals with engineered nanomaterials, which show biocompatibility and low toxicity for treating various diseases. It is challenging to scale up the production of effective nanomedicines in the safety (23). There is an urgent need to facilitate the production and monitoring of the therapeutic consequences of nanomedicines. In this regard, ML and DL techniques can be applied to predict interactions and potential side effects to determine optimum dose concentration and treatment time before drug delivery to the patients. Regulatory frameworks are also required to ensure the safety and effectiveness of AI-developed nanomedicines as they help set up guidelines and standard procedures in healthcare systems. AI-based digital therapies combined with conventional drug treatment are being investigated for various health issues like obesity, diabetes, dementia, and cancer. AI-based combination therapy using multidrugs has yet to overcome major difficulties in killing cancer cells (24). Therefore, modifying and improving these drugs for the sake of reduced toxicity and improved viability is a primary goal in the nanomedicine (25).

A study conducted on nanodiamond (ND) drug delivery platforms driven by AI-assisted feedback control system (FSC) has shown promising results for combinational drug delivery (26). This study used FSC, an AI-driven ND platform, to optimize combination nanomedicine drug delivery. In the first stage, three categories of FND drugs (doxorubicin, mitoxantrone, and bleomycin) were prepared. In the next stage, 57 groups of varying concentrations of these three groups of FND drugs were synthesized and tested on breast cancer cell lines (DA-MB-231, BT20, and MCF7) and control cells (MCF10A, IMR-90, and H9C2) using viability assay schemes to evaluate drug efficacy and safety. The toxicity and efficacy were deduced by using the relationship between FND drugs and the available therapeutic window to determine the difference between viability assays for cancerous cells. The study (26) found that the FSC platform could optimize drug delivery, improving efficacy and reducing toxicity. The FSC-optimized FND performed better than all other drugs with a therapeutic window of 51.50%.

12.6 Convolutional Neural Network Enhances Fluorescence Imaging

The tracking and visualization of drug delivery carriers and biomarkers is an important task for the reliability and screening of biomedicines. Generally, biomarkers are excited by optical sources due to fluorescence emission within the visible band. Biological tissues strongly absorb and respond to the optical window producing cell autofluorescence. Due to the presence of a variety of proteins, enzymes, and nucleic acids, the autofluorescence background is mixed with that of diseased biomarkers; this mixing creates difficulty in detecting biomarker fluorescence at very low concentrations. The problem of background fluorescence can be resolved by introducing novel fluorescent biomarkers that do not overlap with cell autofluorescence. Alternatively, experimental techniques can be modified so that a minimum background level does not interfere with the required signals. Organic dyes, with a broad spectral range and high quantum yield, offer an alternate solution. However, they suffer from photobleaching and uncertain cytotoxicity. Some other available options for fluorescence imaging are quantum dots and FNDs. Besides having high quantum yield and photostability, FNDs allow functionalization with various biomolecules. However, currently, the combination of AI algorithms to solve the inverse problem of separation of cellular background fluorescence is being tested, and promising results have been obtained.

The theranostic monitoring of FNDs in human urine has been reported using the fluorescence spectroscopy of NV centers. The procedure outlined here is vital to understanding nanomedicine drug delivery and its excretion from the human body. Accurate knowledge of the excretion efficiency of nanomedicine is pivotal to determining drug efficacy. In this regard, an ANN algorithm has been implemented to accurately determine the concentration of excreted FNDs from human urine samples (27). The diluted FNDs (100 nm) in a urine sample (from healthy adults aged 19–50 years) with varying concentrations of 1 mg/l up to 20 mg/l were tested. The concentrations of excreted FNDs were evaluated using the fluorescence and Raman spectroscopy of the samples, where the integral spectral intensity (640–800 nm) of the samples without FNDs was subtracted from a sample carrying FNDs. It was found that the fluorescence intensity of the FNDs has a linear relationship to the concentration of nanoparticles, whereas the accuracy of this measurement was 0.3 mg/l. In this experiment, the undesired background autofluorescence of urine was removed from the required data by an ANN algorithm. By using the ANNs, an experimental approach solved the inverse problem of optical spectroscopy. The accuracy of the solution was greatly influenced by appropriate noise estimation and the analytical model of the target. The experimental training approach (using the real data) of the samples (315 spectra) realistically incorporated the experimental noise content and the associated molecular interactions. The method reported here could potentially be extended for analysis of skin, blood vessels, and different biofluids. Furthermore, this approach can be implemented with different fluorescent nanosensors for *in vivo* applications like the detection of proteins, enzymes, and intracellular metabolism. Another study (17) reported the detection of FNDs using CNNs with a low concentration of 3 μg/ml from chicken egg white. It was predicted that with the inverse problem approach, the detection limit of FNDs could be extended by a factor of 1.5.

12.7 Challenges and Conclusion

The integration of AI and FNDs in biosensing is associated with numerous challenges in various applications. The major barrier to AI biosensing device fabrication is the need to train the AI to predict signals generated by the biosensors accurately. The difficulty is heightened by the emergence of mutations with unexpected biological interactions, which cause abnormal symptoms. The physical model of the disease must be incorporated, along with all its uncertainties. Moreover, the selection of ingredients for synthesis and surface modification of FNDs for the target biological system can also be complex and time-consuming. These developments require trained manpower with expertise in multidisciplinary aspects of materials science, chemistry, physics, and biology.

Despite these difficulties, combining AI and FNDs would lead to developing highly sensitive biosensors capable of investigating intercellular metabolic activity at the individual cell level. The consequences would endeavor medical diagnosis, food safety, and general healthcare. The role of AI in smart FND sensor design could also enable real-time monitoring of free radicals, intracellular signaling, and heat exchange. Their scope would be further extended to the synthesis of personalized nanomedicine with optimum dose concentration and duration of treatment. By analyzing patient health profiles and medical history, effective treatment plans would be clinically devised and adopted. Overall, progressive research in biosensing using AI and FNDs for the detection of disease biomarkers provides an opportunity for the early-stage detection and diagnosis of a variety of malicious diseases and infections.

Acknowledgments

This work was supported by the National Science and Technology Council (NSTC), Taiwan, under Grant No. 112-2221-E-011-031.

References

1 Qureshi SA, Raza SEA, Hussain L, Malibari AA, Nour MK, Rehman Au, *et al.* Intelligent ultra-light deep learning model for multi-class brain tumor detection. *Appl Sci.* 2022;12(8):3715.

2 Qureshi SA, Hsiao WW-W, Hussain L, Aman H, Le T-N, Rafique M. Recent development of fluorescent nanodiamonds for optical biosensing and disease diagnosis. *Biosensors.* 2022;12(12):1181.

3 Alpaydin E. *Introduction to machine learning.* MIT Press; 2020.

4 Cunningham P, Delany SJ. k-Nearest neighbour classifiers – A Tutorial. *ACM Comput Surv.* 2021;54(6):1–25.

5 Liaw A, Wiener M. Classification and regression by random forest. *R News.* 2002;2(3):18–22.

6 Fausett LV. *Fundamentals of neural networks: Architectures, algorithms and applications.* Pearson Education India; 2006.

7 Qureshi SA, Hussain L, Ibrar U, Alabdulkreem E, Nour MK, Alqahtani MS, *et al.* Radiogenomic classification for MGMT promoter methylation status using multi-omics fused feature space for least invasive diagnosis through mpMRI scans. *Sci Rep.* 2023;13(1):1–29.

8 Nair V, Hinton GE, editors. Rectified linear units improve restricted Boltzmann machines. *Proceedings of the 27th international conference on machine learning (ICML-10)*; 2010.

9 He K, Zhang X, Ren S, Sun J. Spatial pyramid pooling in deep convolutional networks for visual recognition. *IEEE Trans Pattern Anal Mach Intell.* 2015;37(9):1904–16.

10 Scherer D, Müller A, Behnke S, editors. Evaluation of pooling operations in convolutional architectures for object recognition. *Artificial Neural Networks–ICANN 2010: 20th International Conference*, Thessaloniki, Greece, September 15–18, 2010, Proceedings, Part III 20; 2010: Springer.

11 Nagi J, Ducatelle F, Di Caro GA, Cireşan D, Meier U, Giusti A, *et al.*, editors. Max-pooling convolutional neural networks for vision-based hand gesture recognition. *2011 IEEE international conference on signal and image processing applications (ICSIPA)*; 2011: IEEE.

12 Srivastava N, Hinton G, Krizhevsky A, Sutskever I, Salakhutdinov R. Dropout: A simple way to prevent neural networks from overfitting. *J Mach Learn Res.* 2014;15(1):1929–58.

13 Bro R, Smilde AK. Principal component analysis. *Anal Methods.* 2014;6(9):2812–31.

14 Leong YX, Tan EX, Leong SX, Lin Koh CS, Thanh Nguyen LB, Ting Chen JR, *et al.* Where nanosensors meet machine learning: Prospects and challenges in detecting disease X. *Acs Nano.* 2022;16(9):13279–93.

15 Jin X, Liu C, Xu T, Su L, Zhang X. Artificial intelligence biosensors: Challenges and prospects. *Biosens Bioelectron.* 2020;165:112412.

16 Cui F, Yue Y, Zhang Y, Zhang Z, Zhou HS. Advancing biosensors with machine learning. *ACS Sensors.* 2020;5(11):3346–64.

17 Dolenko TA, Burikov SA, Vervald AM, Vlasov II, Dolenko SA, Laptinskiy KA, *et al.* Optical imaging of fluorescent carbon biomarkers using artificial neural networks. *J Biomed Opt.* 2014;19(11):117007.

18 Tsukamoto M, Ito S, Ogawa K, Ashida Y, Sasaki K, Kobayashi K. Accurate magnetic field imaging using nanodiamond quantum sensors enhanced by machine learning. *Sci Rep.* 2022;12(1):13942.

19 Dushenko S, Ambal K, McMichael RD. Sequential Bayesian experiment design for optically detected magnetic resonance of nitrogen-vacancy centers. *Phys Rev Appl.* 2020;14(5):054036.

20 Perevedentseva E, Lin Y-C, Cheng C-L. A review of recent advances in nanodiamond-mediated drug delivery in cancer. *Expert Opin Drug Del.* 2021;18(3):369–82.

21 Ho D. Nanodiamond-based chemotherapy and imaging. In: Chad A. Mirkin, Thomas J. Meade, Sarah Hurst Petrosko, Alexander H. Stegh (Eds.), *Nanotechnology-Based Precision Tools for the Detection and Treatment of Cancer.*

Springer Link; 2015. pp. 85–102, https://link.springer.com/book/10.1007/978-3-319-16555-4.

22 Chen S, Li W, Zheng X, Yu P, Wang P, Sun Z, *et al.* Immunomagnetic microscopy of tumor tissues using quantum sensors in diamond. *Proc Natl Acad Sci USA.* 2022;119(5):e2118876119.

23 Ho D, Wang P, Kee T. Artificial intelligence in nanomedicine. *Nanoscale Horiz.* 2019;4(2):365–77.

24 Serov N, Vinogradov V. Artificial intelligence to bring nanomedicine to life. *Adv Drug Deliv Rev.* 2022;184:114194.

25 Wang H, Lee D-K, Chen K-Y, Chen J-Y, Zhang K, Silva A, *et al.* Mechanism-independent optimization of combinatorial nanodiamond and unmodified drug delivery using a phenotypically driven platform technology. *Acs Nano.* 2015;9(3):3332–44.

26 Khong J, Wang P, Gan TR, Ng J, Anh TTL, Blasiak A, *et al.* The role of artificial intelligence in scaling nanomedicine toward broad clinical impact. In: Eun Ji Chung, Lorraine Leon, Carlos Rinaldi (Eds.), *Nanoparticles for Biomedical Applications.* Elsevier; 2020. pp. 385–407, https://www.sciencedirect.com/science/article/pii/B9780128166628010018.

27 Laptinskiy K, Burikov S, Dolenko S, Efitorov A, Sarmanova O, Shenderova O, *et al.* Monitoring of nanodiamonds in human urine using artificial neural networks. *Phys Status Solidi A.* 2016;213(10):2614–22.

13

Perspective and Outlook on Nanodiamond Research

Trong-Nghia Le, Shu-Wei Chang, Joshua Ko, Yu-Jou Lin, Wei-Hung Chiang, and Wesley Wei-Wen Hsiao**

** Corresponding authors*

The chemical inertness, nontoxic nature, and ease of surface modification are shared by both nanodiamonds (NDs) and fluorescent nanodiamonds (FNDs), with the added benefits of high brightness, excellent photostability, and non-photobleaching for FNDs. As a result, these materials have garnered significant interest for applications such as bioimaging, drug delivery, quantum sensing, and biomedical analysis. FNDs have been extensively utilized for *in vitro* and *in vivo* imaging to investigate biological processes, while also demonstrating potential in anti-cancer therapeutic delivery, magnetic field sensing, and nanothermometric sensing. The use of FNDs as a fluorescent probe for detecting disease biomarkers has seen development in recent years. Although challenges in FND research are inevitable, this chapter explores the strengths and challenges of FNDs in recent applications, as well as the emerging topic of integrating artificial intelligence algorithms with FND nanosensors.

13.1 Strengths of Fluorescent Nanodiamonds

Fluorescent nanodiamonds (FNDs) possess particular strengths in comparison with other nanocarbon materials, as shown in Table 13.1 (1–3). FNDs have inherited many desirable characteristics of bulk diamonds, including optical refraction and film transparency, exceptional hardness and mechanical stability, and chemical inertness in oxidizing media. The fluorescence of FNDs emanates from color point defects (centers) that exhibit superior photostability with emission at room

Nanodiamonds in Analytical and Biological Sciences: Principles and Applications, First Edition.
Edited by Wesley Wei-Wen Hsiao and Trong-Nghia Le.
© 2023 John Wiley & Sons Ltd. Published 2023 by John Wiley & Sons Ltd.

Table 13.1 The strengths and challenges of various nanocarbon materials used in bioapplications.

Kind of nanocarbon materials	Strengths	Challenges
NDs/FNDs	Excellent mechanical properties, chemically inert, inherent fluorescence, ODMR for biosensing applications, rich and tailorable surface chemistry, or dense functional groups	Unspecified degradability, insulator, irregular-particle morphology
Graphene	Excellent mechanical properties, tailorable size, easy preparation, photothermal effect	Hydrophobic, easily forms agglomerates irreversibly, no fluorescence itself, low functionalization degree
Graphene oxide	Excellent water dispersibility, tailorable size, easy preparation, dense functional groups, photothermal effect	Low fluorescence, broad size distribution, batch-to-batch variability, inflammatory response
Fullerenes	Well-established chemical functionalization, determined molecular structure, radical scavenging behavior	Hydrophobic, low functionalization degree, very small size, extremely large interfacial area, challenging physical and mechanical properties
Carbon nanotubes	Excellent mechanical properties, tailorable size, good cellular uptake, near-infrared fluorescence, photothermal effect	Hydrophobic, inflammatory response, size-dependent toxicity, organ accumulation potential

temperature. Notably, the distinctive magneto-optical properties of nitrogen-vacancy (NV) centers result in optically detected magnetic resonance (ODMR), a phenomenon that is not observed in other nanocarbon materials. While FNDs may not be the thermodynamically favored allotrope under standard conditions, they can be formed in conditions of high temperatures and high pressures. FNDs are also highly biocompatible, nontoxic, and can be easily functionalized by all biomolecules. These characteristics make FNDs a robust platform for various applications, including diagnostic imaging or the development of therapeutic agents for use in nanobiomedicine.

The growth of nanobioscience and nanotechnology has resulted in the discovery and synthesis of a wide range of nanomaterials. Owing to their unique mechanical, chemical, optical, and biological properties, the usefulness of FNDs has been tested within various applications. FNDs contain a high concentration of negative nitrogen-vacancy (NV^-) defect centers, which serve as fluorophores. Formed by ion beam irradiation and thermal annealing processes, these point defects possess a number of remarkable features, including their high biocompatibility and ability to project bright photoluminescence in the extended red region. These FNDs can also emit strong and stable fluorescence from NV^- color centers when excited by visible light without photobleaching and photoblinking. This is true even under continuous high-power laser excitation at the single-molecule level, allowing for deeper penetration and better spectral separation from autofluorescence. Compared with other nanoscale carbon allotropes, FNDs are made of solely sp^3 carbon atoms, leading to an optically transparent and biologically inert material. Furthermore, the surface of nanodiamonds (NDs) can be flexibly modified with either carboxyl or other functional groups for specific binding with proteins and nucleic acids, thereby expanding their potential uses in biosensing, diagnostics, and drug delivery systems. FNDs can also be considered as potential alternative fluorescent probes over organic dyes, fluorescent proteins, and quantum dots for long-term *in vitro* and *in vivo* imaging and tracking applications. As a result of these properties, FNDs have the potential to revolutionize disease detection for the benefit of both patients and society.

13.2 Nanodiamond's Surface Chemistry

NDs' unique nanosized carbon building blocks are "captivating" crystals. This is because their mechanical, chemical, optical, and biological properties make them useful for variety of applications in biomedicine (1, 2). Their size, structural properties, and surface attributes allow them to be used in a wide variety of fields, including bioanalysis, drug delivery, cancer therapy, and bioimaging. In comparison with other nanocarriers, NDs provide an all-encompassing drug delivery system that is both safe and effective. A large number of research studies on the structure, surface chemistry, colloidal properties, and surface modification all point to the feasibility of designing an effective ND-based platform for applications in bioimaging, tracking, and drug delivery system. Along with the extensive studies, NDs' prospective applications in biology and nanoscale medicine will continue to attract the interest of researchers in these fields, ensuring a continual flow of advancement.

Despite its potential across many biomedical applications, effectively mobilizing this nanomaterial requires better knowledge of its structure and surface chemistry. Increased understanding of FNDs properties could improve manufacturing quantities, possibly exceeding those of fullerenes, carbon nanotubes, graphene, and other carbon nanomaterials. The resulting growth in supply will continue to fuel novel applications. Thus, NDs are expected to surge forward and find widespread use in research, technology, and medicine.

13.3 Nanodiamond in Bioimaging

ND-based chemistry and biomedicine is a blossoming field of research. A notable development in this research area has been the conversion of ND to FND, which can host a variety of vacancy-related color centers after proper radiation damage and work as a nontoxic and photostable biomarker for high-resolution imaging, long-term tracking, and deep tissue imaging in both *in vitro* and *in vivo* systems. The detection of FNDs is wieldy with a broad range of techniques that suit this purpose, including multiphoton excitation, super-resolution, cathodoluminescence, and magneto-optical imaging.

FNDs are not without their limitations. Real-time 3-D tracking with good fluorescence imaging contrast remains underdeveloped, since scientists are still trying to successfully execute deep tissue imaging with FNDs at suitable levels of sensitivity and resolution. The key to overcoming this issue likely lies in the longer-wavelength and higher-order nonlinear excitation of FNDs as well as their links with magnetic resonance imaging. We expect that FND-enabled technologies will gain wider recognition for their unique surface and diamond core properties. When this happens, more experts will be attracted to this area of study, fueling new developments in the field of bioimaging. The end result will be the profound ability to perceive the molecular mechanisms of life.

13.4 Nanodiamond in Quantum Sensing

An NV^- center in diamond is a solid-state artificial atom with optical detectable spin states. It has long coherence time in ambient environments, and its ODMR spectrum is sensitive to variations in magnetic fields, electric fields, temperature, and mechanical stress. This has allowed for many breakthroughs in quantum sensing and imaging. NV centers used in temperature sensing techniques have unprecedented precision and sensitivity. Despite the NV center's unique properties, diamond nanothermometry still faces challenges. One ongoing issue is the size of the temperature sensor. Sub-20-nm FNDs are required to get sustained fluorescence and detectable zero-phonon line (ZPL) and ODMR signals within intracellular organelles. However, when FNDs are 20 nm or smaller, the average number of NV centers per particle is close to 1, and the fluorescence quantum yield drops to 20%. Surface modification, size, shape, crystal strain, impurity concentration, and number of NV centers in nanocrystals needs additional study. With further developments to these technologies, a future "all-in-one" treatment with nanoscale tracking, heating, and sensing capabilities is expected to be produced. We anticipate that studies of the outstanding coherence properties of NV^- centers will be awarded a Nobel Prize in Physics in the very near future.

13.5 Nanodiamond as Therapeutic Carrier

A vast range of nanomaterials have been synthesized and applied for drug delivery as a result of rapid advancements in nanoscience and nanotechnology. ND is currently being tested for use in several medical fields. The mechanical, chemical, optical, and biological qualities of ND set it apart from its competitors. These features indicate NDs potential in drug delivery.

NDs can be used to combine targeting, imaging, and drug delivery in a nanoplatform. As a result, the ND platform has emerged as a viable agent for translationally relevant treatment and diagnostics. One recent breakthrough in this sector is the use of NDs in combination with doxorubicin (DOX) to treat drug-resistant cancers (4). NDs increase DOX's circulatory life span and allow deadly amounts of the medication to be administered in a less hazardous manner. In the context of glioblastoma therapy, researchers discovered that localized drug delivery via convection-enhanced administration is possible with NDs adsorbed with DOX (CED) (5). This treatment strategy permits medicinal compounds to penetrate the blood–brain barrier (BBB) without causing uncontrolled drug distribution or toxicity increases. In addition, NDs have risen to prominence in sectors unrelated to cancer. Drug-eluting contact lenses with NDs can help deliver medicine to specific areas of the eye (6). Preclinical experiments have further validated ND-mediated drug delivery strategies. This necessitates additional investigation into the clinical applications of ND-based imaging and medicinal medicines. The ease with which NDs can be functionalized suggests a wide range of possibilities for the regulated delivery of tiny and large compounds (e.g., proteins or genes).

Despite these advancements, it is still difficult to deliver ND drugs in a realistic manner. Despite some success with fluorescent tagging, research on the use of NDs in drug delivery is still in its early stages. More studies are needed to determine the safety of NDs in people before determining that they are clinically appropriate, innocuous, and effective drug delivery vehicles. NDs are so inert that they cannot be broken down, making the substance difficult to rid from the human body. Although multiple compelling findings have shown that in culture cells and even animals, NDs can carry active anticancer drugs require more studies to be approved in human clinically.

13.6 Nanodiamond for Biomedical Analysis

Recently, researchers have pioneered the development of FND-based diagnostic systems for the detection of infectious and chronic diseases. Chang *et al.* demonstrated that, compared with conventional gold nanoparticle-based lateral flow immunoassays (LFIAs), FND-based spin-enhanced lateral flow immunoassay

(SELFIA) exhibited a greater sensitivity by 100 times via magnetic modulation (7). The potential of SELFIA have been demonstrated by the detection of SARS-CoV-2 antigens from different variants and dengue virus nonstructural protein (NS1) with ultrasensitivity (8, 9). SELFIA also surpassed enzyme-linked immunosorbent assay (ELISA) in terms of efficiency and flexibility. Miller *et al.* were able to detect a single copy of HIV-1 RNA via microwave-modulated LFIA (10). These ultrasensitive quantum diagnostics platforms are applicable to a wide range of diagnostic test formats.

On another prospect, the application of FND in cancer diagnosis is both promising and challenging. It is promising that FND could efficiently bind with biomolecules, such as proteins, which can be disease biomarkers or antibodies for targeting. In addition, FND is highly biocompatible and has unique fluorescence properties due to the NV centers, as discussed throughout the book. However, cancer diagnosis is still challenging due to the need for early-stage detection, the complexity of tumor types and conditions, as well as many unpredictable variables in the human body naturally. Most of the research focuses on tracking and monitoring the cancer cell to provide more information on cancer progression. There is still room for further research in the knowledge of image analysis used for early cancer detection and various types of tumors. Integration with artificial intelligence (AI) is one of them. AI is used in a variety of applications, such as voice synthesis, image recognition, and image interpretation. There are many other successful developments by AI integration, such as a deep learning system that can outperform pathologists when giving a Gleason score and deep neural networks that can detect abnormal lymph nodes or polyps. AI can also be utilized with the technology of FNDs to track cancer cells, improving the precision of cell recognition by computers and cameras. Cells can be tagged and tracked by FNDs, so monitoring cell activity and identifying abnormalities can be a much easier job. By incorporating FNDs and AI image scanning, the tracked cells within images will make recognition more efficient and evident. It may unleash a whole new potential for AI and FNDs.

13.7 Challenges of Fluorescent Nanodiamond Research

New technological advancements aim to solve more problems than they create. NDs have already been commercialized in diverse fields, often for abrasives and lubricants. However, to expand their use to biomedicine, different questions need to be addressed. FNDs constituted by sp^3 are rendered biocompatible by inertness. Indeed, owing to their limited reactivity in biological surroundings, FNDs are usually considered safe nanomaterials. So, compared with other sp^2 nanocarbon materials, the superior stability of FNDs may influence their biodegradability in the human body or natural environment. To the best of our knowledge, no studies yet exist on the biodegradability of FNDs. To prevent unwanted bioaccumulation and ensure safety in the long term, FNDs need to be excreted from the body. Distinct *in vivo* approaches have been carried out to help resolve this issue. Because of their high surface energy (electrostatic interactions) and highly specific surface area, FND particles can aggregate into larger clusters. They also tend to adsorb plasma proteins upon intravenous administration, which can alter their performance in biological applications. Even so, both the coronation and the aggregation of FNDs can be modified through surface functionalization approaches.

In terms of *in vivo* fluorescence imaging, the insufficient number of luminescent centers in single-crystal FNDs gives rise to weak emission intensity that hardly penetrates the skin and tissues. To form NV center-enriched FNDs, irradiation with high-energy helium ion beams following the high-temperature annealing process is necessary. However, this significantly raises the cost of FNDs and ineluctably breaks their intrinsic surface functionality. Furthermore, enhancing the colloidal stability of synthesized FND particles to fulfill the needs of *in vivo* applications remains challenging. In response to these concerns, many research groups are actively developing approaches to generate highly doped FNDs. Advanced knowledge of FNDs' structure and surface chemistry can help scientists design better systems that consider the synergistic properties of the core and surface of FNDs as well as their relevant applications.

References

1 Chang H-C, Hsiao WW-W, Su M-C. *Fluorescent Nanodiamonds*. Hoboken, NJ, USA: John Wiley & Sons; 2018.

2 Hsiao WW-W, Le T-N, Chang H-C. Applications of fluorescent nanodiamond in biology. In: Meyers, R.A. (Ed.), *Encyclopedia of Analytical Chemistry*. Hoboken, NJ, USA: John Wiley & Sons; 2022. pp. 1–43, https://doi.org/10.1002/9780470027318.a9776.

3 Reina G, González-Domínguez JM, Criado A, Vázquez E, Bianco A, Prato M. Promises, facts and challenges for graphene in biomedical applications. *Chem Soc Rev*. 2017;46(15):4400–16.

4 Chow EK, Zhang XQ, Chen M, Lam R, Robinson E, Huang HJ, *et al*. Nanodiamond therapeutic delivery agents mediate enhanced chemoresistant tumor treatment. *Sci Transl Med*. 2011;3(73):73ra21.

5 Xi G, Robinson E, Mania-Farnell B, Vanin EF, Shim KW, Takao T, *et al*. Convection-enhanced delivery of nanodiamond drug delivery platforms for intracranial tumor treatment. *Nanomedicine-Uk*. 2014;10(2):381–91.

6 Kim H-J, Zhang K, Moore L, Ho D. Diamond nanogel-embedded contact lenses mediate lysozyme-dependent therapeutic release. *ACS Nano*. 2014;8(3):2998–3005.
7 Hui YY, Chen OJ, Lin H-H, Su Y-K, Chen KY, Wang C-Y, *et al*. Magnetically modulated fluorescence of nitrogen-vacancy centers in nanodiamonds for ultrasensitive biomedical analysis. *Anal Chem*. 2021;93(18):7140–7.
8 Hsiao WW-W, Sharma N, Le T-N, Cheng Y-Y, Lee C-C, Vo D-T, *et al*. Fluorescent nanodiamond-based spin-enhanced lateral flow immunoassay for detection of SARS-CoV-2 nucleocapsid protein and spike protein from different variants. *Anal Chim Acta*. 2022;1230:340389.
9 Le T-N, Hsiao WW-W, Cheng Y-Y, Lee C-C, Huynh T-T, Pham DM, *et al*. Spin-enhanced lateral flow immunoassay for high-sensitivity detection of nonstructural protein NS1 serotypes of dengue virus. *Anal Chem*. 2022;94(51):17819–26.
10 Miller BS, Bezinge L, Gliddon HD, Huang D, Dold G, Gray ER, *et al*. Spin-enhanced nanodiamond biosensing for ultrasensitive diagnostics. *Nature*. 2020;587(7835):588–93.

Index

Note: figures are indecated by *italics* and tables by **bold** page numbers.

Nanodiamonds in Analytical and Biological Sciences: Principles and Applications, First Edition.
Edited by Wesley Wei-Wen Hsiao and Trong-Nghia Le.
© 2023 John Wiley & Sons Ltd. Published 2023 by John Wiley & Sons Ltd.

b

c

n

o